T5-AFQ-204

INSECT NEUROCHEMISTRY and NEUROPHYSIOLOGY 1993

595.70188
I59
36,308

INSECT NEUROCHEMISTRY and NEUROPHYSIOLOGY 1993

Edited by

Alexej B. Bořkovec
Marcia J. Loeb

CRC Press 1994
Boca Raton   Ann Arbor   London   Tokyo

Cover design: Schematic illustration of the CNS neurons that express the big PTTH phenotype during larval development of *Manduca sexta*. (Reprinted from Westbrook, A.L., Regan, S.A., and Bollenbacher, W.E., *The Journal of Comparative Neurology,* 327, 1–16, 1993. With permission.)

Library of Congress Cataloging-in-Publication Data

Insect neurochemistry and neurophysiology, 1993 / edited by Alexej Bořkovec,
    Marcia J. Loeb.
       p. cm.
      Includes bibliographical references and index.
      ISBN 0-8493-4591-X
        1. Insects--Nervous system--Congress. 2. Insects--Physiology--Congress
    3. Neurochemistry--Congresses, 4. Neurophysiology--Congress.
    I. Bořkovec, A.B. (Alexej B.), 1925-   II. Loeb, M.J.

QL495.I49645 1993   93-38493
595.7′.0188—dc20   CIP

This book contains information obtained from authentic and highly regarded sources. Reprinted material is quoted with permission, and sources are indicated. A wide variety of references are listed. Reasonable efforts have been made to publish reliable data and information, but the author and the publisher cannot assume responsibility for the validity of all materials or for the consequences of their use.

Neither this book nor any part may be reproduced or transmitted in any form or by any means, electronic or mechanical, including photocopying, microfilming, and recording, or by any information storage or retrieval system, without prior permission in writing from the publisher.

All rights reserved. Authorization to photocopy items for internal or personal use, or the personal or internal use of specific clients, may be granted by CRC Press, Inc., provided that $.50 per page photocopied is paid directly to Copyright Clearance Center, 27 Congress Street, Salem, MA 01970 USA. The fee code for users of the Transactional Reporting Service is ISBN 0-8493-4591-X/94/$0.00+$.50. The fee is subject to change without notice. For organizations that have been granted a photocopy license by the CCC, a separate system of payment has been arranged.

CRC Press, Inc.'s consent does not extend to copying for general distribution, for promotion, for creating new works, or for resale. Specific permission must be obtained in writing from CRC Press for such copying.

Direct all inquiries to CRC Press, Inc., 2000 Corporate Blvd., N.W., Boca Raton, Florida 33431.

© 1994 by CRC Press, Inc.

No claim to original U.S. Government works
International Standard Book Number 0-8493-4591-X
Library of Congress Card Number 93-38493
Printed in the United States of America 1 2 3 4 5 6 7 8 9 0
Printed on acid-free paper

Acknowledgement:

Preparation of this book was made possible by the diligence of I. Bedard, S. Cohen, M. Weinrich and L. Villa.

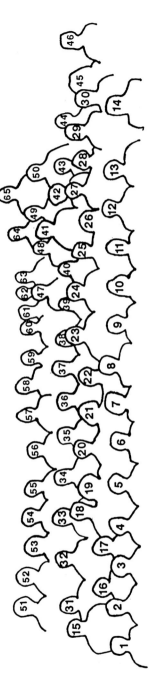

Key to the photo:

1.Rosemary Gray; 2. Peter Harvie; 3. Jon Hayashi; 4. Marcia Loeb; 5. Peter Evans; 6. Ken Davey; 7. Dale Gelman; 8. Sasha Borkovec; 9. Lawrence Gilbert; 10. Miriam Altstein; 11. Ada Rafaeli; 12. Roger Downer; 13. Paul Pener; 14. Vladimir Novak; 15. Pete Masler; 16. Jan Zdarek; 17. P. Sivasubramanian; 18. Radislav Kotera; 19.Linda Hall; 20. Barry Ganetsky; 21. Blanka Kalinova; 22. Florence Tribut; 23. Bruno Lapied; 24. Karel Slama; 25. Danuta Konopinska; 26. Alice Huckova 27. Ron Nachman; 28. Jum Chung 29.Wieslav Sobotka; 30. Hiromu Akai; 31.Jim Truman; 32. Angela Lange; 33. Ian Orchard; 34. Malcolm Burrows; 35. Vladimir Landa; 36. Yasuhiso Endo; 37. David Muehleisen; 38. ⸮ ; 39. Henk Vullings; 40.⸮ ; 41. Hans Laufer; 42. Joanna Michalik; 43. Ivan Gelbic ; 44. Maciej Pszczolkowski; 45. Renee Wagner; 46. Tom Kelly; 47. Veeresh Sevala; 48. Vaclav Nemec; 49 Dalibor Kodrik; 50. Tomas Soldan; 51. Richard Baines; 52. John Hildebrand; 53. Richard Rayne; 54. Maurice Elphick; 55. Steven Warburton; 56. Anne Westbrook; 57. Barry Loughton; 58. John Stout; 59. Klaus Richter; 60. Peter Verhaert; 61. Klaus Hoffmann; 62. Mirko Slovak; 63. Laszlo Hiripi 64. Karl Dahm; 65. Hans Smid

Not Identified, or not in the photo:
Blanka Bennettova, Yang Chansun, Ornella Cusinato, Valerij Filippov, Maria Filippova, Adrien Girardie, Josianne Girardie, Graham Goldsworthy, Magda Hodkova, Jacques Puiroux, Sandra Robb, Frantisek Sehnal, John Steele, Richard Tykva, Li Wei Wei.

Preface

The first International Conference on Insect Neurochemistry and Neurophysiology, ICINN'83, was held almost exactly ten years ago in the United States. At that time, only two neurosecretory peptides, proctolin and adipokinetic hormone, were fully characterized but a large number of other neuropeptides were being targeted for isolation and identification. The focus on these extremely highly active substances appeared justified because their possible exploitation in insect control and management seemed reasonable. However, the following two Conferences in 1986 and 1989 have introduced some skepticism about that assumption. Although the number of newly isolated and identified neuropeptides increased dramatically, research in insect neurobiology broadened and diversified well beyond the original range. The fourth Conference, ICINN'93, the subject of this volume, confirms the broad pattern of contemporary research in this field. As in the previous Conferences, the ICINN'93 program contained state-of-the-art plenary lecture reviews by invited speakers as well as shorter research communications presented orally or as posters. In this volume the longer and shorter articles are grouped in two corresponding sections.

The decision to stage the fourth Conference in Europe rather than in the United States was influenced primarily by two factors: one was the desire to increase the participation of European scientists in general and the other was to include, for the first time, a significant contribution from those who could not attend the previous conferences because of their geographic location behind the Iron Curtain. We are grateful to our industrial sponsors, Agrochemical Division of Schering AG and Rhone-Poulenc AG Company, for contributions that facilitated the participation of a large number of our East-European colleagues. The Program Chairman of the Conference, Dr. Dale B. Gelman, deserves our sincere gratitude. We also wish to thank the Organizing Commitee and the American and Czech Executive Committees for making ICINN'93, and thus this book, possible.

Marianske Lazne, Czech Republic
July 1993

Alexej B. Bořkovec
Marcia J. Loeb

Introduction

In the first Conference, **ICINN 1983**, papers were presented on nerve and synapse morphology, physiology, and pharmacology, the roles of catecholaminergic, acetylcholinergic and GABAergic neurotransmitters and their receptors, particularly in mediating neuropeptide-controlled events, cAMP as a second messenger and its role in the post-receptor cascade of events, and the effects of toxins. We noted a gradual transition from studies of insect neuropeptides based on morphological and physiological evidence to physical isolation and thus better characterization of the peptides. Papers presented at the first Conference provided evidence for physiological activity or reported the partial isolation of adipokinetic hormone, prothoracicotropic hormone, proctolin, oostatic hormone, neurophysin, egg development neurosecretory hormone, diuretic hormone, pheromone biosynthesis activating neurohormone, cardioaccelleratory peptides, myotropins, allatotropins, allatostatins and brain factors regulating juvenile hormone esterase. In the following Conference, **ICINN 1986**, better chemical and physiological characterization of these same peptides was provided; testis ecdysiotropin, FMRFamide, hypertrehalosemic hormone, and leucomyosuppressin were added to the list of peptides discussed. Horizons were expanded to insect relatives, the Crustacea, by recognizing that insect adipokinetic hormone was identical to the crustacean red pigment concentrating hormone. Pericardial sinuses and calcium channels were explored. At **ICINN 1989** we saw significant changes as methods developed for studying relatively large amounts of more abundant vertebrate tissue were adapted to small scale insect work, and use of molecular genetics became more widespread. Peptide isolation and characterization as well as gene expression were recurrent themes, while knowledge of the physiological action of many of the peptides discussed in earlier years was expanded; leucomyosupressin, hypolipaemic hormone, human atrial natriuretic peptide, and vasopressin were added to the list of recognized insect peptides. The subjects discussed at this Conference included studies of neural development, sodium and calcium channels, mediation of hormonal signals by the inositol phosphate messenger system, calcium and protein kinase C, and the always pertinent physiology of prothoracic glands, corpora allata, and reproductive organs. Toxins or pesticides such as baculovirus, azadirachtin, endosulfan, precocene, and other chemical regulators, were discussed.

In this book which represents **ICINN 1993**, we see continued interest in nerve function, neurotransmitters, ion channels, second messengers and neuropeptides. By using a variety of techniques, combining aspects of neurophysiology, pharmacology, immunology, peptide separation and sequencing or molecular biology, it has become possible to study systems in greater detail and complexity than before. Regulators of the functions of the predominant insect hormones, juvenile hormones and ecdysteroids, are discussed. Peptide hormone receptors and G proteins in the post-receptor cascade have become hot topics. Signalling within the nervous system by means of light, biogenic amines, nitric oxide, phosphokinases, calcium, calmodulin, cGMP and the now old standard, cyclic AMP, is examined. Not surprisingly, updates on the regulation of synthesis, secretion or action of many of the same neuropeptides discussed in previous **ICINNs** are seen in the present one, **ICINN 1993**, but the list of peptides has been expanded to include hindgut ecdysiotropin, egg prothoracicotropic hormone, silk gland regulators, insulin, peptide-P-like material, retinoic acid binding protein, neuropeptide F, peptides involved in reproduction (several isolated and characterized), antagonists and agonists of juvenile hormone synthesis in insects and crustacea, and peptides controlling circadian rhythm and mating. The classic interest in eliminating harmful insects now reflects our current environmental consciousness as **ICINN 93** reports on natural toxins, peptide hormone and neurotransmitter mimics, peptide hormone antagonists, and the insect hormone-disrupting agent, KK42. Mechanisms developed by parasites to prevent release of development-controlling neuropeptides are also discussed.

    ICINN 1993 presents two pathways of study: in one, we see an interest in understanding complex systems such as the interaction of nerve networks, regulation of overall motor behavior by genetic modification of muscle ion channels, and the effect of one peptide hormone on several critical portions of the whole animal. At the same time, we see a gradual shift of interest from the identification of neuropeptides and neurotransmitters *in situ* or isolated from neural tissue or reproductive organs, to the minutiae of mechanisms of action at the level of target cells. The quantity of peptide needed to activate receptor cascades may be only a few molecules, since peptides and non-peptidic transmitters operate *in vivo* at concentrations approaching 10^{-10} M, causing bioassays and specific detection to be more precise than before. Eclosion hormone effects appear in milliseconds in some tissues, and in hours or days in others. Instead of one-peptide-one-action, there is evidence to show that, for example, PBAN, has a coloration effect in larvae and controls pheromone synthesis in adults of the same species. Families of similar neurohormones (proctolin-like) or multiple non-chemically related peptide hormones, such as prothoracicotropic hormones, have been shown to act at the same cell types, or to produce different effects in different tissues; in the future, the message may be shown to be in a cocktail of substances rather than in one factor. Although there is now great interest in identifying specific receptors for specific peptides, there is no surety that reports at **ICINNs** to come will show that an identified receptor will respond in the same fashion in every instance, or in different locations in the body. Will the dimers and tetramers of steroid-like receptors alter the ultimate message as they bind to gene receptor sequences? Will a genetically engineered change in a receptor make it respond to a different ligand?

    Papers in this book explore and promote understanding of specific second messenger systems which mediate insect transmitter and peptide hormone responses. However, evidence is also presented to show that both cAMP and inositol phosphate-derived second messengers can have the same ultimate effect in two separate systems. These apparent contradictions may be resolved some day when different receptor sites for the same peptide are found, as is the case for some catecholamine receptor systems. We may anticipate future **ICINN** papers describing transmitter and neuropeptide cascades that utilize some of the multiple isoforms of phosphokinases, families of phosphodiesterases, and the several G_s, G_i or other G proteins. Some of the work described in this volume has explored methods by which peptides are broken down to stop a message once it has been delivered. It would be interesting to know if a specific hormone-receptor pathway determines the method by which a message will be terminated.

Ultrastructural studies provide knowledge of the organelles within and between cells, and are vital in aiding our understanding of how secreting and transmitting cells function intrinsically, and how they may interact. They may allow us to understand how messages are modulated. However, we also need methods which will allow us to listen to chemical conversations between cells and systems.

To obtain precise data, we are forced to be reductionists and to work in stripped down *in vitro* systems. However, these methods do not ordinarily lead to an understanding of how minute events are translated in the whole organism. Some progress in this regard is reported in this volume. However, we look to future reports for more insight in this area.

M. J. Loeb
A. B. Bořkovec

Beltsville, MD
September, 1993

CONTENTS

Reviews

Reports of Current Research

Neuroanatomy and Neural Function

Signalling Within The Nervous System

Insect Neuropeptide Research

Adipokinetic Hormone

Allatoregulatory Peptides

FMRFamide

Pheromone Biosynthesis Activating Neuropeptide

Proctolin

Peptides Affecting Development

Neuropeptide Metabolism

Receptor Studies

Reviews

Neural networks and the control of locomotion in insects

M. Burrows

Zoology Department, University of Cambridge
Downing Street, Cambridge CB2 3EJ, England

This paper analyses some of the neural mechanisms by which adaptive locomotory behaviour of an insect is produced, in order to learn about the general properties of the neurones involved, and the circuits that they form. This vast topic is reduced to more manageable proportions by concentrating on a specific question: how are sensory signals from mechanoreceptors on a leg processed in the central nervous system so that an appropriate and fine adjustment of posture or locomotion is produced? What are the networks that are involved and how are they designed for this processing? How are the constituent interneurones connected to perform these tasks and in particular what are the integrative properties of the essential local interneurones?

The leg and the nervous system chosen for study belong to a locust (*Schistocerca gregaria* (Forskål)). This insect is chosen for the design of its nervous system and for the range of behaviour that it shows. Moreover, intracellular recordings can be made from neurones whilst the animal is alert, so that the action of a particular neurone can be related directly to the behaviour of the animal. Individual neurones can be identified according to their physiological actions and properties, their morphology at the level of the light microscope, and the distribution of their synapses as seen in the electron microscope.

Local reflexes

Some 10,000 sensory axons from one hind leg converge onto the segmental, metathoracic ganglion that controls the movements of the hind legs. Many of these afferents are from mechanoreceptors that provide information about touch (exteroceptors) and about movement of the joints (proprioceptors), but many others are from chemoreceptors. By contrast, only some 100 motor neurones are responsible for generating the complex patterns of muscular contractions that are required of this leg in locomotion and in adjusting posture. Considerable convergence and integration is therefore indicated. Moreover, the spatial information provided by the receptors must be preserved because specific local reflexes can be evoked by stimulation of restricted arrays of receptors (Siegler and Burrows, 1986). For example, touching hairs on the ventral surface of the tibia causes the trochanter to levate and move the femur forwards, the tibia to extend and

0-8493-4591-X/94/$0.00 + $.50
© 1994 by CRC Press, Inc.

the tarsus to depress, whereas touching hairs a few mm away on the dorsal tibia causes a different sequence of movements of these joints. These are adaptive compensatory reflexes ensuring that the leg moves to avoid an object into which it might bump during the course of locomotion. These reflexes define the nature of the integrative problems that must be solved in the nervous system. Similarly, movements of a joint evoke reflex responses in the motor neurones innervating muscles of that and adjacent joints. In quiescent animals these inputs result in resistance reflexes that oppose the imposed movements, but in more active animals assistance reflexes result which reinforce the movement.

The neural components

Each half of the metathoracic ganglion contains all the motor neurones that move a hind leg and perhaps as many as 1,000 interneurones, of which a few hundred are involved with the control of each leg. Many are local interneurones with branches restricted entirely to this ganglion. There are 2 classes of these local interneurones, some which normally generate action potentials (spiking local interneurones) and some which normally do not (non spiking local interneurones). Others are intersegmental interneurones with axons that project to adjacent ganglia in the segmental chain. Axons from interneurones that receive sensory stimuli from other legs and other parts of the body also project to this ganglion. These are the neurones that provide the basic framework from which the local networks must be organised.

Processing the afferent signals

Tactile hairs are stimulated when a leg bumps into an external obstacle and their afferents elicit an appropriate adaptive response. Each of these hairs is innervated by a single sensory neurone whose spikes can be recorded from the cut end of a hair shaft. The terminals of these sensory neurones form a three dimensional map of the surface of the leg within an area of ventral neuropil (Newland, 1991). Which neurones process this initial barrage of sensory signals? Spiking local interneurones with cell bodies in a ventral midline (Siegler and Burrows, 1984) or antero-medial group (Nagayama, 1989) receive direct excitatory inputs from these and other afferents (Siegler and Burrows, 1983; Nagayama and Burrows, 1990). The gain of this first synapse between the afferents and the interneurones can be high so that each afferent spike can evoke a spike in the interneurone. Each interneurone receives inputs from a particular array of receptors, and these comprise its receptive field. Not all the receptors in a field contribute equally, some forming high gain synapses, others of lower gain (Burrows, 1992). These interneurones form a map of the surface of the leg, by virtue of the arrays of direct connections formed by the particular afferents. The interneurons have two fields of branches within the ganglion: on in ventral neuropil to which the afferent neurones project, and one in more dorsal neuropil where motor neurons and other interneurons have branches. The ventral branches correspond to the afferent map and to the receptive field of the interneuron (Burrows and Newland, 1993). Thus for example, an interneuron with a receptive field on the tarsus has branches only in a posterior region of neuropil to which the tarsal hair afferents project. Two important features of these receptive fields emerge; first one region of the leg is represented by several interneurones so that there is parallel processing within this class of interneurone. Second, the specificity of the connections ensures that spatial information provided by the receptors is preserved in separate channels.

The exteroceptive afferents do not, however, simply connect with these interneurones. Many also connect with intersegmental interneurones so that information is conveyed to ganglia that control the movements of the other legs (Laurent, 1988; Laurent and Burrows, 1988). A few also connect directly with non spiking interneurones (Burrows, Laurent and Field, 1988; Laurent and Burrows, 1988), but most of the input to these interneurones is first processed by the spiking local interneurones. Some afferents bypass the circuitry formed by the local interneurones and synapse directly with motor neurones (Weeks and Jacobs, 1987; Laurent and Hustert, 1988). There is therefore divergence of the afferent signals to several classes of neurone and parallel distributed processing.

Many of the same neurones and the same pathways are involved in processing the input from the proprioceptors. A major difference is that many of these afferents synapse directly on the motor neurones themselves (Burrows, 1987b). In addition to these direct pathways, parallel pathways involving the spiking and the non spiking interneurones also exist. Considerable processing also occurs in the terminals of these afferents (Burrows and Laurent, 1993). Afferents that respond to a particular movement of the joint excite an unidentified group of interneurones that then evoke depolarising inhibitory synaptic potentials in the terminals of other afferents responding to the same movement. The result is a reduced efficacy of transmission by the afferent to its postsynaptic motor neurone (Burrows and Matheson, 1993).

Output connections of the local interneurones

What happens to the signals after they have been processed by the local interneurones? The midline spiking local interneurones make inhibitory output connections (Burrows and Siegler, 1982), whereas the antero-medial interneurones make excitatory connections (Nagayama and Burrows, 1990). All the effects revealed so far are mediated by spikes, so that each spike in an interneurone is followed by either an IPSP or an EPSP in the postsynaptic neurone. Members of the midline group make a restricted number of inhibitory connections with motor neurones (Burrows and

Siegler, 1982), but more widespread connections with non spiking interneurones (Burrows, 1987a), with intersegmental interneurones (Laurent 1987b) and with antero-medial spiking interneurones (Nagayama and Burrows, 1990). Many of the members of this group of spiking local interneurones stain with an antibody raised against GABA and their inhibitory actions can be blocked by a GABA antagonist (Watson and Burrows, 1987).

Non spiking interneurones make either excitatory or inhibitory output connections, but they do so without the intervention of spikes. Their effects on postsynaptic neurones are mediated by the graded release of chemical transmitter (Burrows and Siegler, 1978). They exert profound effects on the spiking patterns of motor neurones which they organise into sets that are appropriate for the execution of normal movements of a leg (Burrows, 1980). They also make inhibitory connections with other non spiking interneurones in lateral inhibitory networks (Burrows, 1979).

Properties of the local interneurones

From this information about the two physiological classes of local interneurones the following conclusions can be drawn.

Midline spiking local interneurones

1. Collate afferent information from arrays of receptors whilst preserving spatial information.
2. Reverse the sign of the afferent signal from excitatory to inhibitory in their output connections with many other types of neurone.
3. Limit the receptive fields of non spiking and intersegmental interneurones and enhance the borders through lateral inhibition.
4. Exclude the action of non spiking interneurones whose motor effects would be inappropriate, and disinhibit those whose action would be appropriate.

Non spiking interneurones

1. Release chemical transmitter in a graded manner in response to small changes in their voltage. This means that they exert a precise control over the membrane potential and spike frequency of their postsynaptic neurones.
2. Exclude the action of interneurones with inappropriate actions for a particular movement through lateral inhibition.
3. Control groups of motor neurones in sets appropriate for normal locomotion. Each interneurone synapses on several motor neurones, and in turn each motor neurone receives inputs from several non spiking interneurones.

Adjusting the motor output

What are the key elements that have been identified so far in the circuits that are responsible for adjusting the motor output so that it is appropriate to the prevailing behavioural circumstances? Non spiking interneurones, by virtue of their connections with pools of motor neurones, their inputs from afferents and their interactions with local interneurones appear crucial for the execution of a local reflex. Manipulating the membrane potential of one of these interneurones alters the effectiveness of a local reflex in a graded manner. Therefore, despite the parallel and distributed processing, a single interneurone can play a substantial role in a reflex pathway. From this observation it would be expected that any synaptic inputs to a non spiking interneurone might be able to change a reflex, and that this might be the way that intersegmental inputs place a local response in the appropriate behavioural context. Members of a population of intersegmental interneurones with cell bodies in the mesothoracic ganglion receive inputs from a middle leg and have axons that project to the metathoracic ganglion (Laurent, 1987a). The receptive fields of these interneurones are shaped by the same series of connections as for the local interneurones: direct excitation from afferents, inhibition by spiking local interneurones (Laurent, 1987b). In the ganglion controlling the hind legs these intersegmental interneurones connect with non spiking interneurones and with some motor neurones (Laurent and Burrows, 1989a). The connections are specific and are related to the receptive field of the intersegmental interneurone and to the output connections of the non spiking interneurone. The effect of these connections is to alter the output of a non spiking interneurone and thus its participation in a local reflex (Laurent and Burrows, 1989b).

Reflex pathways

The pathways used for local reflex movements of a hind leg can thus be defined in detail that is sufficient to allow a sensory signal to be followed through its various integrative stages to its emergence as an adaptive motor response. Moreover, it is possible to pin-point some of the crucial elements in these pathways where modifications can occur. For example, the role of non spiking interneurones in adjusting the motor output and acting as the summing points for both intra- and intersegmental effects has been highlighted. The following characteristics of the networks can be recognised:

1. There is considerable convergence from the afferents to the first interneuronal layer, but spatial information is nevertheless preserved.
2. The afferent signals diverge to make excitatory connections with different classes of neurones. Therefore, parallel distributed processing of the same signals by neurones with different integrative properties occurs.

3. The signals from exteroceptive afferents also diverge to make connections with several interneurones in the same class, resulting in overlapping receptive fields. Thus the position of an afferent on the leg is not represented by one specific central neurone, but by all the interneurones whose receptive fields overlap.
4. An afferent does not connect with all the interneurones in the local network, and similarly a single interneurone does not connect with all the output elements. The connections are specific and can be understood in functional terms by their role in organising specific movements.
5. Lateral inhibitory interactions predominate between the local interneurones that have so far been identified as elements of the networks. Disinhibition contributes to the excitation of the motor neurones.
6. The flow of information is in one direction in which each interneurone has a well defined place. Feedback connections between the elements that have are so far been found are limited to those that regulate the gain of the output synapses of the proprioceptive afferents. This feedback acts like an automatic gain control that limits the gain of a sensory response in the context of the activity of other afferents responding to the same movement. The effect of a sensory signal from one afferent therefore depends on the network response of the other active afferents. Feedback amongst the local interneurones has not been found. For example, non spiking neurones presynaptic to intersegmental interneurones or to spiking local interneurones have not been found.
7. The complex receptive fields of the interneurones are understood only if their output connections are known by reference to the behaviour.

While these experiments have provided a good understanding of the types of processing performed in these local networks, the next step must be to understand how the pathways operate during a voluntary movement and to identify the elements in the networks and the processes that are responsible for this level of integration.

Acknowledgement
The work described in this paper is supported by NIH grant NS16058, by a grant from the SERC (UK) and by the Japanese Human Frontier Science Program. Many of the experiments described have been carried out jointly with Drs M.V.S. Siegler, G.J. Laurent, A.H.D. Watson, P.L. Newland and T. Matheson.

REFERENCES

Burrows, M. Graded synaptic transmission between local premotor interneurons of the locust. *J. Neurophysiol.*, 42, 1108-1123, 1979.

Burrows, M. The control of sets of motoneurones by local interneurones in the locust. *J. Physiol.*, 298, 213-233, 1980.

Burrows, M. Inhibitory interactions between spiking and non spiking local interneurones in the locust. *J. Neurosci.*, 7, 3282-3292, 1987a.

Burrows, M. Parallel processing of proprioceptive signals by spiking local interneurones and motor neurones in the locust. *J. Neurosci.*, 7, 1064-1080, 1987b.

Burrows, M. Reliability and effectiveness of transmission from exteroceptive sensory neurones and spiking local interneurons in the locust. *J. Neurosci.* 12, 1477-1489, 1992.

Burrows, M. and Laurent, G. Synaptic potentials in the central terminals of locust proprioceptive afferents generated by other afferents form the same sense organ. *J. Neurosci.*, 13, 808-819, 1993.

Burrows, M. and Matheson, T. A presynaptic gain control mechanism among sensory neurones of a locust leg proprioceptor. *J. Neurosci.*, (in press) 1993.

Burrows, M. and Newland, P. L. Correlation between the receptive fields of locust interneurons, their dendritic morphology and the central projections of mechanosensory neurones. *J. Comp. Neurol.*, 329, 412-440, 1993.

Burrows, M., Laurent, G. J. and Field, L. H. Proprioceptive inputs to non spiking local interneurones contribute to local reflexes of a locust hindleg. *J. Neurosci.*, 8, 3085-3093, 1988.

Burrows, M. and Siegler, M. V. S. Graded synaptic transmission between local interneurones and motoneurones in the metathoracic ganglion of the locust. *J. Physiol.*, 385, 231-255, 1978.

Burrows, M. and Siegler, M. V. S. Spiking local interneurons mediate local reflexes. *Science (N. Y.)*, 217, 650-652, 1982.

Laurent, G. The morphology of a population of thoracic intersegmental interneurones in the locust. *J. Comp. Neurol.*, 256, 412-429, 1987a.

Laurent, G. The role of spiking local interneurones in shaping the receptive fields of intersegmental interneurones in the locust. *J. Neurosci.*, 7, 2977-2989, 1987b.

Laurent, G. Local circuits underlying excitation and inhibition of intersegmental interneurones in the locust. *J. Comp. Physiol.* 162, 145-157, 1988.

Laurent, G. J. and Burrows, M. Direct excitation of nonspiking local interneurones by exteroceptors underlies tactile reflexes in the locust. *J. Comp. Physiol.*, 162, 563-572, 1988.

Laurent, G. J. and Hustert, R. Motor neuronal receptive fields delimit patterns of activity during locomotion of the locust. *J. Neurosci.*, 8, 4349-4366, 1988.

Laurent, G. and Burrows, M. Distribution of intersegmental inputs to non spiking local interneurones and motor neurones in the locust. *J. Neurosci.*, 9, 3019-3029, 1989a.

Laurent, G. and Burrows, M. Intersegmental interneurones can control the gain of reflexes in adjacent segments by their action on non spiking local interneurones. *J. Neurosci.*, 9, 3030-3039, 1989b.

Nagayama, T. Morphology of a new population of spiking local interneurones in the locust metathoracic ganglion. *J. Comp. Neurol.*, 283, 189-211, 1989.

Nagayama, T. and Burrows, M. Input and output connections of an autero-medial group of spiking local interneurones in the metathoracic ganglion of the locust. *J. Neurosci.* 10, 785-794, 1990.

Newland, P. L. Morphology and somatotopic organisation of the central projections of afferents from tactile hairs on the hind leg of the locust. *J. Comp. Neurol.*, 312, 493-508, 1991.

Siegler, M. V. S. and Burrows, M. Spiking local interneurons as primary integrators of mechanosensory information in the locust. *J. Neurophysiol.*, 50, 1281-1295, 1983.

Siegler, M. V. S. and Burrows, M. The morphology of two groups of spiking local interneurones in the metathoracic ganglion of the locust. *J. Comp. Neurol.*, 224, 463-482, 1984.
Siegler, M. V. S. and Burrows, M. Receptive fields of motor neurones underlying local tactile reflexes in the locust. *J. Neurosci.*, 6, 507-513, 1986.
Watson, A. H. D. and Burrows, M. Immunocytochemical and pharmacological evidence for GABAergic spiking local interneurones in the locust. *J. Neurosci.*, 7, 1741-1751, 1987.
Weeks, J. C. and Jacobs, G. A. A reflex behavior mediated by monosynaptic connections between hair afferents and motoneurons in the larval tobacco hornworm. *J. Comp. Physiol.*, 160, 315-329, 1987.

GENETIC AND MOLECULAR ANALYSIS OF POTASSIUM CHANNELS IN *DROSOPHILA*

Barry Ganetzky, Jeffrey W. Warmke, Gail Robertson, Nigel Atkinson and Rachel Drysdale

Laboratory of Genetics, University of Wisconsin, Madison, WI 53706

I. INTRODUCTION

A. A MUTATIONAL APPROACH TO ION CHANNEL GENES IN *DROSOPHILA*

The behavior of insects, like other higher organisms, requires the function of the nervous system for sensory input, integration, and motor output. The function of the nervous system in turn depends upon the membrane properties of individual neurons enabling them to receive, process and transmit information in the form of electrical impulses. Ultimately, the distinctive signalling capabilities of neurons are determined by the activity of ion channels (Hille 1991). These transmembrane proteins form gated, aqueous pores in the cell membrane and mediate the selective diffusion of various ions across the lipid bilayer. The regulated ion fluxes underlie the propagation of action potentials and synaptic transmission. Hence, to understand behavior and the function of the nervous system at the molecular level, it is necessary to understand the molecular properties of ion channels, including their structure, function and regulation.

We have used *Drosophila melanogaster* as an experimental system for molecular studies of ion channels (Wu and Ganetzky 1992). In addition to lending itself to various kinds of electrophysiological analyses, *Drosophila* is particularly well-suited for such studies because of the powerful genetic and molecular techniques that can be applied to this organism (Rubin 1988). Thus, in our studies we utilize a multidisciplinary approach combining genetics, electrophysiology, and molecular biology. The basic premise of our approach is that in a genetically amenable organism such as *Drosophila*, it should be possible to identify the genes that encode ion channels or that otherwise affect their function or regulation by obtaining mutations affecting these genes. The main advantage of this type of approach is that it enables identification of the relevant genes and their encoded polypeptides in the absence of any prior biochemical information, which is often lacking or difficult to obtain for ion channels. This strategy has proven to be quite successful and has led to the molecular identification of a sodium channel structural gene (Loughney et al. 1989) as well as several potassium channel structural genes (Temple et al. 1987; Warmke et al. 1991; Atkinson et al. 1991). Here, we review our recent work on potassium channel genes.

B. AN EXTENSIVE VARIETY OF POTASSIUM CHANNELS SHAPE NEURONAL ACTIVITIES

It is known from electrophysiological studies that potassium channels

0-8493-4591-X/94/$0.00 + $.50

© 1994 by CRC Press, Inc.

comprise a large and diverse group that differ in their kinetics, gating, pharmacology and other properties (Rudy 1988). Potassium channels are involved in membrane repolarization and thereby largely determine the rate and pattern of action potentials generated by different neurons, the duration of individual action potentials and the amount and time course of transmitter release. The modification of potassium channel activity in response to various hormones, neurotransmitters, and second messengers can modify the output of a given neuron or network. Potassium channels and their regulation thus exert a strong influence on synaptic plasticity, including the mechanisms involved in learning and memory.

C. THE *SHAKER* LOCUS PROVIDED THE FIRST MOLECULAR HANDLE FOR POTASSIUM CHANNEL GENES

Until recently, the lack of suitable affinity ligands precluded the biochemical purification of potassium channels for molecular studies. The molecular analysis of potassium channels was initiated with the cloning of the *Drosophila Shaker* (*Sh*) locus, which was shown to encode a family of rapidly inactivating potassium channels (Timpe et al. 1988). Subsequently, three additional potassium channel genes, *Shaw*, *Shab* and *Shal*, were cloned from *Drosophila* on the basis of their homology with *Sh* (Salkoff et al. 1992). The polypeptides encoded by all of these genes (Figure 1) contain a hydrophobic core comprising six putative membrane-spanning domains (S1-S6) and a segment (P) between S5 and S6 that forms a hairpin loop within the membrane (reviewed by Miller 1991). The S4 segment is involved in the voltage-sensing mechanism of the channel and the P segment lines the channel pore and is involved in ion selectivity and permeation. The polypeptides encoded by these genes have been functionally expressed in *Xenopus* oocytes where they apparently assemble as homotetramers to form voltage-activated, potassium-selective channels. Homologues of each of these potassium channel genes in *Drosophila* have now been cloned from a wide variety of organisms including other invertebrates as well as vertebrates (Rudy et al. 1991).

D. RELATIVES OF *SHAKER* MISSED IN HOMOLOGY SCREENS MAY STILL BE FOUND VIA GENETIC STRATEGIES

Because the cloning of potassium channel genes by homology screens has favored the isolation of those with substantial similarity to members of the *Sh* family, other potassium channels that are only distant relatives would likely be missed in such screens. For example, genes encoding calcium-activated potassium channels were not among those recovered as *Sh* homologues. For this reason, as an alternative strategy that made no assumptions about sequence conservation, we continued to pursue the molecular analysis of other mutations in *Drosophila* that we had identified as having defects in potassium channels. The analysis of two of these mutants, *slowpoke* (*slo*) and *ether `a go-go* (*eag*) has led to the identification of two previously uncharacterized potassium channel structural genes.

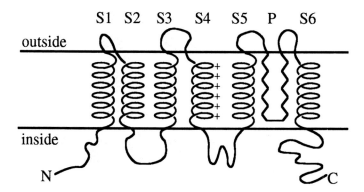

Figure 1. Diagrammatic representation of a proposed transmembrane-folding model for voltage-activated potassium channel polypeptides. Seven hydrophobic segments are shown, six of which (S1-S6) are thought to form alpha-helices that span the membrane and one of which (P) is thought to form a beta-hairpin loop within the membrane. The S4 segment has a repeating motif consisting of a positively charged amino acid at every third position spaced by two hydrophobic amino acids and functions as the voltage sensor. The P segment corresponds to the lining of the channel pore. Functional voltage-activated potassium channels are thought to be a tetrameric assembly of such subunits. (Based on a figure in Miller 1991)

II.   RESULTS

A.   GENETIC AND MOLECULAR ANALYSIS OF CALCIUM-ACTIVATED POTASSIUM CHANNELS

1.   Potassium currents in *Drosophila* muscles are affected by different mutations

Voltage-clamp studies of adult flight muscles and larval body wall muscles revealed the presence of at least four different potassium currents (Salkoff and Wyman 1983; Singh and Wu 1989). These include two voltage-activated potassium currents, the fast I_A and the delayed I_K, and two calcium-activated potassium currents, the fast I_{CF} and the delayed I_{CS}. Although mutations of the *Sh* locus completely eliminated the I_A current, they had no effect on I_{CF} indicating that the channels mediating these currents were encoded by different genes (Salkoff 1983; Wu and Haugland 1985). The discovery and analysis of mutations of the *slowpoke* locus (*slo*) identified it as a candidate for encoding a component of calcium-activated potassium channels.

2.   The *slo* mutation eliminates a calcium-activated potassium current

The *slo* mutation was recognized on the basis of the sluggish and uncoordinated phenotype that it conferred (Elkins et al. 1986). Intracellular recordings indicated that the repolarization of action potentials in *slo* adult and larval muscles was about ten times slower than normal (Figure 2A). Voltage-

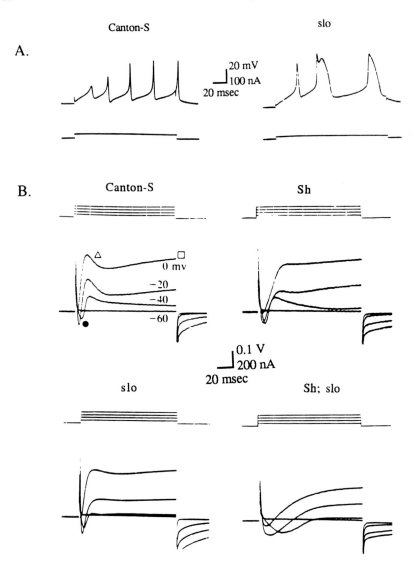

Figure 2. Electrophysiological analysis of *slo*. A. Action potentials evoked in adult flight muscle fibers of wild-type (left) and *slo* (right) flies in response to constant current injection. Note the prolonged duration of action potentials in *slo* mutants. (Reproduced with permission from Elkins et al. 1986). B. Membrane currents in wild-type (Canton-S) and mutant adult flight muscle fibers. The lower traces in each panel represent currents elicited by voltage steps from a holding potential of -80mV to the indicated potentials (upper traces). In wild-type (upper left), an inward calcium current (filled circle) is evoked at -40mV followed by an inactivating outward potassium current (open triangle). At -20mV a late outward potassium current (open square) that does not inactivate during the voltage pulse

first appears. The early inactivating potassium current is separable into the voltage dependent I_A, which is removed by the Sh^{KS133} mutation (upper right), and the calcium dependent I_{CF}, which is removed by the *slo* mutation (lower left). Thus, in *Sh; slo* double mutants (lower right) no early outward current is present. Temperature for these recordings was 4°C. (Adapted from Elkins et al. 1986).

clamp studies demonstrated that the underlying defect was a complete and specific elimination of I_{CF} in these cells (Figure 2B). Because of the specificity of its defect, the *slo* mutation was useful in dissecting the functional contributions of the various potassium currents in *Drosophila* to repolarization of muscles and presynaptic terminals. For example, I_{CF}, rather than I_A, was found to have the major role in repolarization of action potentials in adult flight muscles (Elkins and Ganetzky 1988). I_{CF} also is important in repolarization of presynaptic motor terminals as indicated by the alteration in evoked synaptic currents in *slo* mutants (Gho and Ganetzky 1992).

Although the electrophysiological experiments clearly indicated that the function of calcium-activated potassium channels was defective in *slo* mutants, it was not possible to determine from these experiments whether *slo* encoded a structural component of these channels or affected their expression or function in some other way. To address this question it was necessary to carry out a molecular analysis of *slo* and to identify its encoded polypeptide.

3. Cytological mapping and cloning of *slo*

The gene was first pinpointed cytologically on the polytene chromosomes by generating new gamma-ray induced mutations on the basis of their failure to complement the original allele (Atkinson et al. 1991). Three of these new mutations were found by cytological examination to be associated with chromosome rearrangements that shared a breakpoint at polytene chromosome band 96A17, identifying it as the location of *slo*. One of these rearrangments, *slo⁴*, was an inversion with one breakpoint in *slo* and the other near another gene that had been previously cloned.

Beginning with DNA probes from the cloned region, a chromosome walk was carried out to reach the neighboring *slo⁴* breakpoint. A cloned DNA fragment that spanned the breakpoint was then used to isolate genomic clones from the *slo* locus at the other end of the inversion. Additional genomic clones from the *slo* region were obtained by chromosome walking. The *slo⁴* breakpoint was mapped on the genomic DNA by Southern blot analysis and chromosomal in situ hybridization to provide an approximate location of the *slo* transcription unit. Subsequently, genomic probes from this region were used to screen a head cDNA library. Several incomplete but overlapping cDNAs were isolated and shown to represent the *slo* transcript by chromosomal in situ hybridization and Northern blot analysis. The composite cDNA represented a spliced transcript that encompassed over 40kb of genomic DNA. The composite cDNA contained a single long open reading frame (ORF) encoding a polypeptide 1184 amino acids in length (Atkinson et al. 1991). Fifteen amino acids at the amino terminal end of the *slo* polypeptide were not included in this composite cDNA (Adelman et al. 1992).

4. Sequence analysis of *slo*

Although initial examination of the deduced *slo* polypeptide did not detect any striking similarity with other protein sequences in computer databases, hydropathy analysis indicated the presence of seven hydrophobic peaks near the amino terminus, as found previously for members of the *Sh* family (Atkinson et al. 1991). Within these hydrophobic segments, significant amino acid sequence similarity could be found between *slo* and a *Sh*-family consensus sequence. The most highly conserved region is the sixth hydrophobic segment, corresponding to the pore domain, in which there are 11 identities and five conservative substitutions in a region of 25 amino acids (Figure 3). Within this region is a segment of 12 amino acids that contains nine identities and three conservative substitutions. There is also good conservation of the fourth hydrophobic segment, corresponding to the S4 domain (Figure 3). Across all the hydrophobic segments taken together, there are 25% amino acid identities and 20% conservative substitutions when the *slo* and *Sh* sequences are aligned. The similarity between the *slo* polypeptide and members of the *Sh* family of voltage-activated potassium channels clearly indicates that they are derived from a common ancestor. This similarity together with the observation that *slo* mutations eliminate a calcium-activated potassium current in vivo led to the conclusion that *slo* encodes a structural component of calcium-activated potassium channels. The functional expression of calcium-activated potassium channels in *Xenopus* oocytes following the injection of *slo* mRNA transcribed in vitro has confirmed this conclusion (Adelman et al. 1992).

5. The *slo* polypeptide defines a new class of potassium channels

Despite the similarity of *slo* to members of the *Sh* family of potassium channels, it does not appear to be simply another member of this family. Known members of this family in *Drosophila*, including *Sh*, *Shaw*, *Shab*, and *Shal* share about 40% amino acid identities across their hydrophobic cores (Salkoff et al. 1992). In contrast, *slo* shares less than 20% identity with any of these polypeptides. These results indicate that the evolutionary split between calcium-activated and voltage-activated potassium channel genes preceded the expansion of the latter group. *slo* thus represents a distinct type of potassium channel polypeptide not previously cloned in any other species.

B. GENETIC AND MOLECULAR ANALYSIS OF A NOVEL FAMILY OF POTASSIUM CHANNELS

1. Discovery of *eag* as a membrane hyeprexcitability mutation

eag was originally discovered on the basis of its ether-sensitive leg-shaking phenotype (Kaplan and Trout 1969). More than ten years later, *eag* was rediscovered on the basis of a striking synergistic interaction with *Sh* in double mutants (Ganetzky and Wu 1983). Electrophysiological experiments revealed that *eag* mutations cause a high frequency of spontaneous action potentials in larval nerves that triggered synaptic potentials whose amplitude and duration were longer than normal (Figure 4). Long trains of action potentials, associated with synaptic potentials of very large amplitude and whose duration was tenfold greater than in *eag* or *Sh* alone, were observed in *eag Sh* double mutants. These results indicated that loss of *eag* function resulted in a phenotype of neuronal hyperexcitability. This phenotype and the synergistic interaction of *eag* with *Sh* in double mutants could be explained by

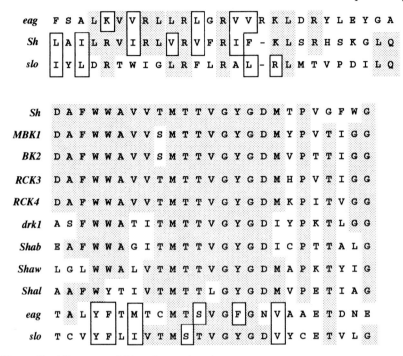

Figure 3. Alignment of S4 and pore domains encoded by *slo* and *eag* cDNAs with the corresponding segments of members of the *Sh* family of potassium channels. Top: Alignment of S4 domains with *Sh*. Amino acid identities with the *Sh* sequence are shaded. Conservative substitutions relative to the *Sh* sequence are boxed. A single amino acid gap has been introduced into the *slo* and *Sh* sequence relative to the *eag* sequence to maximize alignment. Bottom: Alignment of the pore domains with members of the *Sh* family cloned from *Drosophila* and mammals. If four or more members of the *Sh* family contained the same amino acid at a given position, that amino acid was defined as the consensus residue at that position and shaded. The *eag* and *slo* sequences are shaded where there is an identity with the consenus sequence. Boxed amino acids correspond to conservative substitutions relative to the consensus sequence.

supposing that *eag* mutations perturb one or more potassium currents besides the one affected by *Sh*. Voltage-clamp studies showing that the amplitudes of four different potassium currents were reduced in larval muscles of various *eag* mutations provided direct evidence for an effect of this gene on potassium channels (Zhong and Wu 1991).

2. Cloning and sequence analysis of *eag*

Molecular analysis of *eag* was undertaken to identify the encoded polypeptide and to elucidate its effect on potassium currents. Following the

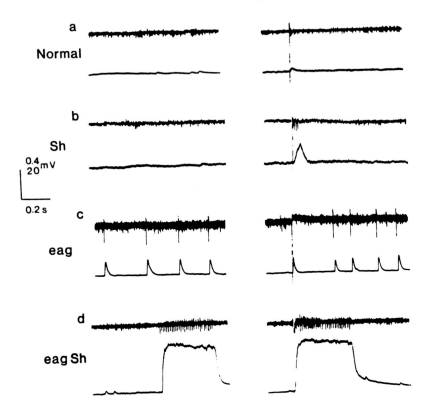

Figure 4. Simultaneous nerve and muscle recordings from normal (Canton-S), *Sh, eag* and *eag Sh* double mutant larvae. Activities evoked by nerve stimulation (right panels) and occurring spontaneously in the absence of stimulation (left panels) are shown. The motor axon spikes are the most prominent units that can be picked up by the suction electrode from segmental nerves. At low external calcium (0.2mM for a and c; 0.1mM for b and d), nerve stimulation evokes enhanced neuromuscular transmission in mutants as compared with normal larvae. In *Sh* larvae, a single nerve stimulus causes the motor axon to fire several extra spikes. Repetitive firing of motor axons, correlated with synaptic potentials, occurs spontaneously in *eag. eag Sh* double mutants display striking synergistic effects, resulting in greatly prolonged synaptic potentials associated with bursts of motor axon spikes. (Adapted from Ganetzky and Wu 1985. With permission).

cytological mapping of *eag* to polytene region 13A, genomic DNA from this region was cloned by chromosome jumping and walking (Drysdale et al. 1991). Genomic probes from this region were then used to isolate cDNAs corresponding to the *eag* transcript. Sequence analysis of a 4.0kb cDNA containing a complete ORF indicated that the encoded polypeptide was 1174 amino acids in length (Warmke et al. 1991).

Although computer analysis of this amino acid sequence did not initially reveal any strong similarities with known polypeptide sequences, hydropathy analysis again indicated the presence of seven hydrophobic domains. Each hydrophobic domain showed significant similarity when aligned with the corresponding segment of a consensus sequence from the *Sh* family. As in the case of *slo*, the alignment was particularly striking in S4 (nine identities and five conservative substitutions out of 26 amino acids) and in the putative pore region (seven identities and six conserved substitutions out of 25 amino acids) (Figure 3). Overall, the membrane-spanning regions of *eag* shared 25% amino acid identities and 20% conserved substitutions with a *Sh*-family consensus sequence. The sequence similarity of the *eag* polypeptide to other known potassium channel subunits, along with the electrophysiological phenotype of *eag* mutations led to the conclusion that *eag* encodes a structural component of potassium channels.

3. Relationship of *eag* to cyclic nucleotide-gated channels

Despite the similarities of *eag* with members of the *Sh* family, the extent of its divergence from all known members of this family indicates that *eag* represents a distinct type of channel polypeptide. In fact, subsequent analysis of the amino acid sequence revealed that the hydrophobic core of *eag* is slightly more similar to the cyclic nucleotide-gated cation channels in vertebrate photoreceptors and olfactory neurons ($\approx$21% identities) than to voltage-gated potassium channels in the *Sh* family (<20% identities). Moreover, a putative cyclic-nucleotide-binding domain (cNBD), not found in any member of the *Sh* family of potassium channel polypeptides, is present in the carboxy terminal half of *eag* (Guy et al. 1991). Thus, *eag* appears to be a novel channel polypeptide related both to voltage-gated potassium channels in the *Sh* family and to cyclic nucleotide-gated cation channels.

4. Expression of *eag* in *Xenopus* oocytes

To confirm our prediction that *eag* encoded a channel polypeptide and to characterize the properties of these channels, we injected *eag* RNA transcribed in vitro from a cDNA into *Xenopus* oocytes (Robertson et al. 1993). Two electrode voltage-clamp recordings from injected oocytes demonstrated expression of a voltage-dependent outward current that activated quickly and decayed slowly (Figure 5). The channels are selective for potassium over sodium as judged by measurements of the reversal potential of tail currents with ion substitution. Experiments to determine whether cyclic nucleotides are important in the activation or modulation of these channels are underway. So far, at least one site-directed mutation that alters an amino acid in the cNBD thought to be essential for binding cyclic nucleotides appears to eliminate channel activity. Additional experiments with excised inside-out patches are necessary to determine directly how the level of exogenous cyclic nucleotides affects the activity of wild-type channels. Although cyclic-nucleotide-gated potassium channels have been identified in vivo in patch-clamp studies of larval muscles (Delgado et al. 1991), the properties of these channels do not appear to correspond precisely with the *eag* channels expressed in oocytes. It is possible that *eag* channels have not yet been detected in vivo because they have not been looked for in the appropriate cells or by using the correct protocols. Alternatively, *eag* may not assemble as a homomultimeric channel in vivo, as it appears to in *Xenopus* oocytes, and may thus form channels whose properties differ from those expressed in oocytes. Subsequent experiments are needed to distinguish these possibilities.

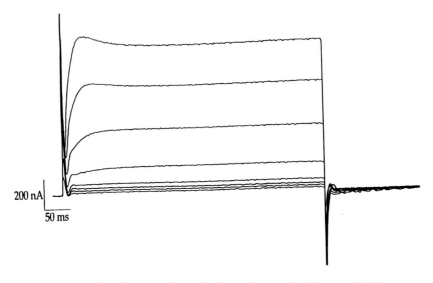

Figure 5. Functional expression of channels encoded by *eag* in *Xenopus* oocytes. Oocytes were injected with 50nl of RNA (0.05mg/ml) transcribed in vitro from an *eag* cDNA template. Currents were recorded after incubating the oocytes for 2 days. The traces show the outward currents recorded by standard two electrode voltage clamp in response to 10mV depolarizing pulses between -70 and 0mV.

5. A conserved family of *eag*-related potassium channel genes

By analogy with the identification of a family of *Sh*-related potassium channel genes, we reasoned that *eag* could be the prototype of a new family of potassium channel genes. Screens of head-specific cDNA libraries with *eag* probes resulted in the identification of another *Drosophila* gene, *elk* (*eag*-like K channel), with homology to *eag*. cDNAs containing a complete ORF (1284 amino acids) for *elk* have been sequenced in full. Across the entire sequence *elk* shares over 30% amino acid identities and another 32% conserved substitutions with *eag*. Within the region encompassing the hydrophobic core and cNBD of these polypeptides the amino acid identity is about 43%. These results demonstrate that the similarity between *eag* and *elk* is much greater than that between either of these genes and any member of the *Sh* family. It thus appears that at least two different families of potassium channel genes exist in *Drosophila*. It will be of interest to determine the size of the *eag* family in *Drosophila* as well as the respective functions of the different family members.

C. USE OF DROSOPHILA PROBES TO ISOLATE CHANNEL GENES FROM OTHER SPECIES

As the results with *eag* and *slo* demonstrate, the genetic tools available in

Drosophila can enable the cloning of ion channel genes that were not previously identified or accessible in other organisms. Sequence comparisons of channel genes isolated from a variety of species reveal that the amino acid sequences of channel polypeptides often exhibit significant evolutionary conservation. Consequently, it is possible to use probes from channel genes cloned in *Drosophila* to isolate homologous counterparts from other species, including vertebrates as well as invertebrates. For example, after the *para* sodium channel gene was cloned in *Drosophila*, primers based on the sequence of this gene were used to clone the corresponding gene from several other insects including *Musca* and *Heliothis*, which has made it possible to test whether pyrethroid resistance maps to the sodium channel gene in these species (Doyle and Knipple 1991; Taylor et al. 1993).

Similarly, we have used probes from the *slo* cDNA to isolate the *Aedes* homologue of this gene (L. Pallanck and B. Ganetzky unpublished). These two genes were found to share greater than 90% amino acid identities from the amino terminus through the pore region. Recently, we have been able to clone *slo* homologues from mouse and human as well (L. Pallanck and B. Ganetzky unpublished). The *Drosophila* and mammalian genes share about 70% amino acid identities indicating strong conservation of the sequence from insects to mammals.

The *eag* sequence is also highly conserved from insects to mammals. Homologues of *eag* have been cloned from mouse and humans (Warmke and Ganetzky 1993). Sequence comparisons among these genes suggest the existence of several distinct subfamilies of *eag*-related genes in both insects and mammals. For example, the greatest similarity in amino acid sequence in any pairwise combination is found between *eag* the mouse homologue (*m-eag*) rather than between *eag* and *elk* or between the mouse and human genes. Thus, *eag* and *m-eag* appear to define members of one subfamily. *elk* may represent the founding member of another subfamily and is likely to have mammalian counterparts more closely related to it than to *eag*. Finally, *h-erg*, which is equally diverged from both *eag* and *elk* may define a third subfamily.

III. PERSPECTIVES

The work reviewed here demonstrates that the application of a genetic strategy in *Drosophila* can be incisive in studies of ion channels and related polypeptides involved in neural signalling. Beginning with the observation of a mutant behavioral phenotype it is possible to proceed in systematic fashion to the molecular isolation of the affected gene and the identification of the encoded polypeptide. This has enabled the isolation of several genes specifying ion channels that have not previously been isolated from any other organism. The availability of these clones in *Drosophila* has now made it possible to isolate the corresponding genes from other insects and and also from mammals. Thus, such studies in *Drosophila* can provide the information and molecular tools that can be used in neurobiological studies in other insects as well as help elucidate basic neurobiological principles that are common to all higher organisms.

An increasing number of channel genes are now being identified on the basis of sequence similarity to previously cloned channel genes. Examples include *elk* as well as members of the *Sh* family, *Shaw*, *Shal*, and *Shab*. Although much can be learned about the properties of these channel proteins from in vitro mutagenesis and expression in heterologous systems, many questions remain about the particular in vivo functions of these polypeptides. Phenotypic characterization of mutations in these genes will be of critical importance in addressing these questions. Fortunately, techniques for site-selected mutagenesis have been developed in *Drosophila* that enable the selective recovery of mutations in target genes of interest (Ballinger and Benzer 1989; Kaiser and Goodwin 1990). Thus, it is now possible in *Drosophila* to utilize a "molecules to mutants" approach to complement the traditional "mutants to molecules" strategy. This will further enhance the power and importance of *Drosophila* as an experimental system for studies of ion channels.

Finally, although much has been learned in recent years about the structure and function of an increasing number of ion channels, many questions remain concerning the molecular mechanisms that regulate expression, assembly, cellular localization, and modulation of channels in vivo. By continuing to screen for appropriate behavioral and electrophysiological phenotypes, it should be possible to isolate mutations in *Drosophila* that affect these processes. These mutations will provide starting points to elucidate essential aspects of channel regulation. It is therefore reasonable to expect that *Drosophila* will continue to play a leading role in molecular studies of neural signalling for the forseeable future.

ACKNOWLEDGMENTS

Support for work describe here was provided by grants from the National Institutes of Health, the Markey Charitable Trust, the Muscular Dystrophy Association and by a Klingenstein Fellowship and a McKnight Neuroscience Award to B. G.

REFERENCES

Adelman, J. P., Shen, K.-Z., Kavanaugh, M. P., Warren, R. A., Wu, Y.-N., Lagrutta, A., Bond, C. T. and North, A., Calcium-activated potassium channels expressed from cloned complemnetary DNAs, *Neuron*, 9, 209-216, 1992. .

Atkinson, N., Robertson, G. and Ganetzky, B., A structural component of calcium-activated potassium channels encoded by the *Drosophila slo* locus, *Science* 253, 551-555, 1991.

Ballinger, D. G. and Benzer, S., Targeted gene mutations in *Drosophila*, *Proc. Natl. Acad. Sci. USA*, 86, 9402-9406, 1989.

Delgado, R., Hidalgo, P., Diaz, F., Latorre, R. and Labarca, P., A cyclic AMP-activated K^+ channel in *Drosophila* larval muscle is persistently activated in dunce. *Proc. Natl. Acad. Sci. USA*, 88, 557-560, 1991.

Doyle, K. E. and Knipple, D. C., PCR-based phylogenetic walking: isolation of *para*-homologous sodium channel gene sequences from seven insect species and an arachnid, *Insect Biochem.*, 21, 689-696, 1991.

Drysdale, R. A., Warmke, J. W., Kreber, R. and Ganetzky, B., Molecular characterization of *eag*, a gene affecting potassium channels in *Drosophila, Genetics* 127, 497-505, 1991.

Elkins, T., Ganetzky, B. and Wu, C. F., A *Drosophila* mutation that eliminates a calcium-activated potassium current, *Proc. Natl. Acad. Sci. USA* 83,8415-8419, 1986.

Elkins, T. and Ganetzky, B., The roles of potassium currents in *Drosophila* flight muscles, *J. Neurosci.* 8, 428-434, 1988.

Ganetzky, B. and Wu, C.-F., Neurogenetic analysis of potassium currents in *Drosophila*: Synergistic effects on neuromuscular transmission in double mutants, *J. Neurogenet.*, 1, 17-28, 1983.

Ganetzky, B. and Wu, C.-F., Genes and membrane excitability in *Drosophila, Trends NeuroSci.*, 8, 322-326, 1985.

Gho, M., McDonald, K., Ganetzky, B. and Saxton, W., Effect of kinesin mutations on neuronal functions, *Science*, 258, 313-316, 1992.

Guy, H. R., Durell, S. R., Warmke, J., Drysdale, R. and Ganetzky, B., Similarities in amino acid sequences of *Drosophila eag* and cyclic nucleotide-gated channels, *Science*, 254, 730, 1991.

Hille, B., *Ionic Channels of Excitable Membranes*, 2nd. Ed., Sinaur, Sunderland, 1991.

Kaiser, K. and Goodwin, S. F., Site-selected transposon mutagenesis of *Drosophila. Proc. Natl. Acad. Sci. USA*, 87, 1686-1690, 1990.

Kaplan, W. D. and Trout, W. E. III., The behavior of four neurological mutants of *Drosophila, Genetics*, 61, 399-409, 1969.

Loughney, K., Kreber, R. and Ganetzky, B., Molecular analysis of the *para* locus, a sodium channel gene in *Drosophila, Cell*, 58, 1143-1154, 1989.

Miller, C., Annus mirabilis of potassium channels, *Science*, 252, 1092-1096, 1991.

Robertson, G. A., Warmke, J. W. and Ganetzky, B., Functional expression of the Drosophila *EAG* K$^+$ channel gene, *Biophys. J.*, 64(2), 340 (abstr.), 1993.

Rubin, G., *Drosophila* as an experimental organism, *Science*, 240, 1453-1459.

Rudy, B., Diversity and ubiquity of K channels, *Neuroscience*, 25, 729-749, 1988.

Rudy, B., Kentros, C. and Vega-Saenz de Miera, E., Families of potassium channel genes in mammals: Toward an understaniding of the molecular basis of potassium channel diversity, *Mol. Cell. Neurosci.*, 2, 89-102, 1991.

Salkoff, L., *Drosophila* mutations reveal two components of fast outward current. *Nature Lond.*, 302, 249-251, 1983.

Salkoff, L., Baker, K., Butler, A., Covarrubias, M., Pak, M. D. and Wei, A., An essential "set" of K$^+$ channels conserved in flies, mice and humans, *Trends Neurosci.*, 15, 161-166, 1992.

Salkoff, L. and Wyman, R., Ion currents in *Drosophila* flight muscles. *J. Physiol. Lond.*, 337, 687-708, 1983.

Singh, S. and Wu, C.-F., Complete separation of four potassium currents in *Drosophila, Neuron*, 2, 1325-1329, 1989.

Taylor, M. F. J., Heckel, D. G., Brown, T. M., Krietman, M. E. and Black, B., Linkage of pyrethroid insecticide resistance to a sodium channel locus in the tobacco budworm, Insect Biochem. Mol. Biol., In press, 1993.

Temple, B. L., Papazian, D. M., Schwarz, T. L., Jan, Y. N. and Jan, L. Y., Sequence of a probable potassium channel component encoded at *Shaker* locus of *Drosophila, Science*, 237, 770-775, 1987.

Timpe, L. C., Jan, Y. N. and Jan, L. Y., Four cDNA clones from the *Sh* locus of *Drosophila* induce kinetically distinct A-type potassium currents in *Xenopus* oocytes, *Neuron*, 1, 659-667, 1988.

Warmke, J., Drysdale, R. and Ganetzky, B., A distinct potassium channel polypeptide encoded by the *Drosophila eag* locus, *Science*, 252, 1560-1562, 1991.

Warmke, J. W. and Ganetzky, B. A novel potassium channel gene family: *EAG* homologs in *Drosophila*, mouse and human. *Biophys. J.*, 64(2), 340, 1993.

Wu, C.-F. and Ganetzky, B., Neurogenetic studies of ion channels in *Drosophila*, in, *Ion Channels*, Vol. 3, T. Narahashi, Ed., Plenum Press, New York, 1992, chap. 9.

Wu, C.-F. and Haugland, F. N., Voltage-clamp analysis of membrane currents in *Shaker* mutants, *J. Neurosci.*, 5, 2626-2640, 1985.

Zhong, Y. and Wu, C.-F., Alteration of four identified K^+ currents in *Drosophila* muscle by mutations in *eag*, *Science*, 252, 1562-1564, 1991.

BIOGENIC AMINES IN INSECTS

Roger G.H. Downer and Laszlo Hiripi

Univ. of Waterloo, Waterloo, Ontario, Canada N2L 3G1

I. INTRODUCTION

Octopamine (AO), dopamine (DA), tyramine (TA) and 5-hydroxytryptamine (5HT) have been identified and proposed as putative neurotransmitters, neurohormones and/or neuromodulators in the central nervous system of insects. The current chapter provides a brief overview of the metabolism, physiological roles and pharmacology of these biogenic amines with particular emphasis on recent studies from our respective laboratories. Aspects of the molecular biology of the aminergic system in insects will be covered in other contributions to the conference and, therefore, this important component of current research on biogenic amines will not be included in the present chapter. Much background information, early literature and experimental detail are available in several reviews and the reader is referred to these for a more inclusive consideration of the field [Brown and Nestler, 1985; Evans, 1985; David and Coulon, 1985; Downer, 1990].

II. DISTRIBUTION AND METABOLISM

Monoamine levels have been determined throughout the nervous system of several species using a variety of analytical procedures [Downer et al., 1985]. Typical values for *Periplaneta americana* and *Locusta migratoria* were obtained using high performance liquid chromatography with electrochemical detection and demonstrate clearly that the ratios of specific monoamines to each other vary between tissues (Table 1) [Evans, 1978; Martin et al., 1984; Downer et al., 1993]. These values are supported by results obtained using the more sophisticated analytical technique of gas chromatography-negative ion chemical ionization mass spectroscopy [Shafi et al., 1989]. Immunohistochemical studies using specific antibodies against DA [Viellemaringe et al., 1984; Milton et al., 1991], 5HT [Davis, 1987] and OA [Konlags et al., 1988; Eckert et al., 1992; Stevenson et al., 1992] have provided tentative identification of dopaminergic, serotonergic and octopaminergic neurons in the central nervous system. Peripheral serotonergic [Sloley et al., 1987; Baines et al., 1990] and octopaminergic [Evans and O'Shea, 1977, 1978] neurons have also been identified.

0-8493-4591-X/94/$0.00 + $.50
© 1994 by CRC Press, Inc.

Table 1.   Levels of Octopamine, Dopamine, 5-Hydroxytryptamine and Tyramine in Nervous Tissues of *Periplaneta americana* and *Locusta migratoria migratoides*

| TISSUE | BIOGENIC AMINE (pmol/organ) | | | | | | |
| | P. americana | | | L. migratoria | | | |
	OA	DA	5HT	OA	TA	DA	5HT
Cerebral Ganglion	14.7-21.6	40.8	15.0	46.0	6.6	40.8	32.2
Optic Lobes (pair)	7.2	--	--	87.1	15.8	69.7	60.6
Suboesophageal Ganglion	5.6	--	--	15.1	3.1	13.3	13.5
Thoracic Nerve Cord	13.8-15.0	7.8	5.4	37.2	7.8	29.9	23.6
Abdominal Nerve Cord	7.8	4.8	0.7	14.6	5.3	56.5	15.7

-- indicates that tissue was not assayed

OA octopamine ; DA dopamine ; 5HT 5-hydroxytryptamine ; TA tyramine

III. METABOLISM

A. BIOSYNTHESIS

It is generally assumed that the synthesis of OA, DA, TA and 5HT in the insect nervous system proceeds according to the metabolic pathways that have been described in vertebrates [Brown and Nestler, 1985; Downer, 1990]. This conclusion is supported by studies on the incorporation of radio-labelled precursors into specific monoamines [Owen and Bouquillon, 1992], characterisation of enzymes involved in biosynthetic pathways [Lake et al., 1970; Sloley and Yu, 1987] and genetic studies on mutant insects with lesions at the genetic locus of key biosynthetic enzymes [Livingstone, 1981; Livingstone and Tempel, 1983]. The proposed biosynthetic pathways and the enzymes involved are identified in Figure 1. The decarboxylase enzyme which catalyses the conversion of tyrosine to TA appears to be different from that which is responsible for the decarboxylation of dihydroxyphenylalanine (DOPA) to DA and 5-hydroxytryptophan to 5HT [Livingstone and Tempel, 1983; Yu and Sloley, 1987]. Thus the insect decarboxylase differs from the general L-aromatic amino acid decarboxylase of vertebrates and, therefore, represents a potential insect-specific target for insecticide development.

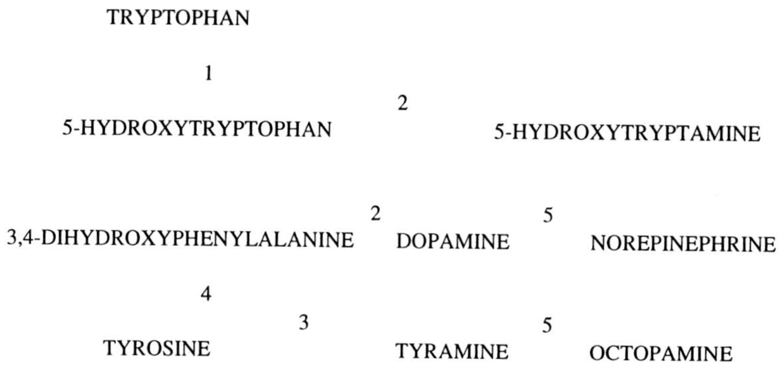

TRYPTOPHAN

1

2

5-HYDROXYTRYPTOPHAN   5-HYDROXYTRYPTAMINE

2   5

3,4-DIHYDROXYPHENYLALANINE   DOPAMINE   NOREPINEPHRINE

4

3   5

TYROSINE   TYRAMINE   OCTOPAMINE

Figure 1. **Proposed Biosynthetic Pathways for Biogenic Amines in Insects**

1.	TRYPTOPHAN HYDROXYLASE (Sloley and Yu, 1987)
2.	AROMATIC AMINO ACID DECARBOXYLASE
	(Livingstone, 1981; Livingstone and Tempel, 1983)
3.	TYROSINE DECARBOXYLASE
	(Livingstone and Tempel, 1983; Yu and Sloley, 1987)
4.	TYROSINE HYDROXYLASE (Brown and Nestler, 1985)
5.	DOPAMINE (TYRAMINE) HYDROXYLASE (Lake et al., 1970)

B.   CATABOLISM

Sodium-dependent and sodium-insensitive uptake mechanisms for octopamine have been reported in the cockroach nerve cord with the latter system possibly located in the surrounding connective tissue sheath [Evans, 1978]. The adrenergic uptake mechanism of vertebrates is associated with monoamine degradation and it is likely that an inactivation process occurs also in insects. Several possible routes of monoamine inactivation have been proposed including N-acetylation [Sekeris and Karlson, 1966], oxidative deamination [Scott et al., 1985] and conjugation with amino acid, sugar, sulphate and/or phosphate [Maxwell et al., 1980, Downer, 1990]. N-acetylation has been studied extensively in several species and shown to occur with OA, DA, TA and 5HT [Dewhurst et al., 1972; Evans and Fox, 1975; Vaughan and Neuhoff, 1976; Hayashi et al., 1977; Martin and Downer, 1989 a.b.c.; Martin et al, 1989]. N-acetyltransferase (NAT), the enzyme which catalyses the acetylation of the amino group of the monoamine using acetyl CoA as the acetyl donor, is widely distributed throughout insect tissues and differences have been reported in the properties of NAT from nervous and peripheral tissues [Martin and Downer, 1989 b; Martin et al., 1989]. N-acetylation is clearly the primary route of monoamine inactivation in the insect nervous system and, therefore, NAT has been identified as a potential target for insecticide development [Downer and Martin, 1987]. Studies to identify possible inhibitors of NAT indicate that the presence of a 4-hydroxyl group on the phenol ring enhances the ability of phenol and catechol derivatives to inhibit the enzyme [Martin and Downer, 1989]. Sulphate conjugation may also be involved in the inactivation of DA in the nervous system particularly around the time of ecdysis [Sloley and Downer, 1987] when N-acetylation of DA is of pivotal importance in cuticular tanning. A two-step inactivation of 5HT has been described in the mosquito with the N-acetyl 5-HT, resulting from N-acetylation, undergoing subsequent sulphate conjugation in the presence of phenosulphotransferase [Khoo and Wong, 1993].

IV. PHYSIOLOGICAL ROLES AND PHARMACOLOGY

A.   OCTOPAMINE
1.   Physiological Roles

OA has been reported to effect or modulate many physiological processes in insects (Table 2). In some cases the response results from direct action of OA at an OA-specific receptor located on a target tissue whereas others are indirect consequences of OA-mediated effects on the neuroendocrine system. In general, OA appears to be central in eliciting the overall arousal response [Corbet, 1991] and the complex sequence of biochemical/physiological processes associated with the onset and prolongation of flight activity in some species [Orchard et al., 1993]. These proposed roles for OA are consistent with the "orchestration hypothesis" [Sombati and Hoyle, 1984] which suggests that particular behaviour patterns can be triggered by activation of specific octopaminergic neurons. Further evidence for a behaviour-elicitation role for octopamine is provided by the marked increase in haemolymph octopamine levels that occurs immediately following flight or excitation [Bailey et al., 1983] and the demonstrated activation of certain octopaminergic nerve cells before flight activity commences [Ramirez and Orchard, 1990]. In addition to the neuromodulatory roles that have been identified for OA, the monoamine functions also as a neurohormone and neurotransmitter [Evans, 1985].

Table 2.　Physiological Processes Influenced by Octopamine (after Corbet, 1991)

PROCESS	REFERENCE
RESPONSIVENESS AND BEHAVIOUR	
pheromonal sensitivity	Linn and Roelofs 1986
pheromone production	Christensen et al. 1991
antennal hair erection	Nijhout 1977
sensitivity to olfactory stimuli	Mercer and Menzel 1982
enhancement of propioceptive pathways	Ramirez and Pearson 1991
circadian eclosion, activity, courtship	Livingstone 1981
excitability of respiratory neurones	Ramirez and Pearson 1991
light production in firefly lantern	Carlson 1968
tracheal ventilation and spiracular closure	Sombati and Hoyle 1984
MUSCULATURE	
induction of flight rhythm activity	Classen and Kammer, 1986
flight muscle activity	Whim and Evans 1988
myogenic rhythmicity	Evans and O'Shea 1977
gut muscle contraction	Huddart 1985
oviduct muscle contraction	Orchard and Lange 1987
METABOLISM	
stimulation of glycogenolysis	Robertson and Steele 1972
substrate utilisation by muscle	Candy 1978
hypertrehalosemia	Downer 1979a
diacylglycerol release from fat body	Orchard 1986
SECRETION	
malpighian tubule secretion	Coast 1989
salivary gland secretion	House and Ginsburg 1985
Neuropeptide release	Orchard et al. 1983, Downer et al. 1984
OTHER PHYSIOLOGICAL PROCESSES	
increased rate of respiration	Bellah et al. 1984
cardioacceleration	Collins and Miller 1977
phagocytosis	Baines and Downer 1993

2. Pharmacology

The initial classification of OA receptors was based upon the physiological responses of OA receptors in the extensor tibiae preparation of locusts to a variety of OA agonists and antagonists [Evans, 1981]. Subsequent studies with a broad spectrum of species and tissues have employed OA-mediated stimulation of cyclic AMP production [Nathanson and Hunnicutt, 1979; Bodnaryk, 1982; Downer, 1988] and radiolabelled ligand binding [Minhas et al., 1987] to further characterise OA receptors. Although some interspecific differences have been identified with regard to the relative potencies of particular agonists and antagonists, the original scheme proposed by Evans [1981] prevails as the basis for OA receptor classification.

According to this scheme, OA receptors show pharmacological similarities to the α-adrenergic receptors of vertebrates, although the greater affinity of OA receptors for monohydroxyphenolamines rather than their dihydroxy equivalents indicates their distinctiveness from catecholamine receptors. Evans [1981] distinguishes OA_1 receptors from OA_2 receptors on the basis of their response to the agonists clonidine and naphazoline with the former demonstrating greater potency against OA_2 receptors and the latter being most active against OA receptors. Also chlorpromazine and yohimbine are more effective inhibitors of OA_1 receptors than OA_2 receptors whereas the reverse is true for metachlopromide. OA_2 receptors can be further subdivided with OA_{2A} receptors responding better to naphazoline whereas OA_{2B} receptors are more responsive to tolazoline. OA_{2A} receptors can also be distinguished from OA_{2B} receptors on the basis of their greater sensitivity to the inhibitors cyproheptadine, mianserin and metachlopromide and lesser sensitivity to chlorpromazine.

A central concern in current classifications of insect aminergic receptors is that pharmacological characterisations are based on drugs that were developed for use with vertebrate receptors. Thus the potent inhibitors of OA_{2A} receptors, mianserin and cyproheptadine are recognised in the vertebrate nervous system as inhibitors of 5HT and histamine; similarly, another octopamine inhibitor, gramine is a serotonergic blocker in vertebrates. Precise classification of OA receptors and their relationship to other aminergic receptor classes must await the development of more specific pharmacological agents and, ultimately, the isolation and sequencing of individual receptors.

B. DOPAMINE

1. Physiological Roles

DA is the most abundant monoamine in the insect nervous system (Table 1) and several immunohistochemical studies have localised and examined the distribution of dopaminergic neurons in the central nervous system [Klemm, 1976; Dymond and Evans, 1979; Milton et al.,

1991]. However, in spite of the obvious importance of DA in the nervous system of insects, there are relatively few data on the physiological role(s) that it plays. Furthermore, in that DA can act as a partial agonist of octopaminergic receptors, it is not always possible to ensure that an action which has been ascribed to DA results from direct interaction of DA with a DA-specific receptor. The situation is complicated further by the pivotal role that DA and N-acetyl DA play in cuticular tanning and the impact that this may have on DA levels in the central nervous system [Owen and Fisher, 1988].

There is strong evidence to suggest that DA acts as a neurotransmitter to effect hyperpolarisation of the acinar cells of the salivary glands in the cockroach *Nauphoeta cinerea* [Bowser-Riley and House, 1976; Ginsborg and House, 1976] and to stimulate salivary secretion in the moth, *Manduca sexta* [Robertson, 1975].

The demonstration of dopaminergic innervation of the corpus cardiacum [Klemm, 1976] may indicate a role in neuropeptide release such as has been demonstrated for OA in the locust [Orchard et al., 1983] and cockroach [Downer et al., 1984]; however, definitive studies in this regard have not yet been completed. Other studies have suggested a modulatory role for DA in circadian rhythmicity and/or the induction or termination of diapause [Houk and Beck, 1977; Bodnaryk, 1979; Gelman and Hayes, 1980]. However, the lack of any circadian rhythmicity in N-acetyl transferase (NAT) activity in *Ostrinia nubialis* [Evans and Soderlund, 1982] tends to eliminate possible parallelism with the NAT pacemaker in the pineal gland of vertebrates. A possible role for DA in cellular differentiation is suggested by the observation that DA promotes the differentiation of epidermal cells into wing buds in the aphid, *Myzus persicae* [Harrewijn, 1976].

2. Pharmacology

Studies on the DA-mediated hyperpolarisation of acinar cells in the salivary gland of *N. cinerea* first suggested that the DA receptor of insects is pharmacologically distinct from the adrenergic receptors that have ben characterised in vertebrates [Ginsborg et al., 1976]. This conclusion has been confirmed by preliminary studies on a dopaminergic inhibitory motor neuron in the cockroach prothorax [Pitman and Fleming, 1985] and detailed pharmacological characterisation of a DA-receptor that is coupled to adenylate cyclase in the cockroach nervous system [Harmar and Horn, 1977; Orr et al., 1987].

The coupling of the insect DA receptor to adenylate cyclase and the stimulatory effects of the D_1 agonists, ADTN and epinine, on cyclic AMP production in a cockroach brain preparation [Orr et al., 1987] indicates similarities between the insect DA receptor and the D_1 receptor of vertebrates. However, the potent D_1 agonist, SKF 38393 does not elicit increases in cyclic AMP production in the insect whereas the D_2 agonist, LY171555 has a pronounced stimulatory effect [Orr et al., 1987]. Specific inhibitors of vertebrate D_1 and D_2 receptors are also inconsistent in their effects on the insect DA receptor. Thus, strong inhibition is obtained with the D_1 antagonists, SCH 23390 and SKF 83566 and with the D_2 blocker, spiperone [Orr et al., 1987]. The most effective blockers of the insect OA receptor are the non-selective DA antagonists, pifluthixol, *cis*-flupenthixol and (+)-butaclamol. Binding studies using radiolabelled pifluthixol as ligand support the pharmacological conclusions derived from DA-mediated cyclic AMP production [Notman and Downer, 1987] and confirm the pronounced influence of monovalent cations on DA-sensitive adenylate cyclase [Orr et al., 1988].

C. 5-HYDROXYTRYPTAMINE
1. Physiological Roles

5HT is widely distributed throughout the insect nervous system and (immuno)histochemical procedures have been used to localise and study the distribution of serotonergic neurons [Klemm, 1976; Maxwell et al., 1978; Sloley et al., 1986; Baines et al., 1990]. Several physiological processes are influenced by 5HT and in some cases there is strong evidence to suggest that the effect results from direct action of 5HT.

The most extensively studied action of 5HT in insects is its regulatory role in salivary secretion. This process involves two separate 5HT receptors, one of which is coupled to adenylate cyclase whereas the other stimulates hydrolysis of phosphatidylinositol and influx of calcium [Berridge et al., 1975; Berridge, 1981]. A 5HT-mediated effect on fluid secretion has been reported also in the Malpighian Tubules of *Rhodnius prolixus* [Maddrell and Phillips, 1975].

Several reports indicate a role for 5HT in the stimulation and/or enhancement of muscle contraction in the heart [Roussel, 1975; Collins and Miller, 1977], midgut [Cook and Holman, 1978; Hukuhara et al., 1981], oviduct [Cook and Meola, 1978] and malpighian tubules [Crowder and Shankland, 1974]. A detailed investigation of the modulation of mandibular closer muscle function in locusts demonstrates serotonergic innervation of the muscle and that 5HT affects muscle contraction by increasing the amplitude, rate of contraction and the rate of relaxation [Baines et al., 1990]. A similar role for 5HT on the mandibular closer muscle of the cricket has also been described [Baines and Downer, 1991].

The survival of cockroaches that have been exposed to a LD_{50} dose of *Staphylococcus aureus* is enhanced in the presence of 5HT. Studies *in vivo* and *in vitro* indicate that this results from 5HT-mediated stimulation of the phagocytic and nodule formation activities of haemocytes [Baines et al., 1992; D. Baines and Downer, 1992].

Other reports have implicated 5HT in the regulation of circadian rhythmicity [Muszynska-Pytel and Cymborowsky, 1978], aggressive behaviour in ants [Kostowski and Tarchalska, 1972] and the inhibition of wing development in apterous aphids [Harrewijn, 1976].

2. Pharmacology

Although several of the physiological studies indicated above used serotonergic antagonists to provide preliminary information on the pharmacological nature of 5HT receptors in insects, there have been few detailed investigations.

5HT-sensitive adenylate cyclase has been described in several tissues and the effects of some potential antagonists of 5HT-mediated cyclic AMP production have been compared in preparations of the nerve cord and corpus cardiacum of the cockroach [Downer et al., 1985; Gole et al., 1987]. The data indicate pharmacologically distinct receptors in the two organs [Downer, 1990].

High and low affinity binding sites for radiolabelled 5HT have been described in whole head homogenates of *Drosophila* [Dudai and Zvi, 1984]. However, as the preparation undoubtedly included 5HT receptors associated with mandibular musculature, salivary glands and nervous tissue, it is not possible to determine the receptor(s) that is (are) present in a particular tissue.

High affinity binding of the $5HT_2$ ligand ketanserin, has been reported in the brain of the house cricket [Pyza et al., 1991] and there are indications of a $5HT_2$ receptor in peripheral tissues of the locust [Banner et al., 1987] and cockroach [Baines and Downer, 1992]. However, no specific, saturable binding of ketanserin could be demonstrated in locust brain [Hiripi and Downer, 1993] and there is clearly need to expand the cricket study to include detailed kinetic and pharmacological analyses.

[3H]-5HT binding to locust brian membrane preparations indicates a single binding site with apparent Kd values of 2.98 ± 0.19nM and a maximal site concentration of 14.45 ± 1.12 pmol/g tissue [Hiripi and Downer, 1993]. The pharmacological properties of 5HT binding suggest similarities with the vertebrate 5HT receptor, although the low affinity of some selective drugs for the receptor suggest differences from the vertebrate $5HT_1$ receptor [Hiripi and Downer, 1993].

D. TYRAMINE

TA has been described as a trace amine in the central nervous system of vertebrates and physiological effects on neurotransmitter release have been proposed [Boulton, 1985]. TA occurs also in the invertebrate nervous system at higher concentrations than in vertebrates [Robertson and Juorio, 1976; McFarlane et al., 1990; Downer et al., 1993] and several physiological effects have been ascribed to TA. These include a proposed role as an antagonist of proctolin [Brown, 1975], involvement in suppression of spontaneous gut muscle contraction [Cook and Holman, 1978, 1979], effector of small spontaneous hyperpolarisations in acinar cells of salivary glands [House, 1973] and suppression of trehalogenesis [Downer, 1979b]. None of these observations provide conclusive evidence that TA is a specific effector of physiological events in insects but they raise the possibility and further support for the hypothesis is provided by the report that cloned Drosophila cDNA encodes for a specific TA receptor [Saudou et al., 1990].

TA has been proposed as a putative neurotransmitter in the central nervous system of the locust [Downer et al., 1993]. This proposal is supported by the demonstration of tyrosine decarboxylase activity in brain tissue, high-and low-affinity TA uptake mechanisms that are independent of those for OA and DA, potassium and reserpine-induced release of TA from incubated ganglia [Downer et al., 1993] and [3H] TA binding to brain membrane preparations that is pharmacologically distinct from [3H]DA binding [Hiripi et al., in this volume].

V. CONCLUSIONS

The aminergic component of the insect nervous system is central to the regulation of many essential physiological processes. Aspects of the metabolism and pharmacology are sufficiently distinctive from equivalent systems in vertebrates to suggest that compounds with high specificity for insects could be developed to interfere with normal aminergic function.

REFERENCES

Bailey, B.A., Martin, R.J. and Downer, R.G.H., Haemolymph octopamine levels during and following flight in the American cockroach, *Periplaneta americana L.*, Can. J. Zool., 62, 19-22, 1983.

Baines, D. and Downer, R.G.H., 5-hydroxytryptamine-sensitive adenylate cyclase affects phagocytosis in cockroach haemocytes, Archs. Insect Biochem. Physiol., 21, 303-316, 1992.

Baines, D., DeSantis, T. and Downer, R.G.H., Octopamine and 5-hydroxytryptamine enhance the phagocytic and nodule formation activities of cockroach (*Periplaneta americana*) haemocytes, J. Insect Physiol., 38, 905-914, 1992.

Baines R.A., and Downer, R.G.H., Pharmacological characterization of a 5-hydroxytryptamine-sensitive receptor/adenylate cyclase complex in the mandibular closer muscles of the cricket, *Gryllus domestica*, Archs. Insect Biochem. Physiol., 16, 153-163, 1991.

Baines, R.A., Tyrer, N.H. and Downer, R.G.H., Serotonergic innervation of the locust mandibular closer muscle modulates contractions through the elevation of cyclic adenosive monophosphate, J. Comp. Neurol., 294, 623-632, 1990.

Banner, S.E., Osborne, R.H. and Cattel, K.J., The pharmacology of the isolated foregut of the locust *Schistocerca gregaria*: II characterisation of a 5HT-2 like receptor, Comp. Biochem. Physiol, 88C, 139-144, 1987.

Bellah, K.L., Fitch, G.K., Kammar, A.E., A central action of octopamine on ventilation frequency in *Corydalus cornutus*, J. Exp. Zool, 231, 289-292, 1984.

Berridge, M.J., Electrophysiological evidence for the existence of separate receptor mechanisms mediating the action of 5-hydroxytoyptamine, Mol. Cell Endocrinol., 23, 91-104, 1981.

Berridge, M.J., Lindley, B.D. and Prince, W.T., Membrane permeability changes during stimulation of isolated salivary glands of *Calliphora* by 5-hydroxytryptamine, J. Physiol., 244, 549-567, 1975.

Bodnaryk, R.P., Identification of specific dopamine-and octopamine-sensitive adenylate cyclases in the brain of *Mamestra configurata,* Insect Biochem., 9, 155-162, 1979.

Bodnaryk, R.P., Biogenic amine-sensitive adenylate cyclases in insects, Insect Biochem., 12, 1-6, 1982.

Boulton, A.A., The trace amines: recent overview and future pointers, in "Neuropsychopharmacology of the trace amines, Boulton, A.A., Bieck, P.R., Maitre, L. and Riederer, P., Eds. Humana Press, Clifton, N.J., 1985, pp. 3-12.

Bowser-Riley, F. and House, C.R., The actoins of some putative neurotransmitters on the cockroach salivary gland, J. Exp. Biol., 64, 665-676, 1976.

Brown, B.E., Proctolin: a peptide transmitter candidate in insects, Life Sci., 17, 1241-1252, 1975.

Brown, C.S. and Nestler, C., Catecholamines and indolalkylamines in "Comprehensive Insect Physiology, Biochemistry and Pharmacology", Vol. 11, Kerkut, G.A. and Gilbert, L.I., Eds. Pergamon Press, Oxford, pp. 435-497.

Candy, D.J., The regulation of locust flight muscle metabolism by octopamine and other compounds, Insect Biochem, 8, 177-181, 1978.

Carlson, A.D., Effect of adrenergic drugs on the lantern of the larval *Photuris* firefly, J. Exp. Biol., 48, 381-387, 1968.

Christensen, T.A., Itagaki, H., Teal, P.E.A., Jasensky, R.D., Tumlinson, J.H. and Hildebrand, J.G., Innervation and neural regulation of the sex phermone gland in female *Heliothis* moths, PNAS, USA, 88, 4971-4975, 1991.

Classen, D.E. and Kammer, A.E., Effects of octopamine, dopamine and serotonin on production of flight motor output by thoracic ganglia of *Manduca sexta*, J. Neurobiol., 17, 1-14, 1986.

Coast, G.M., Stimulation of fluid secretion by single isolated malpighian tubules of the house cricket, *Acheta domesticus*, Physiol. Entomol., 14, 21-30, 1989.

Collins, C. and Miller, T., Studies on the action of biogenic amines on the cockroach heart, J. Exp. Biol., 67, 1-15, 1977.

Cook, B.J. and Holman, G.M., Comparative pharmacological properties of muscle function in the foregut and the hindgut of the cockroach, *Leucophala maderae,* Comp. Biochem. Physiol., 61C, 291-295, 1978.

Cook, B.J. and Holman, G.M., The pharmacology of insect visceral muscle, Comp. Biochem. Physiol., 64C, 183-190, 1979.

Cook, B.J. and Meola, S., The oviduct musculature of the horsefly, *Tabanus sulcifrons*, and its response to 5-hydroxytryptamine and proctolin, Physiol. Entomol., 3, 273-280, 1978.

Corbet, S.A., A fresh look at the arousal syndrome of insects, Adv. Insect Physiol., 23, 81-116.

Crowder, L.A. and Shankland, D.L., Response to 5-hydroxytryptamine and electrophysiology of the Malpighian tubule muscle of the American cockroach, *Periplaneta americana*, Ann. Entomol. Soc. Amer., 67, 281-284, 1974.

David J-C. and Coulon, J-F, Octopamine in invertebrates and vertebrates: a review, Prog. Neurobiol., 24, 141- , 1985.

Davis, N.T., Neurosecretory neurons and their projections to the serotonin neurohaemal system of the cockroach, *Periplaneta americana* (L) and identification of mandibular and maxillary motorneurones associated with this system, J. Comp. Neurol., 259, 604-621, 1987.

Dewhurst, S.A., Croker, S.G., Ikeda, K. and McCaman, R.E., Metabolism of biogenic amines in *Drosophila* nervous tissue, Comp. Biochem. Physiol., 43B, 975-981, 1972.

Downer, R.G.H., Induction of hypertrehalosemia by excitation in *Periplaneta americana*, J. Insect Physiol., 25, 59-63, 1979a.

Downer, R.G.H., Trehalose production in isolated fat body of the American cockroach, *Periplaneta americana*, Comp. Biochem. Physiol., 62C, 31-34, 1979b.

Downer, R.G.H., Octopamine- and dopamine-sensitive receptors and cyclic AMP production in insects, in "Molecular basis of Drug and Pesticide Action", Lunt G.G. Ed., Excerpta Medica, Amsterdam, pp. 255-265, 1988.

Downer, R.G.H., Octopamine, dopamine and 5-hydroxytryptamine in the cockroach nervous system in "Cockroaches as models for Neurobiology: applications in biomedical research", Huber, I., Masler, E.P., and Rao, B.R., Eds. CRC Press, Boca Raton, Fla., pp. 103-124, 1990.

Downer R.G.H. and Martin, R.J., N-acetylation of octopamine: a potential target for insecticide development in "Sites of action for neurotoxic pesticides," Hollingworth, R.M. and Green, M.B., Symposium of American Chemical Society #356, New York, pp. 202-210, 1987.

Downer, R.G.H., Gole, J.W.D., Orr, G.L. and Orchard, I., The role of octopamine and cyclic AMP in regulating hormone release from corpora cardiaca of the American cockroach, J. Insect Physiol., 30, 451-462.

Downer, R.G.H., Bailey, B.A. and Martin, R.J., Estimation of biogenic amines in biological tissues in "Neurobiology", Gilles, R. and Balthazart, J., Eds., Sringer-Verlag, Berlin, pp. 248-263, 1985.

Downer, R.G.H., Hiripi, L. and Juhos, S., Characterisation of the tyraminergic system in the central nervous system of the locust, *Locusta migratoria migratoides*, Neurochem. Res., (in press), 1993.

Dudai, Y. and Zvi, S., High affinity [3H]-octopamine binding sites in *Drosophila melanogaster*, Interaction with ligand and relationship to octopamine receptors, Comp. Biochem. Physiol., 77C, 145-151, 1984.

Dymond, G.R. and Evans, P.D., Biogenic amines in the nervous system of the cockroach, *Periplaneta americana*: association of octopamine with mushroom bodies and dorsal unpaired median (DUM) neurones, Insect Biochem., 9, 535-545, 1979.

Eckert, M., Rapus, J., Nürnberger, A. and Penzlin, H., A new specific antibody reveals octopamine-like immunoreactivity in cockroach ventral nerve cord, J. Comp. Neurol., 322, 1-15, 1992.

Evans, P.D., Octopamine: a high affinity uptake mechanism in the nervous system of the cockroach, J. Neurochem., 30, 1015-1022, 1978.

Evans, P.D., Multiple receptor types for octopamine in the locust, J. Physiol., 318, 99-122, 1981.

Evans, P.D., Octopamine: in "Comprehensive Insect Physiology, Biochemistry and Pharmacology" Vol. II, Kerkut, G.A. and Gilbert, L.I., Eds, Pergamon Press, Oxford, pp. 499-530, 1985.

Evans, P.D. and O'Shea, M., An octopaminergic neurone modulates neuromuscular transmission in the locust, Nature (Lond.), 270, 257-259, 1977.

Evans, P.D. and O'Shea, M., The identification of an octopaminergic neurone and the modulation of a myogenic rhythm in the locust, J. Exp. Biol., 73, 235-260, 1978.

Evans, P.H. and Fox, P.M., Enzymatic N-acetylation of indolealkylamines by brain homogenates of the honeybee *Apis mellifera*, J. Insect Physiol., 21, 343-353, 1975.

Evans, P.H. and Soderlund, D.M., Biogenic amine acetylation: no detectable circadian rhythm in whole brain homogenates of the insect *Ostrinia nubialis*, Experientia 38, 302-303, 1982.

Gelman, D.B. and Hayes, D.K., A survey of adenyl cyclase activity in head homogenates of fith-instar diapansing and non-diapausing larvae of the European corn borer *Ostrinia nubialis* (Hubner), Comp. Biochem. Physiol., 67C, 61-69, 1980.

Ginsborg, B.L. and House, C.R., The response to nerve stimulation of the salivary gland of *Nauphoeta cinerea* Olivier, J. Physiol., 262, 477-487, 1976.

Ginsborg, B.L., House, C.R. and Slinsky, E.M., On the receptors which mediate the hyperpolarisation of salivary gland cells of *Nauphoeta cinerea* Olivier, J. Physiol., 262, 489-500, 1976.

Gole, J.W.D., Orr, G.L. and Downer, R.G.H., Pharmacology of octopamine, dopamine and 5-hydroxytryptamine-sensitive adenylate cyclase on the corpus cardiacum of the American cockroach, *Periplaneta americana* L., Archs. Insect Biochem. Physiol., 5, 119-128.

Harmar, A.J. and Horn, A.S., Octopamine-sensitive adenylate cyclase in cockroach brain: effects of agonists, antagonists and guanylyl nucleotides, Mol. Pharmacol., 13, 512-520, 1977.

Harrewijn, P., Role of monoamine metabolism in wing dimorphism of the aphid, *Myzus persical*, Comp. Biochem. Physiol., 55C, 147-153, 1976.

Hayashi, S., Murdock, L.L. and Florey, E., Octopamine metabolism in invertebrates (*Locusta, Astacus, Helix*): evidence for N-acetylation in arthropod tissues, Comp. Biochem. Physiol., 58C, 183-191, 1977.

Hiripi, L. and Downer, R.G.H., Characterization of serotonin binding sites in insect (*Locusta migratoria*) brain, Insect Biochem. Molec. Biol., 23, 303-307, 1993.

Houk, E.J. and Beck, S.D., Monoamine oxidase in the brain of European cornborer larvae, *Ostrinia nubialis* (Hubner), Insect Biochem., 8, 231-236, 1978.

House, C.R., An electrophysiological study of neuroglandular transmission in the isolated salivary glands of the cockroach, J. Exp. Biol., 58, 29-43, 1973.

House, C.R. and Ginsborg, B.L., Properties of dopamine receptors at a neuroglandular synapse. In "Neuropharmacology of Insects", Ciba Fndn. Symposium #88, Pitman, London, pp. 32-47, 1982.

Huddart, H., Visceral Muscle. In "Comprehensive Insect Physiology, Biochemistry and Pharmacology", Kerkut, G.A. and Gilbert, L.I., Eds. Pergamon Press, Oxford, pp. 131-184, 1985.

Hukuhara, T., Satake, S. and Sato, Y., Rhythmic contractile movements of the larval midgut of the silkworm, *Bombyx mori*, J. Insect Physiol., 27, 469-473, 1981.

Khoo, H.G.N. and Wong, K.P., Sulphate conjugation of serotonin and N-acetylserotonin in the mosquito, *Aedes togoi*, Insect Biochem. Molec. Biol., 23, 507-513, 1993.

Klemm, N., Histochemistry of putative transmitter substances in the insect brain, Progr. Neurobiol., 7, 99-169, 1976.

Konlags, P.N.M., Vullings, H.G.B., Geffard, M., Buljs, R.M., Diederen, J.H.B. and Jensen, W.F., Immunocytochemical demonstration of octopamine-immunoreactive cells in the nervous system of *Locusta migratoria* and *Schistocerca gregaria*, Cell Tissue Res., 231, 371-379, 1988.

Kowstowski, W. and Tarchalska, B., The effects of some drugs affecting brain 5HT on the aggressive behaviour and spontaneous electrical activity of the central nervous system of the ant, *Formica rufa*, Brain Res., 38, 143-149, 1972.

Linn, C.E. and Roeloffs, W.L., Modulatory effects of octopamine and serotonin on male sensitivity and periodicity of response to sex pheromone in the cabbage looper moth, *Trichoplusia ni*, Archs. Insect Biochem. Physiol., 3, 161-171.

Lake, C.R., Mills, R.R., and Brunet, P.C.J., B-hydroxylation of tyramine by cockroach haemolymph, Biochem. Biophys. Acta., 215, 226-228, 1970.

Livingstone, M.S., Two mutations in *Drosophila* differentially affect the synthesis of octopamine, dopamine and serotonin by altering the activities of two different amino acid decarboxylases, Soc. Neurosci. Abstracts, 7, 351, 1981 (cited by Evans, 1985).

Livingstone, M.S. and Tempel, B.L., Genetic dissection of monoamine neurotransmitter synthesis in *Drosophila*, Nature (Land), 303, 67-70, 1983.

Maddrell, S.H.P. and Phillips, J.E., Secretion of hypo-osmotic fluid by the lower Malpighian tubules of *Rhodnius prolixus*, J. Exp. Biol., 62, 671-683, 1975.

Martin, R.J. and Downer, R.G.H., Microassay for N-acetyl transferase activity using high performance liquid chromatography with electrochemical detection, J. Chromatogr., 487, 287-293, 1989a.

Martin, R.J. and Downer, R.G.H., N-acetylation of p-octopamine by Malphghian tubules and other tissues of the American cockroach, *Periplaneta americana* L. in vitro, Can. J. Zool., 67, 1495-1499, 1989b.

Martin, R.J. and Downer, R.G.H., Effects of potential inhibitors on N-acetylation of octopamine by tissue extracts from Malpighian Tubules and cerebral ganglia of *Periplaneta americana*, Archs. Insect Biochem. Physiol., 11, 29-45, 1989c.

Martin, R.J., Bailey, B.A. and Downer, R.G.H., Analysis of octopamine, dopamine, 5-hydroxytryptamine and tryptophan in the brain and nerve cord of the American cockroach. In "Neurobiology of the trace amines", Boulton, A.A., Baker, G.B., Dewhurst, W.G. and Sandler, M., Eds. Humana Press, Clifton, N.J., pp. 91-96, 1984.

Martin, R.J., Jahagirdar, A.P. and Downer, R.G.H., Partial characterisation of N-acetyltransferase activitiy from cerebral ganglia and malpighian tubules of *Periplaneta americana*, Insect Biochem., 19, 351-359, 1989.

MacFarlane, R.G., Midgley, J.M., Watson, D.G. and Evans, P.D., The analysis of biogenic amines in the thoracic nervous system of the locust, *Schistocerea gregaria* by gas chromatography negative ion chemical ionization mass spectrometry (GC-NICIMS), Insect Biochem., 20, 305-311, 1990.

Maxwell, G.D., Tait, J.F. and Hildebrand, J.G., Regional synthesis of neurotransmitter candidates in the CNS of the moth, *Manduca sexta*, Comp. Biochem. Physiol., 61C, 109-119, 1978.

Maxwell, G.D., Morre, M.M. and Hildebrand, J.G., Metabolism of tyramine in the central nervous system of the moth, *Manduca sexta*, Insect Biochem., 10, 657-665, 1980.

Mercer, A.R. and Menzel, R., The effects of biogenic amines on conditioned and unconditioned responses to olfactory stimuli in the honeybee, *Apis mellifera*, J. Comp. Physiol., 145, 363-368, 1982.

Milton, G.W.A., Verhaert, P.D.E.M. and Downer, R.G.H., Immunofluorescent localization of dopamine-like and leucine-enkephalin-like neurons in the supraoesophageal ganglia of the American cockroach, *Periplaneta americana*, Tissue and Cell, 23, 331-340, 1991.

Minhas, N., Gole, J.W.D., Orr, G.L. and Downer, R.G.H., Pharmacology of [3H]-mianserin binding in the nerve cord of the American cockroach, *Periplaneta americana*, Archs. Insect Biochem. Physiol., 61, 191-201, 1987.

Muszynska-Pytel, M. and Cymborowski, B., The role of serotonin in regulation of the circadian rhythms of locomotor activity in the cricket, *Acheta domesticus*, Part 2, Distribution of serotonin and variations in different brain structures, Comp. Biochem. Physiol., 59C, 17-20, 1978.

Nathanson, J.A. and Hunnicutt, E.J., Neural control of light emissionin *Photuris* larvae: identification of octopamine-sensitive adenylate cyclase, J. Exp. Zool., 208, 255-262.

Nijhout, H.F., Control of antennal hair erection in male mosquitoes, Bio. Bull, Woods Hole, 153, 591-603, 1977.

Notman, H.J. and Downer, R.G.H., Binding of [3H]-pifluthixol, a dopamine antagonist, in the brain of the American cockroach, *Periplaneta americana*, Insect Biochem., 17, 587-590, 1987.

Owen, M.D. and Bouquillon, A.I., The synthesis of L-dihydroxyphenylalanine (L-DOPA) in the cerebral ganglia of the cockroach, *Periplaneta americana* L., Insect Biochem. Molec. Biol., 22, 193-198, 1992.

Owen, M.D. and Fisher, J., The association of changes in cerebral ganglion dopamine metabolism with ootheca development in *Periplaneta americana*, Comp. Biochem. Physiol., 91C, 403-409, 1988.

Orchard, I., Adipokinetic hormones - an update, J. Insect Physiol., 33, 451-463, 1987.

Orchard, I. and Lange, A.B., Neuromuscular transmission in an insect visceral muscle, J. Neurobiol., 17, 359-372, 1986.

Orchard, I., Gole, J.W.D. and Downer, R.G.H., Pharmacology of aminergic receptors mediating an elevation in cyclic AMP and release of hormone from locust neurosecretory cells, Brain Res., 288, 349-353, 1983.

Orchard, I., Ramirez, J-M. and Lange, A.B., A multifunctional role for octopamine in locust flight, Ann. Rev. Entomol., 38, 227-249, 1993.

Orr, G.L., Gole, J.W.D., Notman, H.L. and Downer, R.G.H., Pharmacological characterisation of the dopamine-sensitive adenylate cyclase in cockroach brain: evidence for a distinct dopamine receptor, Life Sci., 41, 2705-2715, 1987.

Orr, G.L., Gole, J.W.D., Notman, H.L. and Downer, R.G.H., The regulation of basal and dopamine-sensitive adenylate cyclase by salts of monovalent cations in brain of the American cockroach, *Periplaneta americana*, Insect Biochem, 18, 79-86, 1988.

Pitman, R.M. and Fleming, J.R., The action of dopamine on an identified insect neurone, Pestic. Sci., 16, 447-448, 1985.

Pyza, E., Golembiowska, K. and Antkiewicz-Michaluk, L., Serotonin, dopamine, noradrenaline and their metabolites: levels in the brain of the house cricket (*Acheta domesticus* L) during a 24-hour period and after administration of quepazine - a 5HT receptor agonist, Comp. Biochem. Pysiol., 94C, 365-371, 1991.

Ramirez, J.M. and Orchard, I., Octopaminergic modulation of the forewing stretch receptor in the locust, *Locusta migratoria*, J. Exp. Biol., 149, 255-279, 1990.

Ramirez, J.M. and Pearson, K., Octopaminergic modulation in the flight system of the locust, J. Neurophysiol., 66, 1522-1537, 1991.

Robertson, H.A., The innervation of the salivary gland of the moth, *Manduca sexta*: evidence that dopamine is the transmitter, J. Exp. Biol., 63, 413-419, 1975.

Robertson, H.A. and Juorio, A.V., Octopamine and some related noncatecholic amines in the invertebrate nervous system, Int. Rev. Neurobiol., 19, 173-224, 1975.

Robertson, H.A. and Steele, J.E., Activation of insect nerve cord phosphorylase by octopamine and adenosine 3 ,5 -monophosphate, J. Neurochem., 19, 1603-1606, 1972.

Roussel, J.P., Effect of 5-hydroxytryptamine on cardiac activity by *Locusta migratoria* in vivo, Arch. Int. Physiol. Biochem., 83, 293-298, 1975.

Saudou, F., Amaiky, N., Plassat, J.M., Borelli, E. and Hen, R., Cloning and characterisation of a *Drosophila* tyramine receptor, EMBO. J., 9, 3611-3617, 1990.

Scott, J.A., Johnson, T.L. and Knowles, C.O., Biogenic amine uptake by nerve cords from the American cockroach and the influence of amidines on amine uptake and release, Comp. Biochem. Physiol., 82C, 43-49, 1985.

Sekeris, C.E. and Karlson, P., Biosynthesis of catecholamines in insects, Pharm. Rev., 18, 89-94, 1966.

Shafi, N., Midgley, J.M., Watson, D.G. and Smail, G.A., Analysis of biogenic amines in the brain of the American cockroach (*Periplaneta americana*) by gas chromatography - negative ion chemical ionization mass spectroscopy, J. Chromatogr., 490, 9-19, 1989.

Sloley, B.D. and Downer, R.G.H., Dopamine, N-acetyl dopamine and dopamine-O-sulphate in tissues of newly ecdysed and fully tanned adult cockroaches (*Periplaneta americana*), Insect Biochem., 17, 591-596, 1987.

Sloley, B.D. and Yu, P.H., Apparent differences in tryptophan hydroxylation by cockroach (*Periplaneta americana*) nervous tissue and rat brain, Neurochem. Int., 11, 265-269, 1987.

Sloley, B.D., Downer, R.G.H. and Gillott, C., Levels of tryptophan, 5-hydroxytryptamine and dopamine in some tissues of the cockroach, *Periplaneta americana*, Can. J. Zool., 64, 2669-2674, 1987.

Sombati, S. and Hoyle, G., Generalisations of specific behaviours in a locust by local release into neuropil of the natural neuromodulator octopamine, J. Neurobiol., 15, 481-506, 1984.

Stevenson, P.A., Pfluger, H.J., Eckert, M. and Rapus, J., Octopamine immunoreactive cell populations in the locust thoracic-abdominal nervous system, J. Comp. Neurol., 315, 382-397, 1992.

Vaughan, P.F.T. and Neuhoff, V., The metabolism of tyrosine, tyramine and L-3,4-dihydroxyphenylalanine by cerebral and thoracic ganglia of the locust *Schistocerca gregaria*, Brain Res., 117, 175-180, 1976.

Villemaringe, J., Duris, P., Gefford, M., LeMoal, M., Delange, M., Bensch, C. and Girardie, A., Immunohistochemical localization of dopamine in the brain of the insect locusta migratoria migratoides in comparison with the catecholamine distribution determined by histofluorescence technique, Cell Tissue Res., 237, 391-394, 1984.

Whim, M.D. and Evans, P.D., Octopaminergic modulation of flight muscle in the locust, J. Exp. Biol., 134, 247-266, 1988.

ACTION AND INTERACTION OF PEPTIDES IN REGULATING ECDYSIS BEHAVIOR IN INSECTS

James W. Truman, Randall S. Hewes, and John Ewer

Department of Zoology,
University of Washington, Seattle, WA 98195.

I. INTRODUCTION

Neuropeptides play key roles in regulating the physiology and behavior of insects. These molecules, though, do not work in isolation but are embedded in the ongoing physiology of the animal. Consequently, to elucidate their role in the normal physiology of an organism, we need to understand a number of issues including: 1) the factors that coordinate the timing of release of the neuropeptide so that it is appropriate in the context of ongoing physiology and development, 2) the mechanisms that direct the released peptides to their target organs, and 3) the role of recruited "downstream" peptides in mediating aspects of the overall response. This essay uses recent advances in our understanding of the ecdysis system of insects to illustrate these points.

II. THE PHYSIOLOGY AND BEHAVIOR OF ECDYSIS

The rigid exoskeleton of arthropods has provided this group with a tremendous adaptive advantage and has allowed them to exploit terrestrial and aerial habitats. It has also placed unique physiological and developmental demands on these organisms because a new exoskeleton must be produced periodically to accomodate growth and changes in form. This process of molting requires a precisely timed interaction of developmental, physiological and behavioral processes that are coordinated by a number of hormones. The molt is initiated by the action of the prothoracicotropic hormone which acts on the prothoracic glands to cause the synthesis and secretion of ecdysteroids (Bollenbacher and Granger, 1985), the steroid hormones that drive the molting process (Riddiford, 1985). The rising ecdysteroid titer acts on the epidermis to cause apolysis, secretion of molting fluid, cell division and morphogenetic changes in shape. Steroid titers peak at about 30-40% of the molt, and the subsequent falling phase is associated with the synthesis of new exocuticle (Riddiford, 1985). Late in the molt the activation of enzymes in the molting fluid starts the digestion of the endocuticle of the old exoskeleton. Thus, at the end of the molt the insect is covered by the degraded cuticle of the old stage over the newly developed cuticle of the next stage. The culminating events are the shedding (ecdysis) of the old cuticle and the expansion of the new one.

0-8493-4591-X/94/$0.00 + $.50
© 1994 by CRC Press, Inc.

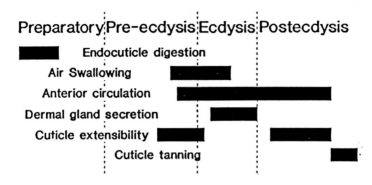

Figure 1. Generalized summary of the physiological events that are typically associated with the various phases of the ecdysis sequence. The black bars show the time when the events typically occur.

The complex processes that occur around ecdysis are reviewed by Reynolds (1980). The behaviors can be divided into 4 phases: 1) the **preparatory** phase involves the search for an appropriate site for ecdysis; 2) during the **pre-ecdysial** phase the insect performs behaviors that break connections between the old and new cuticles; 3) the **ecdysis** phase brings about the actual shedding of the old cuticle; and 4) the **postecdysis** behaviors are involved with the expansion of the new cuticle to its appropriate form. A successful ecdysis requires that these behaviors be coordinated with a number of physiological and developmental changes (Fig. 1). Ecdysis cannot be initiated until the digestion of the old cuticle has progressed to the point that the insect can break through it. Around the onset of ecdysis movements most insects show air swallowing and circulatory adjustments that expand the insect and aid in rupturing the old cuticle and in the expansion of the new one. Prior to ecdysis the new cuticle is very permeable, to allow movement of material across it from the molting fluid, but at ecdysis the cuticle becomes water-proof, in large part due to the secretion by dermal glands. The new cuticle often becomes more extensible at ecdysis to aid in its extraction from the sheath of old cuticle, and this extensibility is further increased after ecdysis to accomodate expansion to its new form (Reynolds, 1977; 1980). After expansion, the cuticle is then rigidly tanned.

This complex coreography of events involves the interactions of at least 4 hormones. The key hormone in this process is the neuropeptide eclosion hormone (EH). This is a 62 amino acid peptide whose actions appear to be dedicated to the triggering of ecdysis behaviors (see Truman, 1992 for a review). Depending on the species, EH is found in one to two pairs of ventromedial brain neurons (the VM cells). The axons from these cells project the length of the ventral CNS, finally exiting into the periphery through the proctodeal branch of the terminal nerve where they form a diffuse neurohemal site. In some species, these cells also send axon collaterals to the corpora cardiaca so that EH is released into the blood from both anterior and posterior sites. We are interested in how the release of EH is timed to the appropriate phase in the molt cycle and how EH, in turn, recruits the release of other neuropeptides involved in the ecdysis response.

III. COORDINATION OF ECDYSIS WITH THE MOLT CYCLE

A. REGULATION OF EH RELEASE

The VM cells appear to release EH only at the end of each molt. For the pupal molt in *Manduca*, for example, about 90% of the stored EH is released over a span of about 20 min, and this is followed by ecdysis approximately 40 to 60 min later (Hewes and Truman, 1991). Injection of EH at various times during the pupal molt show that there is only a narrow window during which the old cuticle is sufficiently digested so that the animal can successfully shed it. Consequently, the timing of EH release is critical and this release appears to be regulated by the interaction of two classes of factors: **molt-related** factors and **extrinsic** factors.

One obvious molt-related cue for EH release is the old cuticle that covers the animal at the end of the molt. This cuticle, however, can be precociously removed manually without interfering with the subsequent timing of EH release showing that its presence is not needed for this release to occur. Moreover, the fact that an implanted brain shows the proper timing of EH release suggests the the information about the progression of the molt is conveyed to the brain via a blood borne signal rather than one carried through neural channels (Truman and Riddiford, 1970). This signal appears to be the falling titer of ecdysteroids (Slama, 1980), which provides a common cue for both endocuticle digestion and EH release. In the case of endocuticle digestion, interruption of the ecdysteroid decline by injection or infusion of 20-hydroxyecdysone (20-HE) delays or blocks endocuticle breakdown (Schwartz and Truman, 1983). Similar 20-HE treatments also cause dose-dependent delays in the subsequent release of EH (Truman et al., 1983) although the critical period for delaying EH release is later than that for blocking endocuticle digestion.

The mechanism by which declining ecdysteroids affect the functioning of the VM cells was assessed for the molt to the 5th larval instar (Hewes and Truman, 1993). When animals were injected with 20-HE at various times during the ecdysteroid decline, those injected 7 hr or more before the normal onset of ecdysis subsequently delayed their ecdysis while those treated at later times ecdysed on schedule. This ecdysis delay reflects an interference with EH release because the steroid-treated animals ecdysed on time if provided with an exogenous source of EH. Consequently, as the ecdysteroid titer drops below a critical level at -7 hr some events begin which "commit" the brain to a time-course that leads to EH release about 6 hr later.

To determine whether this commitment was associated with changes in the VM cells themselves, the electrical properties of these cells were measured by intracellular recordings from their somata (Hewes and Truman, 1993). During the intermolt, the VM cells exhibited a high threshold, and intracellular injection of current generally elicited only a single spike. This relative inexcitability persisted up through about 3 hr before ecdysis. During the next 2 hr, though, the cells showed a marked increase in excitability as evidenced by a severe reduction in threshold in the absence of changes in their resting potential or input resistance. In about half the cases, the cells showed spontaneous firing of 0.6-2.0 Hz at rest. By 4-5 hr after ecdysis, the excitability of the cells had returned to their intermolt values. Thus, the VM cells show an transient increase in membrane excitability that is associated with the release of EH. Intracellular recordings from the cell body showed no evidence of synaptic input at the time of release. Therefore, the excitability change itself may be sufficient to cause the cells to fire. It should be cautioned, though, that recordings from the soma might not detect significant synaptic input if it was electrically distant from the soma.

The transient increase in excitability is regulated by the ecdysteroid decline. When larvae were injected with 20-HE at 11 hr before ecdysis, a time when such treatment delays EH release, the excitability increase in the VM cells was also delayed. The same dosage of steroid given at -4.5 hr, at a time after the CNS was "committed" to release EH but before the excitability changes in the VM cells were evident, had no effect on the subsequent change in excitability. Hence, the presence of high ecdysteroids, *per se*, does not suppress excitability once the cell has become committed. Instead, the high steroid titers appear to regulate the commitment decision itself.

The span of 5-6 hours between the time when ecdysteroids are effective and when the cells subsequently become excitable, suggests that this change is mediated by a transcription dependent mechanism. Indeed, treatment with actinomycin D, which blocks RNA synthesis, serves to block both EH release and the increase in VM cell excitability. Again, this drug is only effective in blocking these events if it is given prior to the commitment of the cells to release EH. From these data, it appears that the decline of ecdysteroids below a critical threshold level approximately 7 hr before ecdysis initiates new transcription which results in the cells becoming more excitable about 5 to 6 hr later. The nature of the genes that are activated at this time is unknown.

The extrinsic factors that influence EH release vary from species to species and from molt to molt. For larval and pupal ecdyses in *Manduca*, for example, the timing of EH release appears to be determined solely by

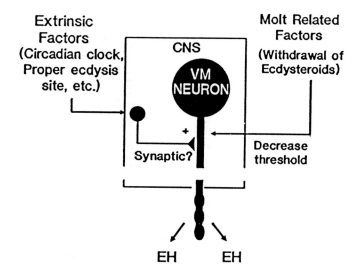

Figure 2. Schematic summary of the factors that act on the VM neurons to cause the release of EH. See text for further details.

the ecdysteroid titer with no extrinsic factors coming into play. In contrast, for adult ecdysis in this and many other species, input from a circadian clock restricts EH release to a temporal "gate" during the day so that animals that become developmentally competent outside of this gate must delay their eclosion until the gate on the next day (Saunders, 1982). In crickets and grasshoppers the absence of a proper site from which the animal can suspend itself during the shedding of the old cuticle can also delay the onset of ecdysis.

Based on the data from larval ecdysis in *Manduca*, which is simplified because of the involvement of only molt-related factors, Figure 2 provides a simple model of how extrinsic factors may then interact with this system in more complex situations. It is most likely that extrinsic factors such as circadian input and information about acquisition of an appropriate ecdysis perch are transmitted synaptically to the VM cells. During most of the molt, though, the high threshold of the VM cells prevents such input, if it occurs, from firing the cells. Only when these cells become highly excitable, due to the decline in ecdysteroids, are they able to respond to these synaptic inputs. Interestingly, insects that require an appropriate perch for ecdysis will eventually attempt ecdysis if such a site is not provided. This suggests that with time, the increasing excitability of the VM cells can become sufficient to cause them to fire on their own.

B. REGULATION OF SENSITIVITY

The coordination of ecdysis with a specific phase of the molt cycle occurs through the control of both the time of EH release and the sensitivity of the insect to the peptide. Indeed, insects are only responsive to EH during a very narrow time window at the end of each molt. Details of the onset of this sensitivity have been recently reviewed (Truman, 1992; Morton and Truman, 1990) and will only be briefly mentioned here. Events that occur in the response to EH involve an increase in cyclic GMP in the target tissue, and this, in turn, results in the phosphorylation of two 54 kD phosphoproteins that are found in the CNS. The manner by which the phosphorylation of these proteins results in ecdysis is not known. During the intermolt, EH treatments evoke neither a behavioral response nor an increase in cGMP, and no 54 kD phosphoproteins are present. With the onset of the molt, ecdysteroids render the CNS competent to show a cGMP increase when challenged with EH, but it is not until the ecdysteroids decline that the CNS becomes behaviorally responsive to EH. Importantly, this same declining phase also causes the appearance of the 54 kD phosphoproteins. Thus, through the action of ecdysteroids, the EH response system becomes assembled at the end of each molt. After this system is activated by EH, the components then disappear and the system is only reassembled when ecdysteroids return at the next molt.

It is interesting to contrast two features of the regulation of nervous system responsiveness versus the regulation of peptide release. In both cases the systems are sensitive to falling ecdysteroid titers and after a critical level, transcription-dependent mechanisms result in the onset of sensitivity to EH (Morton and Truman, unpublished) and in EH release (Hewes and Truman, 1993). The critical titer of ecdysteroids for induction of sensitivity, though, is higher than that which commits the cells to release. Consequently, as the animal is faced with the declining ecdysteroid titers at the end of the molt, the differences in thresholds insure that when hormone is finally released, the CNS will be in a condition to respond to it.

A second point of comparison is in the role of extrinsic factors. Although, these can play a major role in adjusting the time of EH release, the onset of sensitivity seems to be regulated only by the ecdysteroid decline. This is well illustrated by the adult ecdysis of the giant silkmoths, *Hyalophora cecropia*, in which there is a strong circadian influence over the time of EH release, yet the animals become responsive to EH at essentially any time of day (Truman, unpublished).

IV. DELIVERY OF PEPTIDES TO THEIR TARGET TISSUES

Eclosion hormone has diverse targets, both within the CNS and in the periphery, but a striking feature of the EH system is that injection of peptide into the circulation is sufficient to trigger the well-ordered sequence of central and peripheral actions. Since EH is normally found in the circulation prior to ecdysis, it was concluded that the blood was the normal

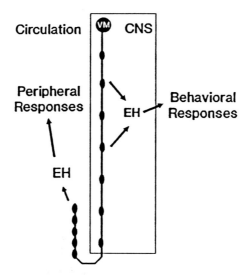

Figure 3. The VM neurons have a dual mode of eclosion hormone release. Peptide is released locally within the CNS to evoke the behavioral responses and into the blood from the proctodeal nerves to act on peripheral tissues.

pathway by which EH reached both its peripheral and central targets (Truman and Riddiford, 1970). The identification of the neurons that release EH, though, presented an alternate mechanism since the axon from these neurons project through the length of the CNS on their way to the proctodeal nerves (Truman and Copenhaver, 1989). Anatomically, these cells are in a position to release peptide locally in each ganglion as well as into the circulation.

The role of peripherally versus locally released EH could be conveniently examined for pupal ecdysis of *Manduca* because in this stage the only peripheral release site was the proctodeal nerves. Manipulations involving the removal of the proctodeal nerves had the expected result of preventing the appearance of circulating EH. Surprisingly, though, these insects nevertheless showed a normal ecdysis behavior (Hewes and Truman, 1991)!

Two pieces of evidence suggest that ecdysis was triggered by the local release of EH from the descending axons of the VM neurons (Hewes and Truman, 1991). Firstly, comparison of immunostained nervous systems from insects a few hours prior to pupal ecdysis versus newly ecdysed pupae showed the expected loss of EH immunoreactivity from the release sites in the proctodeal nerves but also a similar loss of EH-IR from the VM cell axons as they descended through the ventral ganglia. Secondly, antidromic stimulation of the VM axons from the proctodeal nerves resulted in the induction of ecdysis behavior with a latency that was shorter than that seen for the injection of EH. These data support the conclusion that the behavioral responses to EH can be caused by peptide that is released locally within the CNS. This is most likely the normal route by which these behaviors are triggered, although the circulating peptide presumably can act back on the CNS to reinforce the effects of the local release.

What then is the function of the circulating EH? One peripheral target appears to be the Verson's glands, a set of large dermal glands that produce and secrete the outer cement layer over the cuticle. These glands release their products at the time of ecdysis and the secretion is spread over the new cuticle as the old exuvium is shed. Animals that had their terminal segments ligated so that EH was not released into the blood subsequently underwent normal ecdysis, but their Verson's glands failed to secrete (Hewes and Truman, 1991). This blockage of secretion was overcome in these animals by the injection of exogenous EH. Thus, the targets of the circulating peptide appear to include peripheral tissues which must perform functions associated with ecdysis.

Therefore, the VM cells appear to release in two modes: they release locally within the CNS to evoke the behaviors associated with ecdysis and they release systemically into the blood for actions on peripheral tissues (Figure 3). The same cells are involved in both modes of release, and, consequently, the temporal appearance of hormone in the blood is matched by comparable changes within the CNS.

V. PEPTIDE HIERARCHIES

Just as EH release is influenced by ecdysteroids, other peptides are released in a context that is established by EH. Two of these peptides are bursicon, which plasticises and then hardens the cuticle of the new stage (Reynolds, 1977) and the cardioaccelleratory peptides (CAPs) which cause circulatory adjustments that are associated with ecdysis and expansion of the new cuticle (Tublitz and Truman, 1985). The temporal profiles of release for these two peptides have been measured for adult ecdysis in *Manduca*. Both bursicon and the CAPs are secreted by abdominal neurosecretory cells during the postecdysial phase in association with wing expansion. This release can be delayed for a number of hours until the insect finds a suitable site for the inflation and hardening of its new wings (Reynolds *et al.*, 1979; Tublitz and Truman, 1985).

A novel way of identifying neurons associated with EH action resulted from an immunocytochemical search for the neurons responsible for the cGMP increase that occurs at ecdysis in response to EH (Ewer *et al.*, 1993). As indicated above, EH stimulates a rise in cGMP levels which stay elevated for a number of hours (Morton and Truman, 1985). The use of antibodies against cGMP, showed that a significant portion of this increase was due to a network of 50 neurons spread throughout the CNS (Ewer *et al.*, 1993). These include 10 neurons in the subesophageal ganglion (SEG) and 2 pairs of neurons each in ganglion from T1 through A7. The SEG neurons include cells that innervate the foregut musculature and project to the frontal ganglion, a structure associated with the control of swallowing. Also, at least one, and perhaps two, pairs of SEG neurons project through the length of the ventral CNS where they appear to make contact with the two pairs of segmental neurons in each segmental ganglion . Both pairs of the cGMP immunoreactive (cGMP-IR) cells found in the abdominal ganglia are also immunoreactive for Crustacean Cardioactive Peptide (CCAP) which has recently been shown to be one of the insect CAPs (CAP2a, Cheung *et al.*, 1992). One of these neurons is cell 27, an identified projection neuron that extends to the aliary muscles of the heart and to the neurohemal region of the transverse nerve (Taghert and Truman, 1982; Davis *et al.*, 1994). The other CCAP neuron is cell 704, an ascending interneuron (Davis *et al.*, 1994) whose function is unknown. The homologous cells in the thoracic ganglia also show cGMP-IR but in these segments neither are CCAP immunoreactive (Fig. 4; Ewer *et al.*, 1993).

Overall these 50 cells appear to constitute an interconnected network of neurons primarily associated with the foregut and the heart. This relationship is interesting considering that air swallowing and cardiac adjustments are major features associated with ecdysis (Reynolds, 1980). We suggest that this network is responsible for mediating these events.

The timing of the cGMP-IR in this network is interesting considering the time-course of the behavioral events that are triggered by EH. The earliest response to EH is pre-ecdysis behavior which begins about 20 min after EH injection (Copenhaver & Truman, 1982; Miles and Weeks, 1991). By 30 to 35 min, all the neurons in the network showed the rapid appearance of cGMP-IR and levels were just beginning to wane by the time of ecdysis at 40-50 min. Although these cells showed an essentially synchronous activation, the time-course of their loss of cGMP-IR varied dramatically. The descending interneuron was among the first to return to basal levels by 90 min. In both thoracic and abdominal ganglia, detectable levels of cGMP persisted much longer in cell 27 than in 704. In the abdomen, the cGMP-IR in the interneuron returned to basal levels at about the same time as the descending cell but cell 27 maintained elevated levels until about 150 min. The two cells in the thoracic ganglia had longer time-courses than their abdominal counterparts. Indeed, the cell 27s in the thorax still showed detectable levels of cGMP-IR at 6 hr!

The mechanism by which EH activates this network was revealed by experiments in which the larval ventral nerve cord was transected prior to injection of EH. The larvae showed the expected cGMP response in ganglia anterior to the cut but no response in posterior ganglia. This result was observed irrespective of the level of transection in the thoracic or abdominal region of the nervous system. Consequently, cell 27 and 704

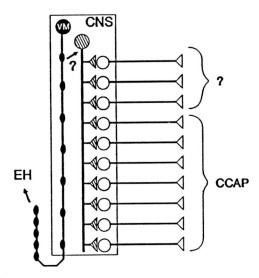

Figure 4. Schematic representation of the relationship of some of the some of the neurons that show a cyclic GMP increase to the VM cells. EH released within the CNS either directly or indirectly activates a descending interneuron in the subesophageal ganglion (cross hatched). This cell in turn apparently makes contact with the cell 27's in each segmental ganglion. The latter cells in the abdomen contain crustacean cardioactive peptide (CCAP); those in the thorax an unknown peptide.

appear not to respond directly to injected EH. In order to activate the network, EH apparently acts on the most anterior ganglia (the brain or SEG) and descending output from these centers then synaptically activates the neurons in the lower ganglia. A good candidate for this descending neuron is the interneuron that extends from the SEG to the terminal ganglion and becomes cGMP-IR after EH treatment. This relationship, though, needs to be proven experimentally.

We assume that this increase in cGMP in these neurons results in an increase in their electrical activity. The fact that CCAP is known to be released at some ecdyses (Tublitz and Truman, 1985) supports this contention. Also, in molluscan neurons, cGMP increases calcium conductances and thus brings about a marked broadening of the action potential (Paupardin-Tritsch *et al.,* 1986).

Thus, the cGMP immunocytochemistry has highlighted a set of cells whose role in ecdysis was not previously appreciated. They are primarily a peptidergic system and over half of them contain CCAP. Their thoracic homologues presumably also contain a peptide whose nature in *Manduca* is unknown. This network is activated well after the onset of pre-ecdysis behavior but prior to the onset of ecdysis. Presumably, through actions on the foregut and through release of CCAP and other peptides this network initiates air swallowing and the circulatory changes that accompany ecdysis. The timing of activation of this network brings up the possibility that actions of these peptides within the CNS might be involved in triggering the ecdysis behavior itself. The subsequent differences in the time-course of activation of the thoracic versus the abdominal cells suggest that the two groups of neurons may subtend different physiological or behavioral functions that continue through the post-ecdysial phase.

VI. CONCLUSIONS

The initial attractiveness of the ecdysis system of insects was its seeming simplicity. The above discussion, though, shows that this simplicity disappears as the system is put in its normal physiological context. Eclosion hormone is a key player in the control of this behavior but its release and action occurs only in the context of shifts in the titer of the ecdysteroids. Likewise, some of the events that occur "down-stream" in the response sequence are not directly due to EH but to peptides that are released as a consequence of EH action. Despite this increasing complexity, the EH system of insects still provides an excellent system for examining the cellular basis of how a neuroendocrine effector system regulates complex physiological processes.

REFERENCES

Bollenbacher, W.E. and Granger, N.A., Endocrinology of the prothoracicotropic hormone, in *Comprehensive Insect Physiology, Biochemistry, and Pharmacology*, Vol. 7, Kerkut, G.A. and Gilbert, L.I. Eds., Pergamon Press, Oxford, 1985, p. 109-151.

Cheung, C.C., Loi, P.K., Sylwester, A.W., Lee, T.D., and Tublitz, N.J., Primary structure of a cardioactive neuropeptide from the tobacco hawkmoth, *Manduca sexta. FEBS Let.*, 313, 165-168, 1992.

Copenhaver, P.F., and Truman, J.W., The role of eclosion hormone in the larval ecdysis of *Manduca sexta. J. Insect Physiol.*, 28, 695-701, 1982.

Davis, N.T., Homberg, U., Dircksen, H., Levine, R.B., and Hildebrand, J.G., Crustacean cardioactive peptide-immunoreactive neurons in the hawkmoth *Manduca sexta* and changes in their immunoreactivity during postembryonic development. *J. Comp. Neurol.*, in press, 1994.

Ewer, J., de Vente, J., and Truman, J.W., Eclosion hormone triggers a long-lasting, synaptically-driven increase in cGMP in identifiable insect neurons. submitted, 1993.

Hewes, R.S., and Truman, J.W., The roles of central and peripheral eclosion hormone release in the control of ecdysis behavior in

Manduca sexta. J. Comp. Physiol. A 168, 697-707, 1991.

Hewes, R.S. and Truman, J.W., Steroid regulation of excitability in identified insect neurosecretory cells. submitted, 1993.

Miles, C.I., and Weeks, J.C., Developmental attenuation of the pre-ecdysis motor pattern in the tobacco hornworm, *Manduca sexta. J. Comp. Physiol. A*, 157, 423, 1985.

Morton, D.B., and Truman, J.W., Steroid regulation of the peptide-mediated increase in cyclic GMP in the nervous system of the hawkmoth, *Manduca sexta. J. Comp. Physiol. A*, 157, 423-432, 1985.

Morton, D.B., and Truman, J.W.,The molecular basis for steroid regulation of eclosion hormone action in the tobacco hornworm, Manduca sexta, in *Comparative Pharmacology of Neuropeptides*, Stefano, G.B., and Florey, E., Eds. Manchester University Press, Manchester, 1990, p. 187-197.

Paupardin-Tritsch, D., Hammond, C., Gerschenfeld, H.M., Nairn, A.C., and Greengard, P., cGMP-dependent protein kinase enhances Ca^{2+} current and potentiates the serotonin-induced Ca^{2+} current increase in snail neurons. *Nature*, 323, 812-814, 1986.

Reynolds, S.E., Control of cuticle extensibility in the wings of adult *Manduca* at the time of eclosion: effects of eclosion hormone and bursicon. *J. Exp. Biol.*, 70, 27-37, 1977.

Reynolds, S.E., Integration of behaviour and physiology in ecdysis. *Adv. Insect Physiol.*, 15, 475-595, 1980.

Reynolds, S.E., Taghert, P.H., and Truman, J.W., Eclosion hormone and bursicon titres and the onset of hormonal responsiveness during the last day of adult development in *Manduca sexta* (L.). *J. Exp. Biol.*, 78, 77-86, 1979.

Riddiford, L.M., Hormone action at the cellular level, in *Comprehensive Insect Physiology, Biochemistry, and Pharmacology*, Vol. 8, Kerkut, G.A. and Gilbert, L.I. Eds., Pergamon Press, Oxford, 1985, p. 37-84.

Saunders, D.S., *Insect Clocks*, Second Edition, Pergamon Press, Oxford, 1982.

Schwartz, L.M., and Truman, J.W., Hormonal control of rates of metamorphic development in the tobacco hornworm *Manduca sexta. Dev. Biol.*, 99, 103-114, 1983.

Slama, K., Homeostatic function of ecdysteroids in ecdysis and oviposition. *Acta. Entomol. Bohemoslovaca*, 77, 145-168, 1980.

Taghert, P.H., and Truman, J.W., Identification of the bursicon-containing neurons in abdominal ganglia of the tobacco hornworm, *Manduca sexta. J. Exp. Biol.*, 98, 385-401, 1982.

Truman, J.W., The eclosion hormone system of insects. *Prog. Brain Res.*,

92,361-372, 1992.

Truman, J.W., and Copenhaver, P.F., The larval eclosion hormone
    neurones in *Manduca sexta*: identification of the brain-proctodeal
    neurosecretory system. *J. Exp. Biol.*, 147, 457-470, 1989.

Truman, J.W. and Riddiford, L.M., Neuroendocrine control of ecdysis
    in silkmoths. *Science*, 167, 1624-1626, 1970.

Truman, J.W., Rountree, D.B., Reiss, S.E., and Schwartz, L.M.,
    Ecdysteroids regulate the release and action of eclosion hormone in
    the tobacco hornworm, *Manduca sexta* (L). *J. Insect Physiol.*, 29,
    895-900, 1983.

Tublitz, N.J., and Truman, J.W. Insect cardioactive peptides:
    II. Neurohormonal control of heart activity by two
    cardioacceleratory peptides (CAPs) in the tobacco hawkmoth.
    Manduca sexta. J. Exp. Biol. 118, 381-395, 1985.

CLONING AND MOLECULAR ANALYSIS OF G-PROTEIN COUPLED RECEPTORS IN INSECTS

Linda M. Hall[1], Frances Hannan[2], Guoping Feng[1], Daniel F. Eberl[3], Yuen Yi Hon[1], and Christopher T. Kousky[1]

[1]Dept. of Biochemical Pharmacology, SUNY at Buffalo, Buffalo, NY 14260
[2]AFRC Lab. of Molecular Signalling, Univ. of Cambridge, Dept. of Zoology, Cambridge, England, CB2 3EJ
[3]Dept. of Genetics, Harvard Medical School, 200 Longwood Ave., Boston, MA 02115

I. GENERAL PROPERTIES OF G-PROTEIN COUPLED RECEPTORS

G-protein coupled receptors are a large family of structurally related transmembrane receptors which share common structural motifs. These receptors consist of a single protein with seven transmembrane domains (TM1 to TM7) joined by alternating extracellular and intracellular loops. They begin with an extracellular amino-terminal domain and end with a cytoplasmic carboxy-terminal tail. The transmembrane regions group together in the membrane to form a "cup" into which the ligand fits to activate the receptor. The ligand may be a neurotransmitter, a peptide, or even light in the case of rhodopsins. Agonist binding to the receptor is thought to induce a conformational change within the transmembrane domains that is transmitted to the cytoplasmic face where interactions with a specific guanine nucleotide binding protein complex (G-protein) is known to occur. The agonist-activated receptor interacts with a G-protein.

"At rest" this G-protein binds GDP and is in an inactive form. The G-protein complex consists of three subunits designated as: alpha(α), beta(β), and gamma (γ). Interaction of the G-protein complex with the activated receptor results in the replacement of GDP by GTP and activates the G-protein complex. When this complex is activated by the binding of GTP, the α subunit with GTP associated with it dissociates from $\beta\gamma$. The activated α subunit interacts with one of the effector systems listed below. At least for some receptors, association with the $\beta\gamma$ complex alone may facilitate phosphorylation by a kinase leading to immediate desensitization of the receptor. When the $\beta\gamma$ complex dissociates from the receptor, it is then free to reassociate with the α subunit or in some cases to interact with other effectors. When the ligand dissociates from the receptor and the phosphate is removed from the receptor, the process is ready to begin again.

Depending on the G-protein involved, these receptors can be coupled to a

0-8493-4591-X/94/$0.00 + $.50
© 1994 by CRC Press, Inc.

variety of effectors including: phosphodiesterase, phospholipase C, adenylate cyclase, phospholipase A2, and direct coupling to ion channels. Whether these interactions are stimulatory or inhibitory is determined by the G-protein involved.

G-protein coupled receptors, also known as heptahelical receptors, include sensory receptors such as the rhodopsins which are activated by light, olfactory receptors activated by odorants, as well as neurotransmitter receptors activated by: adrenaline, noradrenaline, acetylcholine (muscarinic receptors), serotonin, dopamine, and peptides such as Substance K and Neuropeptide Y. This manuscript will emphasize muscarinic acetylcholine receptors and an adrenergic receptor homologue designated as an octopamine/tyramine receptor. These receptors have been cloned and sequenced from the fruitfly *Drosophila melanogaster* based on sequence similarity to previously cloned genes from vertebrates. It should be emphasized that once a successful approach has been developed for a receptor from one insect, it is a relatively simple task to isolate the homologous gene from another insect species because the sequence differences are generally small. Thus, a gene may first be cloned from a well-studied insect such as *Drosophila* where many molecular genetic tools are available. The information gained from the *Drosophila* clone allows for rapid isolation of homologous clones from pest insects or from invertebrates important for physiological studies.

II. APPROACHES FOR CLONING G-PROTEIN COUPLED RECEPTORS

There are two standard approaches in molecular biology which can be used for cloning new members of a gene family or for cloning the homologous gene from a different species once the first cDNA is available. These are: reduced stringency library screening and the polymerase chain reaction (PCR) using degenerate primers. There are many examples of the successful application of the first strategy. (See, for example, Onai et al., 1989; Arakawa et al., 1990.) The PCR based strategy is a newer one that allows more rapid cloning. Successful applications of PCR based cloning appear in the recent literature. [Libert et al., 1989; Buck and Axel, 1991; Murtagh et al., 1993]

A. REDUCED STRINGENCY LIBRARY SCREENING
This method requires that a clone (either genomic or cDNA), partial clone, or oligonucleotide containing protein coding sequence of the gene of interest be available to use for probing a library. The success rate will be improved if the probe is enriched for sequences which are most highly conserved across species (if that is known) or across receptor subtypes. Choice of the probe is a most important aspect of this approach. Intron

bearing regions of genomic clones should be avoided since the sequence of these regions is unlikely to be conserved across species. Similarly, the 5' and 3' untranslated regions of cDNA clones should be avoided.

Once a probe is selected, an appropriate library must be chosen. If the tissues and/or developmental stages which express the receptor of interest are known, then screening a cDNA library prepared from that tissue or stage will be effective. Alternatively, a genomic library may be used since it makes no assumptions about time and place of gene expression. All genes will be represented in a genomic library with approximately equal frequency. Although the presence of noncoding introns in genomic clones may complicate initial sequence analysis, intron regions can generally be recognized by codon usage [Gribskov et al., 1984] or third position codon bias analysis. [Bibb et al., 1984] These methods of analysis take into account the fact that when there is more than one codon for a given amino acid, different species show preferential use of different codons for the same amino acid. A preponderance of rarely used codons in a stretch of genomic DNA is a good indicator of a noncoding region.

Reduced stringency hybridization conditions involve high salt and low temperatures during hybridization and washes. Successful conditions have been defined in the literature and many variations are possible. Hybridization in 5XSSC, 5XDenhardts, and 0.1 mg/ml sheared, denatured salmon sperm DNA at 42°C followed by washes ending with 2X SSC, 0.1%SDS at 42°C is a good starting point. Increasing the wash temperature 5° at a time starting with 42° allows the identification of a series of clones with increasing similarity to the probe. Once candidate clones are available, a Southern blot of each clone hybridized with the original probe and washed with increasing stringency will identify the most conserved regions of the candidate clones. These regions of conserved DNA should be sequenced. The deduced amino acid sequence will reveal whether the clone encodes a G-protein coupled receptor. This approach was used in the successful cloning of the first muscarinic acetylcholine receptor from an insect [Onai et al., 1989; Shapiro et al., 1989], for the cloning of the first insect adrenergic receptor homologue [Arakawa et al., 1990; Saudou et al., 1990], and for cloning a tachykinin-like peptide receptor. [Li et al., 1991]

B. THE POLYMERASE CHAIN REACTION (PCR)

The disadvantage of the reduced stringency library screening approach described above is that substantial effort must be put into plaque purifying clones prior to the sequencing in order to determine the clones of interest. To speed up the cloning process, the polymerase chain reaction (PCR) can be employed. One advantage of PCR is that primers from several different conserved regions can be tested simultaneously and the region to be amplified can be selected so that it includes definitive regions of the G-protein coupled receptor of interest. Two different regions have been used successfully to amplify insect G-protein coupled receptors based on sequence information available from the mammalian literature. To account for cross species differences, it is often necessary to use degenerate primers and to reduce the stringency of the annealing conditions by lowering the annealing temperature or increasing the magnesium concentration (from 1.5 to 2 or 2.5 mM) in the PCR buffer to allow for sequence mismatch. In all cases for PCR amplification, sequence mismatch is more readily tolerated at the 5' end than at the 3' end of the primers. This is important for primer design.

1. Primers from TM3 and TM6

New members of the heptahelical receptor class from *Drosophila* have been cloned in our laboratory using primers to highly conserved regions in transmembrane domains 3 and 6. We have used *Drosophila* genomic DNA as the template to avoid assumptions about when and where the receptors will be expressed. As discussed in a later section, primers designed within conserved regions of the octopamine/tyramine receptor allowed us to clone a closely related, but genetically distinct receptor, Gpcr68. This approach will also yield more distantly related members of the G-protein coupled receptor class as evidenced by the fact that we also isolated a peptide-type receptor using this same primer set. Others [Li et al., 1992] have used a primer pair from these regions to clone a neuropeptide Y receptor from *Drosophila*. Sequencing, comparisons with other receptor sequences in the database, and ultimately expression studies are required to definitively identify the specificity of the newly cloned receptor regardless of which cloning method is used because of the similarities among different members of each receptor class.

Despite the difficulties in identifying the specificity of each receptor, some clues can be obtained simply from the size of the amplification product when PCR primers from TM3 and TM6 are used. These primers amplify a region that includes the third cytoplasmic loop between TM4 and TM5. This loop varies in size among the different classes of heptahelical receptors with the smallest generally being the peptide receptors and the largest being the muscarinic acetylcholine receptors. If cDNA is used as the PCR template, introns will not complicate the interpretation. The largest products will be enriched for muscarinic receptors while the smallest ones will be enriched for peptide receptors.

Recognizing PCR products of interest can be challenging in cross species cloning. Except for the portions closest to the transmembrane domains, the third cytoplasmic loop varies dramatically in sequence among the different receptor types. Even within a given receptor type, it varies dramatically across species. Therefore, this loop itself is not a good diagnostic sequence for determining whether the amplified product represents a receptor. However, the amplified product will also include the sequence encoding two full transmembrane domains. The presence of the hydrophobic TM4 and TM5 domains in the deduced amino acid sequence is an easily recognizable diagnostic feature that aids in the identification of products which encode heptahelical receptors.

This primer set (spanning TM3 and TM6) is very useful when several different receptor classes are sought simultaneously. Products of different sizes will be generated because of the variation in size of cytoplasmic loop 3 and these different size products will be easily separated by agarose gel electrophoresis. Frequently, the size difference among the PCR products will be large enough so that amplification products can be separated cleanly and sequenced directly. This method is very effective for muscarinic receptors which have a larger than average third cytoplasmic loop and for peptide receptors which have very small loops.

2. Primers from TM6 and TM7

Another PCR primer design that has worked successfully for the cloning of a tachykinin (NKD) receptor [Monnier et al., 1992] from *Drosophila* uses degenerate, receptor-specific primers from regions TM6 and TM7. These regions, at the carboxy end of the receptor, are highly conserved in a given receptor type. Because these primers span a shorter sequence than the TM3/6 pair, the product is "easier" to amplify and sequence. The short, amplified sequence contains two highly diagnostic hydrophobic domains separated by a short extracellular loop.

C. ENRICHING FOR PRODUCTS OF INTEREST

A difficulty for both the reduced stringency library screening or the PCR methods described above is that many of the products isolated will be sequences unrelated to G-protein coupled receptors. There are several strategies to enrich for products of interest. For clones isolated by reduced stringency library screening, PCR can be used to see if any can be amplified by receptor specific primers. For PCR products, nested primers can be used to determine whether there is an expected match lying within the amplified region. Alternatively, a cDNA segment from a related receptor clone of interest can be used at reduced stringency to probe a Southern blot followed by sequential washes at increasing stringency (increasing wash temperature, decreasing salt concentration) to identify those products which hybridize the probe at the highest stringency. These approaches can be used to focus initially on the most promising products.

III. MUSCARINIC ACETYLCHOLINE RECEPTORS

Two groups have used reduced stringency library screening with a mammalian muscarinic cDNA probe to clone the same muscarinic receptor from a *Drosophila* head cDNA library. [Onai et al., 1989; Shapiro et al., 1989] This receptor from *Drosophila* has all the features expected of a classical G-protein coupled receptor and shows greater than 60% overall sequence similarity with vertebrate muscarinic receptors.

A.   GENERAL STRUCTURAL FEATURES

As mentioned above, all muscarinic receptors contain a very large third cytoplasmic loop and this is true to an even greater extent for Dromace, the *Drosophila* receptor. In the vertebrate receptors this loop is 157-240 amino acids in length while this loop in the *Drosophila* receptor is even longer with 412-429 amino acids depending on which splice variant is considered. [Hannan and Hall, 1993]

In vertebrate muscarinic receptors, the acetylcholine binding pocket is thought to be lined with amino acid side chains from transmembrane domains 3, 5, 6, and 7. A completely conserved aspartic acid in TM3 is thought to interact with the positively charged amino head group of acetylcholine and that aspartic acid is also conserved in the *Drosophila* receptor. This residue is also conserved in all receptors that bind biogenic amines since it plays a similar role in binding the amine moiety of other ligands as well. Thus, additional amino acid side chains in the ligand binding pocket must be involved in determining ligand specificity for a given receptor.   Sequence comparisons have shown that the transmembrane regions of all muscarinic receptors contain a series of tyrosine and threonine residues which are not conserved among other G-protein coupled receptors. It is thought that the side chains of these amino acids interact specifically with the acetylcholine ester moiety by hydrogen bonding to provide the agonist specificity. [Wess, 1993] Consistent with this suggestion is the observation that all of these tyrosines and threonines are conserved in the *Drosophila* receptor. Also conserved in the *Drosophila* receptor are two cysteine residues in extracellular loops 2 and 3. These cysteine residues are thought to form a disulfide bond important for proper receptor conformation. [Wess, 1993]

Although there is a high degree of conservation between the insect receptor and vertebrate receptors, there are regions in the transmembrane and the extracellular domains which are strikingly different. These regions might serve as sites for insect-specific antagonist action in the development of new insecticides.

B.   RECEPTOR SUBTYPE SPECIFIC COUPLING

Five genetically distinct subtypes have been found for mammalian muscarinic receptors: m1, m2, m3, m4, and m5. Generally m1, m3, and m5 are grouped together because they all activate phospholipase C and increase phosphoinositol hydrolysis whereas the m2 and m4 subtypes inhibit adenylate cyclase. [Richards, 1991] A comparison of the different domains of the *Drosophila* muscarinic receptor with corresponding regions of the vertebrate receptor subtypes indicates that the *Drosophila* receptor is most similar to the vertebrate m1, m3, and m5 receptors. [Hannan and Hall, 1993] This is consistent with the studies of Shapiro et al. [1989] which demonstrated that *Drosophila* muscarinic receptors expressed in mammalian cells in culture increase phosphoinositol (PI) turnover in response to agonist stimulation. This coupling specificity is thought to reside in two groups of ~20 amino acids each lying at either end of the third cytoplasmic loop in the regions closest to TM5 and TM6. Indeed, in these regions the *Drosophila* receptor shows very little sequence similarity to the m2 and m4 receptors, but shows considerable similarity to the m1, m3, and m5 subtypes. [Hannan and Hall, 1993]

Do insects also have genetically distinct classes of muscarinic receptors? This would be an ideal question to address with a PCR based screening strategy. Such work is in progress in our laboratory. There is considerable evidence for muscarinic receptor subtypes in invertebrates from both ligand binding studies and electrophysiological analysis. [reviewed by Hannan and Hall, 1993] It seems likely that further molecular biological screening will uncover additional subtypes in insects.

C. PATTERN OF EXPRESSION

In order to determine the role that the cloned muscarinic receptor plays in the insect, *in situ* hybridization to tissue sections was used to determine the temporal and spatial patterns of expression in *Drosophila*. [Hannan and Hall, submitted for publication] The patterns of expression as determined by Northern blotting of mRNA from body parts and different developmental stages (Table 1) and by *in situ* hybridization to sections with ^{35}S labeled antisense probes (Table 2) are summarized and compared with those patterns for other cloned G-protein coupled receptors from *Drosophila*. Quantitative Northern blotting showed that muscarinic receptor mRNA is more abundant in heads than bodies as would be expected for a neuronal receptor. In addition, developmental Northern analysis showed two peaks of expression for this receptor. The first is in late embryos (18-21 hr at 25°C) and the second is late pupae (3-4 days after puparium formation at 25°C). This pattern is seen for many nervous system specific genes.

The nervous system specific expression of this receptor was confirmed by *in situ* hybridization studies to adult tissue sections which showed hot spots of expression in antennal lobes, mushroom bodies, and ventral lateral protocerebrum (Table 2). This pattern of expression suggests a role for muscarinic receptors in processing both olfactory and visual information. Expression in the mushroom bodies suggests that these receptors may also be involved in learning/memory processes since

Table 1

Northern Blot Analysis of Expression Patterns of Cloned G-Protein Coupled Receptors from *Drosophila*

Receptor Name*	Adult		Developmental Profile Peak Expression	Reference
	Heads	Bodies		
Dronkd	++++	+	9-22hr embryos Adult heads	Monnier et al., 1992
Drotkr	++++	–	14-18 hr embryos First instar larvae Pupae, adults	Li et al., 1991
Dronpy	+++	+++	14-18hr embryos Pupae Adult heads, bodies	Li et al., 1992
Dro5ht2a	++++	–	16-24hr embryos, pupae Adult heads	Saudou et al., 1992
Dro5ht2b	++	–	16-24hr embryos Adult heads	ibid.
Droocr	++++	+	18-21 hr embryos 3-4 day pupae Adult heads	Hannan & Hall, submitted
Dro5htr	+++	–	16-24hr embryos Adult heads	Saudou et al., 1992
Gpcr68	+++	–	Not done	–
Dromace	++++	+	18-21hr embryos 3-4 day pupae Adult heads	Hannan & Hall, submitted

*See Table 4 for full length names of receptors.

Table 2

Tissue *In Situ* Hybridization Analysis of Expression Patterns of Cloned G-Protein Coupled Receptors from *Drosophila*

Receptor Name	Expression Pattern	Reference*
Dronkd	**Embryos**: Subsets of neurons in brain and in each segment of the developing ventral ganglia.	
Drotkr	**Embryos**: Subsets of neurons in brain and in each segment of the developing ventral ganglia.	
Dronpy	Not reported	
Dro5ht2a	**Embryos**: Ventral midline motor neurons in abdominal segments (VUM neurons) that innervate larval muscles; plus lateral cells in thoracic segments.	
Dro5ht2b	**Embryos**: Lateral rows of cells on either side of midline. Low level expression in midline cells.	
Droocr	**Embryos**: Central nervous system **Adult**: Antennal lobe, ventrolateral protocerebrum, mushroom bodies, anterior fat body. Low level expression throughout nervous system.	
Dro5htr	**Embryos**: Lateral cells on either side of midline.	
Gpcr68	Not done.	
Dromace	**Embryos**: Central nervous system **Adult**: Antennal lobe, ventrolateral protocerebrum, mushroom bodies, ovaries plus general low level expression throughout brain and thoracic ganglia.	

*Same references as Table 1.

numerous studies have implicated mushroom bodies in these processes in insects. [Davis, 1993] In addition, expression was found in ovaries suggesting a role in female fertility. Finally, there was a general low level of expression in most cell body regions of the central brain, optic lobes and thoracic ganglia.

D. GENETIC ANALYSIS

Although the *in situ* hybridization studies provide some clues about the *in vivo* receptor function, these studies do not address the question of what the consequences are to the organism if this receptor is inactivated. To address this question, we have begun a genetic analysis of muscarinic receptors in *Drosophila*. The general strategy of these studies is to first map the gene by *in situ* hybridization of the cloned gene to polytene chromosomes from larval salivary glands. Once the initial map position is determined by examination of the hybridization signal relative to the chromosome banding pattern, deletions and other chromosome aberrations are used to provide a more precise map position by determining whether the hybridization signal is within or outside of the chromosome aberration. This process is repeated for a number of deletions in the area of interest and then additional studies are done with mutations thought to map to the same area. The object is to identify mutations which show the same deletion mapping pattern as the hybridization signal. After candidate mutations have been identified, transposable element mediated transformation rescue experiments [Spradling, 1986] are done to determine whether a clone carrying the entire coding sequence of the wild-type gene can rescue ("cure") the candidate mutant defect.

Initial studies showed that this muscarinic receptor mapped to 60C5-8 on the second chromosome. [Onai et al., 1989] Deletion mapping showed that the cloned gene comaps with a number of candidate genes including: a female sterile, several lethal genes and several visible mutations. [Eberl and Hall, unpublished observations] Transformants have been constructed which carry the full length coding sequence for the muscarinic receptor and studies are in progress to determine which of the candidate mutations require the intact muscarinic receptor for rescue. With this approach we should soon know the phenotype of mutations in this muscarinic acetylcholine receptor.

IV. ADRENERGIC RECEPTOR HOMOLOGUES

Two groups have cloned the same adrenergic receptor homologue from *Drosophila* using different probes. [Arakawa et al., 1990; Saudou et al., 1990] In this example, Arakawa et al., [1990] used a human β-adrenergic cDNA probe to isolate *Drosophila* genomic clones by reduced stringency screening. A fragment of these genomic clones which contained an exon encoding a region with high homology to the sixth and seventh transmembrane domains of adrenergic receptors was then used as a hybridization probe to isolate the full length cDNA from a *Drosophila* head cDNA library. Saudou et al. [1990] also used reduced stringency hybridization, but used degenerate oligonucleotide probes from the conserved transmembrane domains TM6 and TM7 to probe a genomic library. Again, they used a genomic clone as a probe to isolate a cDNA encoding the full length receptor. It is quite likely that the oligonucleotide probes used by Saudou et al. [1990] for library screening could also have been used for PCR amplification of the segment between the two probes as described in section II.B above.

Although the gene was cloned with a β_2-adrenergic receptor probe, sequence analysis showed that the clone was actually closest to the α_2 type adrenergic receptor. This illustrates that when reduced stringency is introduced into the screening process in order to allow cross-species cloning, this relaxation of stringency also may result in the isolation of a different receptor type.

A. PHARMACOLOGICAL SPECIFICITY

This cloned receptor was originally designated as an octopamine receptor based on the pharmacological profile of the cloned receptor expressed in CHO-K1 cells. [Arakawa et al., 1990] This receptor was shown to inhibit a forskolin-stimulated cAMP response in a dose dependent fashion when activated with octopamine. Thus, it has an inhibitory action on adenylate cyclase. This inhibition by octopamine could be blocked by yohimbine, an α_2-adrenergic receptor antagonist.

The expressed receptor binds yohimbine with high affinity (K_d of 6.2nM). The rank order potency of antagonist inhibition of [^{3}H]yohimbine binding was : chlorpromazine>mianserin>phentolamine$\geq$ cyproheptadine>metoclopramide$\geq$propranolol. The most effective agonist was synephrine, the N-methylated analog of octopamine (K_i= 10.8$\pm$2.5μM). Other effective agonists were: clonidine (K_i=21$\pm$7μM), octopamine (K_i=29$\pm$6μM), serotonin (K_i=75$\pm$18μM) and epinephrine (K_i=139$\pm$41μM). On the basis of the high receptor affinity for yohimbine and chlorpromazine as compared with metoclopramide, Arakawa et al. [1990] suggested that this receptor was an octopamine type 1 receptor. [Evans, 1980, 1981, 1987; Evans et al., 1988]

Subsequent work on an independently cloned and expressed version of this receptor showed that tyramine was an even more effective inhibitor of both [^{3}H]yohimbine binding and adenylate cyclase activity than octopamine leading the authors to designate this receptor as a tyramine receptor. [Saudou et al., 1990] Since this receptor was originally named an octopamine receptor and since the actual *in vivo* transmitter has not yet been determined, we now refer to this as an octopamine/tyramine receptor (oct/tyr) or as Droocr in the tables.

The question was raised as to whether the high affinity for tyramine of this receptor expressed in NIH3T3 cells was an artefact of the expression system or whether it was also seen when the clone was expressed in CHO cells. [Arakawa et al., 1990] Similarly, is tyramine also a more effective inhibitor of [3H]yohimbine binding to the receptor found in *Drosophila* head membrane extracts? New studies have shown that tyramine is an effective inhibitor of [3H]yohimbine binding both *in vivo* and for the receptor expressed in CHO cells so the high affinity for tyramine is real and not an artefact of an expression system. [S. Robb, T.R. Cheek, F.L. Hannan, L.M. Hall, J.M. Midgley, and P.D. Evans, submitted for publication]

Determining the *bona fide* transmitter for this receptor *in vivo* will be a difficult task. The highest affinity compound is not necessarily the endogenous ligand. For example, the agonist nicotine has a much higher affinity for nicotinic acetylcholine receptors in insects than does the endogenous transmitter, acetylcholine and yet nicotine would not be considered the true transmitter. [Schmidt-Nielsen et al., 1977] Additional work will be required to define the true ligand for this receptor.

B. TARGET FOR FORMAMIDINE INSECTICIDES

Insect octopamine receptors are thought to be the target for the formamidine class of insecticides. To determine whether this cloned receptor has high affinity for formamidines, we have tested the ability of two formamidines (desmethylchlordimeform and Amitraz) to inhibit [3H]yohimbine binding to the cloned receptor expressed in CHO cells. Both are better inhibitors than octopamine showing IC_{50}'s about two orders of magnitude lower than octopamine. [Hall, unpublished observations] Regardless of what the endogenous agonist is, this receptor is at least one potential target for the formamidine class of insecticides.

C. PATTERN OF EXPRESSION

As described for the muscarinic receptor, we have used *in situ* hybridization with ^{35}S-labeled antisense probes from the untranslated regions of the oct/tyr receptor cDNA to determine the pattern of expression of this receptor by both Northern blot analysis and *in situ* hybridization to sections. These patterns are summarized in Tables 1 and 2. The oct/tyr receptor resembles the muscarinic receptor discussed in section III.C since the receptor mRNA is much more abundant in adult heads than in bodies as determined by Northern blotting (Table 1). A developmental Northern blot showed peaks of expression in late embryos at 18-21 hr and late pupae (3-4 days after pupariation) just as for the muscarinic receptor. Again the interpretation is that this receptor is likely to be enriched in the nervous system.

This was confirmed by *in situ* hybridization to tissue sections of embryos and adults. In the thoracic ganglia, there was some generalized expression, but there was also a more punctate expression than seen with the muscarinic receptor. It appears that certain cells are showing high expression of the oct/tyr receptor in the thoracic ganglia. Higher resolution methods such as digoxigenin-labeled oligonucleotide probes or use of receptor-specific antibodies will be needed to map the specific cells expressing this receptor.

Like the muscarinic receptor, the oct/tyr receptor is highly expressed in adult heads. Particularly strong expression was observed in distinct groups of cells in the mushroom body calyces and punctate expression was also observed in the ventral lateral protocerebrum and the antennal lobes. The expression of this receptor in the mushroom bodies is interesting in light of the observations that physical or genetic disruption of this area results in a dramatic decrement in learning/memory. [Erber et al., 1980; Heisenberg et al., 1985; Davis, 1993] Octopamine has demonstrated effects on learning ability in *Drosophila*. [Dudai et al., 1987] However, octopamine has been reported to be a potent stimulator rather than an inhibitor of adenylate cyclase. [Uzzan and Dudai, 1982] It is possible that there is more than one type of octopamine-activated receptor in the *Drosophila* brain. *In vitro* studies may measure only one of these either because the conditions of the experiment favor one type over the other or because one form is more abundant than the other and overwhelms the assay. The advantage of expression of individual cloned receptor genes is that it allows the characterization of one genetically distinct receptor subtype without the confounding influence of other subtypes.

One unusual place where this receptor was expressed was in the anterior (but not posterior) fat body suggesting a possible involvement of this receptor in the regulation of lipid and carbohydrate metabolism in the head. High levels of expression of this receptor in the anterior fat body is consistent with studies on octopamine binding in housefly heads which showed that 90% of the binding was to non-neuronal head tissues. [Hashemzadeh-Gargari and Wierenga, 1989]

V. OTHER G-PROTEIN COUPLED RECEPTORS
A. NEW RECEPTORS FROM PCR EXPERIMENTS

Work on locust has demonstrated that there are at least three distinct octopamine receptor subtypes in locust. [Evans, 1981] A similar situation would be expected for other insects including *Drosophila*. To look for other members of this receptor class, we used the strategy described in section I.B using PCR to amplify a segment between TM3 and TM6 which for the oct/tyr cDNA would produce a fragment ~1kb in length. Multiple products came from this reaction when low stringency annealing was allowed and genomic DNA was used as the template. These products included a 1.7 kb fragment corresponding to the oct/tyr gene itself which carries an intron in this region (accounting for the increased size of the genomic fragment compared with the control cDNA amplification product). When a Southern blot of the amplified products probed with the oct/tyr cDNA was washed at increasing stringency, a 0.68 kb fragment retained a signal at higher stringency than any of the other products. This indicates that the 0.68 fragment is more similar to the previously cloned oct/tyr cDNA than the other products and therefore is another amine receptor candidate. This 0.68 kb fragment was used to isolate a cDNA encoding a new G-protein coupled receptor (Gpcr68).

B.   COMPARISON OF CLONED G-PROTEIN COUPLED RECEPTORS FROM *DROSOPHILA*

Table 3 illustrates the relationship among all of the currently cloned G-protein coupled receptors from *Drosophila*. This tree was prepared using the Pileup program of the Wisconsin GCG Package. [Higgins and Sharp, 1987] The new receptor (Gpcr68) is grouped with the serotonin type 1 receptor whereas the original oct/tyr receptor (designated as Droocr in this table) is grouped with the type 2 serotonin receptors. All 5 of these receptors (Dro5ht2a, Dro5ht2b, Dro5htr, Droocr, and Gpcr68) fall into a subgroup indicating their close structural similarity. This emphasizes the molecular basis for the difficulties in distinguishing these receptors pharmacologically. Octopamine and serotonin have similar affinities for the oct/tyr (Droocr) receptor. (See section IV.A.) This tree analysis predicts that a similar situation could exist for Gpcr68 since it is close to the type 1 serotonin receptor. Expression studies will be required to determine whether Gpcr68 shows the pharmacological specificity predicted for one of the octopamine receptor subtypes as characterized in locust.

Referring again to Table 3, it is apparent that the peptide receptors (Dronkd, Drotkr, and Dronpy) fall into a separate grouping. The muscarinic acetylcholine receptor (Dromace in the table) is grouped with the amine receptors, but is distinct from them. It is anticipated that *Drosophila* and other insects will have a wide variety of G-protein coupled receptors just as has been observed in mammals. This summary represents only the beginning of their characterization. Not included in this summary are the rhodopsins which were the first G-protein coupled receptors to be cloned from *Drosophila*. They are light-activated receptors important for transduction in the visual system. This review includes only the ligand activated receptors.

Table 3

Evolutionary Tree for All G-Protein Coupled Receptors Cloned From *Drosophila* [1]

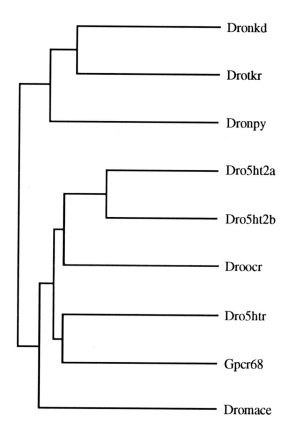

[1]Full length names of receptors and most of the references providing their deduced amino acid sequences are given in Table 4. Dro5htr sequence comes from Witz et al., 1990.

Table 4

Chromosome Mapping of *Drosophila* G-Protein Coupled Receptors

"Tree" Name	Receptor Type	Chromosome Location	Reference
Dro5ht2a	Serotonin receptor 2A	2R [1], 56A-B	Saudou et al., 1992
Dro5ht2b	Serotonin receptor 2B	2R, 56A-B	ibid.
Dromace	Muscarinic acetylcholine receptor	2R, 60C5-8	Onai et al., 1989; Shapiro et al., 1989
Dronkd	Tachykinin (NKD) receptor	3R[2], 86C	Monnier et al., 1992
Dronpy	Neuropeptide Y receptor	3R, 97E1-2	Li et al., 1992
Droocr	Octopamine/tyramine receptor	3R, 99A10-B1	Arakawa et al., 1990; Saudou et al., 1990
Gpcr68	GPCR68	3R, 99B3-5	Feng and Hall, unpublished data
Drotkr	Tachykinin-like peptide receptor	3R, 99D1-2	Li et al., 1992
Dro5htr	Serotonin receptor 1	3R, 100A	Saudou et al., 1992

[1]2R = right arm of the second chromosome
[2]3R = right arm of the third chromosome

C. OVERVIEW OF MAPPING STUDIES

It is obvious from sequence comparisons that G-protein coupled receptors belong to a multigene family that arose very early in evolution. The general components of this signal transduction system including the seven transmembrane segment receptor, the G-protein, and the coupled effector are commonly used in organisms ranging from unicellular eucaryotes (like Tetrahymena, slime mold and yeast) to man. It is commonly accepted that gene families may arise from gene duplication followed by independent evolution of the duplicate genes. Since gene duplication is likely to be a local event, it is interesting to look at the chromosome distribution of the cloned genes to gain insights into a possible path for evolution for the cloned receptors. The map positions of the cloned genes are summarized in Table 4. So far all of the cloned genes lie either on the right arm of chromosome 2 or the right arm of chromosome 3. The two closely related type 2 serotonin receptors map very close together at 56AB on 2R (i.e., the right arm of the second chromosome). Also mapping very close together on the right arm of the third chromosome are the oct/tyr receptor and the newly identified Gpcr68 receptor suggesting they may be recently divergent genes encoding receptors of similar pharmacological specificity.

It is clear that this map distribution is not random. There do appear to be significant clusters of receptor genes. This clustering is not restricted to *Drosophila* since a similar clustering is seen for G-protein coupled receptors in other organisms including humans. [Grandy et al., 1990; Yang-Feng et al., 1990] Given this clustering of genes encoding G-protein coupled receptors, it would be interesting to scan contigs in a region surrounding a cloned receptor gene to look for additional receptor genes. This might provide a rapid way to clone genes encoding closely related receptors. This approach will be facilitated by chromosome walks yielding large contiguous segments from either yeast artificial chromosome libraries (YACS) or P1 clone libraries. [Garza et al., 1989; Smoller et al., 1991] Such large contigs are now increasingly available in *Drosophila* and other species.

REFERENCES

Arakawa, S., Gocayne, J.D., McCombie, W.R., Urquhart, D.A., Hall, L.M., Fraser, C.M., and Venter, J.C., Cloning, localization, and permanent expression of a Drosophila octopamine receptor, *Neuron* 4, 343-354, 1990.
Bibb, M.J., Findlay, P.R., and Johnson, M.W., The relationship between base composition and codon usage in bacterial genes and its use for the simple and reliable identification of protein-coding sequences, *Gene* 30, 157-166, 1984.

Buck, L. and Axel, R., A novel multigene family may encode odorant receptors, *Cell* 65, 175-187, 1991.

Davis, R.L., Mushroom bodies and Drosophila learning, *Cell* 11, 1-14, 1993.

Dudai, Y., Buxbaum, J., Corfas, G., and Ofarim, M., Formamidines interact with *Drosophila* octopamine receptors, alter the flies' behaviour and reduce their learning ability, *J. Comp. Physiol.* [A]161, 739-746, 1987.

Erber, J., Masuhr, T.H., and Menzel, R., Localization of short-term memory in the brain of the bee, *Apis mellifera, Physiol. Entomol.* 5, 343-358, 1980.

Evans, P.D., Biogenic amines in the insect nervous system, *Adv. Insect Physiol.* 15, 317-473, 1980.

Evans, P.D., Multiple receptor types for octopamine in the locust, *J. Physiol.* 318, 99-122, 1981.

Evans, P.D., Phenyliminoimidazolidine derivatives activate both octopamine$_1$ and octopmaine$_2$ receptor subtypes in locust skeletal muscle, *J. Exp. Biol.* 129, 239-250, 1987.

Evans, P.D., Thonoor, M., and Midgley, J.M., Activities of octopamine and synephrine stereoisomers on octopaminergic receptor subtypes in locust skeletal muscle, *J. Pharm. Pharmacol.* 40, 855-861, 1988.

Garza, D., Ajioka, J.W., Burke, D.T., and Hartl, D.L., Mapping the *Drosophila* genome with yeast artificial chromosomes, *Science* 246, 641-646, 1989.

Grandy, D.K., Zhou, Q.Y., Allen, L., Litt, R., Magenis, E., Civelli, O., and Litt, M., A human D1 dopamine receptor gene is located on chromosome 5 at q35.1 and identifies an *Eco* R1 RFLP, *Amer. J. Genet.* 47, 828-834, 1990.

Gribskov, M., Devereux, J., and Burgess, R.R., The codon preference plot: graphic analysis of protein coding sequences and prediction of gene expression, *Nucleic Acids Res.* 12, 539-549, 1984.

Hannan, F. and Hall, L.M., Muscarinic acetylcholine receptors in invertebrates: Comparisons with homologous receptors from vertebrates, in *Comparative Molecular Neurobiology*, Pichon, Y., Ed., Birkhäuser Verlag, Basel, 1993, pp. 98-145.

Hashemzadeh-Gargari, H., and Wierenga, J.M., Binding of octopamine to membranes from the white-eyed and wild-type housefly heads, *Arch. Insect Biochem. Physiol.* 10, 57-71, 1989.

Heisenberg, M., Borst, A., Wagner, S., and Byers, D., *Drosophila* mushroom body mutants are deficient in olfactory learning, *J. Neurogenet.* 2, 1-30, 1985.

Higgins, D.G. and Sharp, P.M., Fast and sensitive multiple sequence alignments on a microcomputer, *Compu. App. Biosci.* 5, 151-153, 1989.

Li, X.-J., Wolfgang, W., Wu, Y.-N., North, R.A., and Forte, M., Cloning, heterologous expression and developmental regulation of a *Drosophila* receptor for tachykinin-like peptides, *EMBO J.* 10, 3221-3229, 1991.

Li, X.-J., Wu, Y.-N., North, R.A., and Forte, M., Cloning, functional expression, and developmental regulation of a neuropeptide Y receptor from *Drosophila melanogaster, J. Biol. Chem.* 267, 9-12, 1992.

Libert, F., Parmentier, M., Lefort, A., Dinsart, C., Van Sande, J., Maenhaut, C., Simons, M.-J., Dumont, J.E., and Vassart, G., Selective amplification and cloning of four new members of the G protein-coupled receptor family, *Science* 244, 569-572, 1989.

Monnier, D., Colas, J.-F., Rosay, P., Hen, R., Borrelli, E., and Maroteaux, L., NKD, a developmentally regulated tachykinin receptor in *Drosophila*, *J. Biol. Chem.* 267, 1298-1302, 1992

Murtagh, Jr., J.J., Lee, F.-J. S., Deak, P., Hall, L.M., Monaco, L., Lee, C.-M., Stevens, L.A., Moss, J., and Vaughan, M., Molecular characterization of a conserved, guanine nucleotide-dependent ADP-ribosylation factor in *Drosophila melanogaster*, *Biochemistry* 32, 6011-6018, 1993.

Onai, T., FitzGerald, M.G., Arakawa, S., Gocayne, J.D., Urquhart, D.A., Hall, L.M., Fraser, C.M., McCombie, W.R., and Venter, J.C., Cloning, sequence analysis and chromosome localization of a *Drosophila* muscarinic acetylcholine receptor, *FEBS Letts.* 255, 219-225, 1989.

Richards, M.H., Pharmacology and second messenger interactions of cloned muscarinic receptors, *Biochem. Pharmacol.* 42, 1645-1653, 1991.

Saudou, F., Amlaiky, N., Plassat, J.L., Borrelli, E., and Hen, R., Cloning and characterization of a Drosophila tyramine receptor, *EMBO J.* 9, 3611-3617, 1990.

Saudou, F. , Boschert, U., Amlaiky, N., Plassat, J.-L., and Hen, R., A family of *Drosophila* serotonin receptors with distinct intracellular signalling properties and expression patterns, *EMBO J.* 11, 7-17, 1992.

Schmidt-Nielsen, B.K., Gepner, J.I., Teng, N.N.H., and Hall, L.M., Characterization of an α-bungarotoxin binding component from *Drosophila melanogaster, J. Neurochem.* 29, 1013-1029, 1977.

Shapiro, R.A., Wakimoto, B.T., Subers, E.B., and Nathanson, N.M., Characterization and functional expression in mammalian cells of genomic and cDNA clones encoding a *Drosophila* muscarinic acetylcholine receptor, *Proc. Natl. Acad. Sci. USA* 80, 9039-9043, 1989.

Smoller, D.A., Petrov, D., and Hartl D.L., Characterization of bacteriophage P1 library containing inserts of Drosophila DNA of 75-100 kilobase pairs, *Chromosoma* 100, 487-494, 1991.

Spradling, A.C., P element-mediated transformation, in *Drosophila a practical approach,* Roberts, D.B., Ed., IRL Press, Washington, DC, 1986, chap. 8.

Uzzan, A., and Dudai, Y., Aminergic receptors in *Drosophila melanogaster:* responsiveness of adenylate cyclase to putative neurotransmitters, *J. Neurochem.* 38, 1542-1550, 1982.

Wess, J., Molecular basis of muscarinic acetylcholine receptor function, *Trends Pharm. Sci.* 14, 308-313, 1993.

Witz, P., Amlaiky, N., Plassat, J.-L., Maroteaux, L., Borrelli, E., and Hen, R., Cloning and characterization of a *Drosophila* serotonin receptor that activates adenylate cyclase, *Proc. Natl. Acad. Sci. USA* 87, 8940-8944, 1990.

Yang-Feng, T.L., Fhong, X.F., Cotecchis, S., Frielle, T., Caron, M.G., and Lefkowitz, R.J., Chromosomal organization of adrenergic receptor genes, *Proc. Natl. Acad. Sci. USA* 87, 1516-1520, 1990.

Insect-Selective Toxins as Biological Control Agents

Jon H. Hayashi and Ivan Gard

American Cyanamid, Insecticide Discovery/Neurobiology,
P.O. Box 400, Princeton, New Jersey, 08543

Many of us, in pursuit of funding, have argued the case for insect neurobiology. We have pointed to the specific advantages that each of our invertebrate preparations has over vertebrate preparations. Advantages run from specific molecular and genetic tools, to fully characterized nervous systems, to elegant studies of circuitry and the mechanisms that underlie their performance and modulation, to behavioral stereotypy, to biophysical studies *in vivo* and *in vitro*, to elegant studies addressing the developmental role of hormones *in vivo* and *in vitro*, to neuronal pathfinding studies, to internal ion fluctuation studies and on and on. And insect neurobiologists have delivered on their promise of productivity.

But in addition to the utility of insect preparations as model systems for vertebrates, insects do, in fact, differ from vertebrates in significant and scientifically revealing ways. Some of these differences might be the result of random genetic drift. Other differences are likely to be the consequence of the different evolutionary pressures acting on vertebrates and on insects. For example, unlike mammals, insects have evolved extensive physiological countermearsures in response to plant chemical defence mechanisms: plants and the insects that graze on them are masters of chemical warfare. Plants evolve toxins that repel, incapacitate, or kill the insects while the insects evolve their own defensive mechanisms that enable them to continue feeding on those plants. One expression of the resulting variability is the observation that different insects have adapted to different food tolerance niches; one insect's food remains another insect's poison. The consequence of both the random and the evolutionarily directed changes is the advent of insect-specific differences that we observe today.

Principle investigators in the agricultural pesticide business, and here I'm referring to those involved in the protection of row crops, as opposed to ornamental crops, know that the surviving insect species have evolved sophisticated and elaborate mechanisms for defence against plant toxins. Insecticides, even ones that are quite toxic to vertebrates, can exhibit toxicity to a limited subset of insects.

Venomous animals that prey on herbivorous insects have evolved toxins

0-8493-4591-X/94/$0.00 + $.50
© 1994 by CRC Press, Inc.

that track the evolution of the prey's nervous system. It is likely that the binding sites for toxins are not purely fortuitous, but that there was selective pressure on the toxin to bind to an essential element on the target channel. If the toxin mimicked a normally occurring functional element, then the toxin's receptor would not be free to drift randomly to evade the toxin.

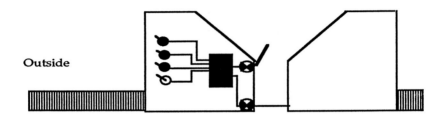

Inside

A neurobiologist's view of an ion channel. The activation gate (upper arm) and the inactivation gate (lower arm) are under the control of a black box that directs opening and closing sequence of the gates. The way these gates operate can be modified by a set of switches that are normally operated by particular circulating elements. Toxins might operate these switches and thereby produce physiologically disastrous consequences to the animal.

Indeed plants have evolved toxins that target vital physiological elements. The pyrethrins are well known examples of defensive plant compounds that are toxic to insects. These compounds act on sodium channels.

The point here is that one can reasonably expect substantial variability among the grazing insects because they have survived millennia of active chemical warfare and each has adapted to a food preference niche and has evolved a strategy for coping with that plant's defenses.

Neurobiologists in the industrial setting are investigating the ways that insects differ from vertebrates and the ways that neural components of one particular insect species differ from another. When these differences are understood, the hope is to develop highly specific agents that will selectively target the pest species and only the pest species. Toward this end we at Cyanamid are using the toxins found in the venoms of animals that prey on the pests of row crops as probes for insect-specific differences. We wish to know what these differences are, what these differences tell us about how voltage- and ligand-gated channels operate and how they are modulated.

One likely source of toxins that attack neuronal elements that are specific to insects is the venom of arachnids that prey on insects. These predators are able to evolve venoms and one might expect these venoms to contain toxins to be directed against targets that are insect sepcific. The discovery of insect-specific toxins will provide neurobiologists with probes with which to understand better the functional aspects of neuronal excitability as well as tools to protect crops from pest insect species.

A. Voltage-Gated Channels
1. Sodium Channels

We now know that toxins directed against insect-specific elements of the nervous system exist. In 1971 Zlotkin purified a toxin component from the venom of the scorpion *Androctonus australis* that affected the sodium channels of insects but not of mammals, arachnids or crustaceans (Zlotkin *et al.*, 1972, Tintpulver *et al.*, 1976, Rathmayer *et al.*, 1977, Ruhland *et al.*, 1977). Thus sodium channels can be pharmacologically distinguished from one another with this insect-specific toxin, AaIT. AaIT is a single chain polypeptide composed of 70 amino acids and crosslinked by four disulfide bridges (Darbon *et al.*, 1982).

Walther *et al.* (1976) showed that uncontrolled, rapid, spontaneous firing within the motor neurons of a locust nerve-muscle preparation was the event that initiated the observed tetanic paralysis. Pelhate and Zlotkin (1981) using a double oil-gap single-fiber technique on a cockroach axon concluded that the repetitive activity induced by AaIT was caused by the retardation of the voltage-dependent sodium channel inactivation mechanism coupled with an increase in the magnitude of the sodium current. And Zlotkin (1986) demonstrated that [^{125}I]AaIT bound to insect neuronal membranes but not to non-neural tissue nor muscle membrane. We now know that the frequency of miniature excitatory junctional currents (MEJC) recorded at synaptic sites using a loose patch-clamp recording technique are not altered by a lethal dose of AaIT (50 nM) and that the uncontrolled, repetitive firing of an AaIT poisoned neuron is abolished by tetrodotoxin (Adams, 1989). Furthermore, we know that AaIT is ineffective against the insect potassium channel (Pelhate and Zlotkin, 1982).

High affinity binding was demonstrated by Gordon *et al.* (1984) using saturation equilibrium assays. They incubated locust synaptosomal membranes with [^{125}I]AaIT for 30 min at 20° C. Their calculated K_d was 1.19 nM. The maximum capacity was low, 1.37 pmol/mg membrane protein as would be expected if binding were specific to the sodium channels. The plot also shows that AaIT binds to a single class of non-interacting binding sites located on the insect neuronal membrane.

Taken together, these observations are consistent with the idea that AaIT is an agent whose specific activity is directed against the insect neuronal sodium channel. Thus, the insect sodium channel, but not the vertebrate channel, must possess some structural peculiarity that is both the recognition site for AaIT and a site of functional significance to the channel.

Let us now recall what we know about the sodium channels of vertebrates and insects so that we can understand better the physiological significance of the insect-specific differences we observe. Analysis of protein sequence predicted by the vertebrate sodium channel cDNA clones (Noda *et al*., 1986, Kayano *et al*., 1988, Catterall, 1988) enabled investigators to construct topological models of the channels within the plasma membrane. The channel apparently has four homologous domains containing six (Noda *et al*., 1986, Catterall, 1986) or perhaps eight (Greenblatt *et al*., 1985, Guy and Seetharamulu 1986, Guy and Conti, 1990) transmembrane segments per domain connected by loops of amino acid sequences.

Neurobiologists model the channel differently. By using toxins to probe the surface of the sodium channel, they have described a number of toxin-binding sites. Site 1 binds tetrodotoxin and saxitoxin. In addition, μ-conotoxin binds at an overlapping site. This site is on the extracellular side of the channel and binding there prevents ion flow through the channel by blocking access to the pore. Site 2 binds a number of lipid-soluble toxins including batrachotoxin, veratridine, aconitine and grayanotoxin. These toxins shift the activation point of the channel in the negative direction and prevent channel inactivation. The result is persistent channel activation and uncontrolled firing of the neuron. Site 3 binds to scorpion α–toxins and sea anemone toxins. This results in a disruption of channel inactivation and causes persistent activation (Catterall, 1986, Courand and Jover, 1984). Site 4 binds to scorpion ß–toxins. This results in persistent channel activation (Catterall, 1986, Courand and Jover, 1984). Site 5 is characterized by its ability to bind brevetoxin. Brevetoxin binding enhances the efficacy of toxins bound to site 2 (Catterall and Gainer 1985, Poli *et al*., 1986) and site 6 binds Goniopora toxin that slows channel inactivation but does not compete for binding at site 3 (Gonoi *et al*., 1986).

Insect sodium channels were shown to resemble those of the vertebrates in terms of their primary structure (Salkoff *et al*., 1987, Ramaswami and Tanouye, 1989, Loughney *et al*., 1989), organization (Loughney *et al*., 1989, Gordon *et al*., 1990, 1992), and basic pharmacological properties (Pelhate and Sattelle, 1982). Thus it is somewhat surprising that an insect-specific toxin binding site should exist. Interestingly, AaIT is not the only insect specific sodium channel toxin that has been discovered.

LqhIT$_2$ is another example of an insect-specific toxin. This toxin, found in the venom of the scorpion *Leiurus quinquestriatus hebraeus,* induces a flaccid paralysis following its binding to the insect sodium channel; the symptoms of this toxin contrast with the tetanic paralysis of AaIT (Lester *et al*., 1982). LqhIT$_2$ bath applied (6 nM) to an insect neuromuscular junction causes an increase in spontaneous transmitter release, as measured by an increase in MEJC but very little repetitive activation of muscle contraction (Adams *et al*., 1989). LqhIT$_2$ competitively displaces AaIT from its site on the sodium channel and the LqhIT$_2$ increase in MEJC frequency is blocked by TTX (Zlotkin *et al*., 1985, 1991, Gordon *et al*., 1984,1985). Scatchard analysis of a saturation binding curve of [^{125}I]LqhIT$_2$ by Gordon *et al*. (1992) revealed the existence of 2 distinct binding sites in locust neuronal membranes whose dissociation constants and capacities are separated by 2 orders of magnitude. They were also able to demonstrate in a competitive binding assay that AaIT was able to displace only half of the bound LqhIT and that the LqhIT bound to the high affinity site was apparently distinct from the AaIT site. Gordong *et al.* (1992) showed AaIT and LaHIT$_2$ shared a common binding site and that LqHIT bound to an additional binding site.

The crude venom of scorpions comprises a number of other polypeptide toxins that are cross linked by four disulfide bridges (Miranda *et al.*, 1970). Among there toxins are the α- and ß-scorpion toxins that are active against the vertebrate and insect sodium channel. Work by De Lima *et al.* (1986) has shown that Ts VII,a toxin purified from the venom of the South American scorpion *Tityus serrulatus,* displays vertebrate toxicity by binding ß site as well as insect toxicity by binding to the AaIT site. De Lima (1986) showed that both [125I]AaIT and [125I]Ts VII bind in a specific and saturable manner on a synaptosomal fraction from house fly heads. Scatchard plots were linear, indicating that each toxin binds to a single class of receptor sites. De Lima *et al.* (1986) showed that although AaIT and Ts VII compete for a common site on insect synaptosomes. They also showed that AaIT was unable to displace Ts VII on rat brain synaptosomes. From these observations they concluded that although the insect and vertebrate sites clearly differ from one another, there is also an element of shared similarity. They hypothesize that the scorpion toxin family consists of a series of conformational variants that grants each variant characteristic target recognition attributes. Gordon *et al.* (1992) used site-directed antibodies that recognize different extracellular regions of the rat brain RII sodium channel α-subunit to define the binding sites of AaIT and LqhIT$_2$. These antibodies were directed against five regions (355-371, 382-400, 1429-1449, 1686-1703, and 1729-1748). The *Drosophila para* locus bears a homology of 67, 79, 50, 56, and 40% to the vertebrate counterparts (Lounghney *et al.*, 1989). Inhibition of [125I]LqhIT was achieved by Ab 382-400 (55 ± 17%), Ab 1429-1449 (56 ± 18%), Ab 1729-1748 (42 ± 20%) and Ab 355-371 (23 ± 5%) (Gordon *et al.*, 1992).

To determine whether AaIT and LqhIT shared a common binding site, Gordon *et al.* (1992) compared the extent to which each antibody inhibited the binding of toxin. The membranes were preincubated in the presence of maximal amounts of each antibody and the binding of the two toxins was then measured. The results clearly indicated that each antibody affected the binding of these toxins differently. Antibodies 382-400, 1429-1449, and 1729-1748 which cause 50% inhibition of [125I]LqhIT binding, inhibit [125I]AaIT binding by less than 20%. Ab 355-371, which inhibits LqhIT binding by 23%, did not significantly affect the binding of AaIT.

Gordon *et al.* (1992) also determined that the ability of a given antibody to inhibit toxin binding was not additive. From this they concluded that each antibody was given in a saturable amount; there were no channels that remained unbound by each antibody. They also conclude that the population of sodium channels is homogeneous; if different subsets of channels existed, then different antibodies might have bound to exclusive subsets of the population and would have thereby produced an additive effect.

Gordon *et al*. (1992) subscribe to the conceptualization that the scorpion toxins bind to the sodium channel with multiple attachment sites (Kharrat *et al*., 1989; Loret *et al*., 1990). These investigators believe a partial overlap in these sites of attachment for different toxins could underlie binding competition between toxins with otherwise dissimilar effects. An example of attachment site overlap might be Ts VII referred to previously or the scorpion toxin AaHIT$_4$ (Loret *et al*., 1991) that displaces both α- and ß-toxins from rat brain sodium channels and also displaces the toxin AaIT from its insect-specific site.

These studies show that important functional elements that are specific to insect sodium channels do exist. These naturally occurring toxins can be exploited as tools to probe the mechanisms of channel function as well as tools that can be developed into efficient pesticides.

2. Calcium Channels

There are compelling reasons to investigate insect calcium channels for insect-specific variability. First of all, calcium is a major regulator of many cell processes and its control by the animal is vital to its survival. It is clear that if an insect-specific calcium channel toxin were discovered that this would represent a potential pesticide. However, it is imperative that these pesticides be inactive on vertebrates as well as on beneficial insect species. There is evidence that suggests to us that calcium channels found in insects differ from those of the vertebrates.

The types of vertebrate channel types are well known and distinctive. These channel types were segregated first by Nowycky *et al*. (1985) who proposed the criteria for T-, L-, and N-type calcium channels. This work done on chick dorsal root ganglion (DRG) neurons established a set of standards against which all other calcium channels were measured. This classification system appears not to hold for invertebrate calcium channels. Charlton and Augustine (1990) examined presynaptic calcium channels at the squid giant synapse and found that those channels to be neither T-, L-, nor N-type. Activation of N- and L-channels in DRG neurons occurs at depolarizations to -20 and -10 mV respectively while T-type channels are activated at the more negative value of -70 mV (Tsien *et al*., 1988). Charlton and Augustine reported that squid presynaptic calcium currents activated at -55 mV, a value that is not consistent with L-, nor N-type channel but perhaps somewhat T-like. Squid calcium channels also differed from those found in DRG neurons in terms of their inactivation properties. T-channels show rapid inactivation with a time constant of roughly 50 msec, while N- and L-channels inactivate with time constants of 50–80 msec and greater than 500 msec, respectively (Tsien *et al.*, 1988). Squid presynaptic calcium channels inactivate with time constants much slower than T-channels. Squid calcium channels show only partial inactivation during depolarizations of several seconds (Charlton and Augustine, 1990). The squid calcium channels activate at voltages too negative to be considered N- or L-type channels and inactivate too slowly to qualify as T-channels.

Charlton and Augustine (1990) also found that the squid presynaptic calcium channels were insensitive to ω-conotoxin GVIA (ω-CgTx), a toxin known to strongly affect N- and L-type channels (Nowycky *et al.*, 1985). They indicate that nitrendipine, a dihydropyridine (DHP) known to block L-type channels (Fox *et al.*, 1987a, b, Nowycky *et al.*, 1985) was ineffective on squid presynaptic calcium channels. Also, the DHP agonist BAY K 8644 (Janis and Triggle, 1984) failed to affect squid calcium channels (Charlton and Augustine, 1990). Therefore the squid calcium channels are not L-type as judged by pharmacological criteria. Charlton and Augustine (1990) have shown that the calcium channels at the squid presynaptic terminal are neither T-, L, nor N-type.

Indeed it is now clear that there are more than three types of calcium channel. For example, the P-type channel in Purkinje neurons was shown to be sensitive to a spider toxin by Llinás *et al.*, (1989). These channels are insensitive to ω-conotoxin and DHPs and exhibit a more negative range of activation than N-type and L-type channels (Regan, 1991, Regan *et al.*, 1991). Furthermore, Mintz *et al.* (1992) has shown that P-channel blockade with the spider toxin ω-Aga-IVA fails to block all of the ω-conotoxin and DHP resistant current in some preparations. They conclude that other types of calcium channels must therefore exist. Indeed, it is now known that calcium channels from other invertebrates (Byerly and Hagiwara, 1987, Llinás *et al.*, 1989, Yeager *et al.*, 1987, Hayashi and Levine, 1992) and from other vertebrate preparations (cf Leonard *et al.*, 1987, Suszkiw *et al.*, 1989) do not conform to the conventional vertebrate categories in terms of kinetics or pharmacology.

Hayashi and Levine (1992) examined voltage-gated currents in cultured leg motor neurons from the moth *Manduca sexta* and found evidence for a slowly inactivating and a non-inactivating calcium current. In fact, a number of invertebrate preparations exhibit low threshold, inactivating currents that do not fit the conventional vertebrate categories (Christensen, 1988, cultured cockroach brains; Tazaki and Cooke, 1990, lobster cardiac ganglion motor neurons; Hayashi and Hildebrand, 1990, *Cultured Manduca* antennal lobe neurons; Angstadt and Calabrese, 1991, heart interneurons of the leech; Hayashi and Stuart, 1993, presynaptic terminal of barnacle photoreceptors) and Byerly and Leung (1988) described calcium currents in *Drosophila* with inactivating and sustained components. Leung *et al.* (1989) found that the noninactivating *Drosophila* component was blocked preferentially by a toxin from the spider *Hololena curta*.

Calcium channels display a considerable amount of variability and the invertebrate calcium channels may be distinct from those of vertebrates. A productive approach to characterizing these differences has been to use invertebrate toxins as probes for novel calcium channel types. Bindokas and Adams (1989) studied the effects of ω-agatoxins, derived from the Funnel Web spider, *Agelenopsis aperta* , that attacked vertebrate and invertebrate neuronal calcium channels. Other spider species such as *Hololena curta* (Bowers *et al.*, 1987) and *Plectreurys tristis* (Branton *et al.*, 1987) cause presynaptic blockade at the insect neuromuscular junction but not at the vertebrate junction.

Given the variety of calcium channels and the differences between vertebrate and invertebrate channels, it is reasonable to expect that insect-specific calcium channel toxins exist within arachnid venoms. The opportunity to pursue this line of research is now at hand; Finding these toxins and defining their modes of action is of great interest to the agricultural industry.

B. Ligand-Gated Receptors
1. Glutamate Receptors

Ligand-gated receptors have also been targets for toxins occurring in nature. Toxins directed against the glutamate receptor have evolved because glutamate is commonly the transmitter used at the neuromuscular junction (NMJ) of insects (Usherwood, 1981). Glutamate receptors of the insect NMJ are similar to the Kainate/AMPA (α–amino-3-hydroxy-5-methyl-4-isoxazole propionic acid) subtype which are sensitive to quisqualate (Eldefrawi *et al.*, 1993). The kainate/AMPA receptor is a ligand-gated ion channel that is permeable to sodium and potassium ions but not to divalent cations such as calcium (Watkins *et al.*, 1990, Young *et al.*, 1990, Unwin, 1989, Rathmayer and Miller, 1990). These receptors are thought to underlie fast excitatory transmission at glutamatergic synapses throughout the vertebrate CNS. In studies done on vertebrates, six kainate/AMPA receptors have been identified and are named GluR1 to GluR6 (Boulter *et al.*, 1990, Keinanen *et al.*, 1990, Nakanishi *et al.*, 1990, Bettler *et al.*, 1990, Egebjerg *et al.*, 1991). GluR1 to GluR4 are closely related to one another while GluR5 and GluR6 have apparently diverged more substantially from the original pattern (Bettler *et al.*, 1990, Egebjerg *et all.*, 1991).

To assess the degree of similarity between an insect an the vertebrate glutamate receptors, Schuster *et al.* (1991) cloned a glutamate receptor subunit (DGluR-II) expressed in embryonic *Drosophila* muscle and found 25.9, 28.2, and 28.4% homology with the rat glutamate receptor subtypes GluR1, GluR4 and GluR5, respectively. The DGluR-II receptor is considered a distant relative of the forms native to rat. Further studies will be required on invertebrate glutamate receptors before generalizations can be made with respect to the degree of similarity with vertebrate receptors. In addition, Schuster *et al.*

(1991) expressed this receptor in *Xenopus* oocytes and found that while it did function as a glutamate activated ion channel, the expressed channel lacked the sensitivity typically observed in *in vivo* Drosophila larval muscle (Jan and Jan, 1976) as well as sensitivity to quisqualate, AMPA and kainate. Schuster *et al.* (1991) speculate that the absence of appropriate complementary subunits may have affected channel performance.

Presently the best known toxins that attack glutamate receptors are the polyamine toxins derived from scorpions and spiders. Philanthotoxin (PhTX) is the best known of these agents and apparently functions as a channel blocker (Bruce *et al.*, 1990). However this is not a reasonable pesticide candidate because it also blocks vertebrate glutamate receptors (Ragsdale, 1989). One possible approach might be to alter the structure of PhTX to achieve the required degree of specificity. Another approach might be to search for peptide toxins that might fortuitously exploit an insect-specific aspect of the glutamate receptor.

2. GABA Receptors

The final potential target for an insecticide that I will discuss is the GABA (γ–aminobutyric acid) receptor. Synthetic organic GABA receptor blockers (cyclodienes and other polychlorocycloalkanes) were extensively used throughout the world before their mode of action was known. These compounds have been withdrawn from the market due to their adverse persistence and toxicity to non-pest species. Nevertheless, the efficacy of a GABA receptor blocker is well known and presently the search is on for an insect-specific, environmentally benign GABA receptor blocker.

Picrotoxin   **Bicuculline**

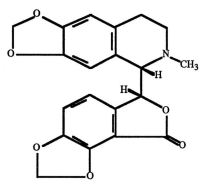

Picrotoxin and bicuculline are toxins that block GABA$_A$ channels in vertebrate preparations.

The question is, do insect-specific differences exist for the GABA receptor and are these sites susceptible to attack by toxins. Like glutamate receptors, GABA receptors are prevalent in the vertebrate CNS. There are two classes of GABA receptor in the vertebrate CNS: $GABA_A$ and $GABA_B$. $GABA_A$ receptors mediate a chloride conductance that is antagonized by picrotoxin and bicuculline while $GABA_B$ receptors are believed to couple calcium and potassium channels via GTP-binding proteins (Enna and Karbon, 1987, Bormann, 1988). GABA receptors on insects neurons most closely resemble the vertebrate type $GABA_A$ (Eldefrawi and Eldefrawi, 1987). In insect brain and ganglia, GABA receptors are inhibited allosterically by picrotoxin and potentiated by benzodiazepines and barbiurtates (Lees *et al.*, 1987); these characteristics are shared with vertebrate $GABA_A$ receptors. However insect GABA receptors differ from the $GABA_A$ classification in their responses to bicuculline. At $GABA_A$ receptors in vertebrates, bicuculline is a competitive antagonist of GABA that blocks GABA binding (Simmonds, 1982, Mann and Enna, 1980, DeFeudis *et al.*, 1980). In some insect preparations bicuculline blocks the responsiveness of the GABA receptor (Walker *et al.*, 1971, Waldrop, *et al.*, 1986) but in other invertebrates, bicuculline antagonism has been shown to be noncompetitive (Takeuchi and Onodera, 1972, Shank *et al.*, 1974, Constanti, 1978). In other insect preparations, Bicuculline fails to inhibit GABA mediated responses (Abalis *et al.*, 1986, Wafford *et al.*, 1987).Therefore the bicuculline binding site may overlap the GABA binding site in some, but not all preparations.

It is clear that vertebrate and insect GABA receptors differ. We know that interference with the normal functioning of the insect GABA receptor is lethal but we have not as yet succeeded in developing an insect-specific GABA receptor toxin. The viewpoint of my company is that such toxins may have been developed by venomous animals that prey on insects and that these toxins could be developed into useful pesticides. Furthermore, the discovery of those highly specific toxins would provide us with tools useful in the exploration of the fundamental mechanisms that underlie channel function.

The search for insect-specific differences is developing rapidly and the expectation is that this development will be enhanced by the discovery of toxin that reside within the venoms of predatory insectivors. It is my hope, and the hope of my company, that industrial and academic labs can find common ground for collaborative studies.

Acknowledgements
We thank Dr Bruce Black, Dr Lynn Brennan, Ms Nancy Fleming, and Ms Nancy Webb for critical reading of the manuscript and for helpful discussions.

REFERENCES

Abalis, I.M. and Eldefrawi, A.T., [3H]Muscimol binding to putative GABA receptor in honey bee brain and its interaction with avermectin B_{1a}. *Pestic. Biochem. Physiol.* 25, 279-287, 1986.

Adams, M.E., Bindokas, V.P., and Zlotkin, E., Synaptic toxins from arachnid venoms: Probes for new insecticide targets, in: *Insecticide Action, from Molecule to Organism.*, Narahashi, T. and Chambers, J.E.Eds., Plemun Press, pp189-203, 1989.

Angstadt, J.D. and Calabrese, R.L., Calcium currents and graded synaptic transmission between heart interneurons of the leech. *J. Neurosci.* 11, 746-759, 1991.

Bettler, B., Boulter, J., Hermans-Borgmeyer I., O'Shea-Greenfield, A., Deneris, E.S., Moll, C., Borgmeyer, U., Hollmann, M., and Heinemann, S., Cloning of a novel glutamate receptor subunit, GluR5: expression in the nervous system during development. *Neuron* 5, 583-595, 1990.

Bindokas, V.P. and Adams, M.E., ω-AGA-I: a presynaptic calcium channel antagonist from venom of the Funnel Web spider, *Agelenopsis aperta. J. Neurobiol.* 20, 171-188, 1989.

Bormann, J., Electrophysiology of $GABA_A$ and $GABA_B$ receptor subtypes. *Trends Neurosci.* 11, 112-116, 1988.

Boulter, J., Hollmann, M., O'Shea-Greenfield, A., Hartley, M., Deneris, E., Maron, C., and Heinemann, S., Molecular cloning and functional expression of glutamate receptor subunit genes. *Science* 249, 1033-10377, 1990.

Bowers, C.W., Phillips, H.S., Lee. P., Jan, Y.N., and Jan, L.Y., Identification and purification of an irreversible presynaptic neurotoxin from the venom of the spider, *Hololena curta. Proc. Natl. Acad. Sci. USA* 84, 3506-3510, 1987.

Branton, W.D., Kolton, L., Jan, Y.N., and Jan, L.Y., Neurotoxins from *Plectreurys* spider venom are potent presynaptic blockers in *Drosophila. J. Neurosci.* 7, 4195-4200, 1987.

Bruce, M., Bukownik, R., Eldefrawi, A.T., Eldefrawi, M.E., Goodnow, R. Jr., Kallimopoulos, T., Konno, K., Nakanishi, K., Niwa, M., and Usherwood, P.N., Structure-activity relationships of analogues of the wasp toxin philanthotoxin: non-competitive antagonists of quisqualate receptors. Toxicon 28, 1333-46, 1990.

Byerly, L. and Leung, H.T., Ionic currents of *Drosophila* neurons in embryonic cultures. *J. Neurosci.* 8, 4379-4393, 1988.

Catterall, W.A., Molecular properties of voltage-sensitive sodium channels. *Ann. Rev. Biochem.* 55, 953-985, 1986.

Catterall, W.A. and Gainer, M., Interaction of brevetoxin A with a new receptor site on the sodium channel. *Toxicon* 23, 497-504, 1985.

Charlton, M.P. and Augustine, G.J., Classification of presynaptic calcium channels at the squid giant synapse: neither T, L- nor N-type. *Brain Res.* 525, 133-139, 1990.

Christensen, B.N., Larmet, Y., Shimahara, T., Beadle, D., and Pichon, Y., Ionic currents in neurones cultured from embryonic cockroach (*periplaneta americana*) brains. *J. Exp. Biol.* 135, 193-214, 1988.

Constanti, A., The 'mixed' effect of picrotoxin on the GABA dose/conductance relation recorded from lobster muscle. *Neuropharmacol.* 17, 159-167, 1978.

Darbon, H., Zlotkin, E., Kopeyan, C., Van Rietschoten, J., and Rochat, H., Covalent structure of the insect toxin of the North African scorpion *androctonus australis* Hector, *Int. J. Peptide Res.* 20, 320-330, 1982.

DeFeudis, F.V., Ossola, L., Schmitt, G., and Mandel, P., Substrate specificity of [3H]muscimol binding to a particulate fraction of a neuron-enriched culture of embryonic rat brain. *J. Neurochem.* 34, 845-849 , 1980.

Egebjerg,.J., Bettler, B., Hermans-Borgmeyer, I., and Heinemann, S., Cloning of a cDNA for a glutamate receptor subunit activated by kainate but not AMPA. *Nature* 351, 745-8, 1991.

Eldefrawi, A.T. and Eldefrawi, M.E., Receptors for γ-aminobutyric acid and voltage-dependent chloride channels as targets for drugs and toxicants. *FASEB J.* 1, 262-271, 1987.

Eldefrawi, M.E., Nabil, A.A., and Eldefrawi, A.T., Glutamate receptor inhibitors as potential insecticides. *Arch. Insect Biochem. and Physiol.* 22, 25-39, 1993.

Enna, S.J. and Karbon, E.W., Receptor regulation: evidence for a relationship between phospholipid metabolism and neurotransmitter receptor-mediated cAMP formation in brain. *Trends Pharmacol. Sci.* 8:21-24, 1987.

Fox, A.P., Nowycky, M.C., and Tsien, R.W., Kinetic and pharmacological properties distinguishing three types of calcium currents in chick sensory neurones. *J. Physiol (London)* 394, 149-172, 1987a.

Fox, A.P., Nowycky, M.C., and Tsien, R.W., Single-channel recordings of three types of calcium channels in chick sensory neurones. *J. Physiol (London)* 394, 173-200, 1987b.

Gonoi, T., Sherman, S.J., and Catterall, W.A., Voltage clamp analysis of tetrodotoxin-sensitive and -insensitive sodium channels in rat muscle cells developing *in vitro*. *J. Neurosci.* 5, 2559-2564, 1985.

Gordon, D., Jover, E., Couraud, F., and Zlotkin, E., The binding of the insect selective neurotoxin (AaIT) from scorpion venom to locust synaptosomal membranes. *Biochim. Biophys. Acta.* 778, 349-358, 1984.

Gordon, D., Moskowitz, H., and Zlotkin, E., Sodium channel polypeptides in central nervous systems of various insects identified with site directed antibodies. *Biochim. Biophys. Acta*, 1026, 80-86, 1990.

Gordon, D., Moskowitz, H., Eitan, M., Warner. C., Catterall, W.A., and Zlotkin, E., Localization of receptor sites for insect-selective toxins on sodium channels by site-directed antibodies. *Biochemistry* 31, 7622-7628, 1992.

Greenblatt, R.E., Blatt, Y., and Montal, M., The structure of the voltage-sensitive sodium channel. Inferences derived from computer-aided analysis of the *Electrophorus electricus* channel primary structure. *FEBS Lett.* 193, 125-134, 1985.

Guy, H.R. and Seetharamulu, P., Molecular model of the action potential sodium channel. *Proc. Natl. Acad. Sci. USA* 83, 508-512, 1986.

Guy, H.R. and Conti, R., Pursuing the structure and function of voltage-gated channels. *Trends Neurosci.* 13, 201-206, 1990.

Hayashi, J.H. and Levine, R.B., Calcium and potassium currents in leg motoneurons during postembryonic development in the hawkmoth *Manduca sexta. J. Exp. Biol.* 171, 15-42, 1992.

Hayashi, J.H. and Stuart, A.E., Currents in the presynaptic terminal arbors of barnacle photoreceptors. *Visual Neuroscience* 10, 261-270, 1993.

Jan, L.Y. and Jan, Y.N., L-glutamate as an excitatory transmitter at the *Drosophila* larval neuromuscular junction. *J. Physiol. (Lond)* 262, 215-36, 1976.

Janis, R.A. and Triggle, D.J., 1,4-Dihydropyridine Ca2+ channel antagonists and activators: a comparison of binding characteristics with pharmacology. *Drug Dev. Res.*, 4, 257-274, 1984.

Keinanen, K., Wisden, W., Sommer, B., Werner, P., Herb, A., Verdoorn, T.A., Sakmann, B., and Seeburg, P.H., A family of AMPA-selective glutamate receptors. *Science* 249, 556-560, 1990.

Kharrat, R., Darbon, H., Rochat, H., and Granier, C., Structure/activity relationships of scorpion alpha-toxins. Multiple residues contribute to the interaction with receptors. *Eur. J. Biochem.* 181, 381-390, 1989.

Kayano, T., Noda, M., Flockerzi, B., Takahashi, H., and Numa, S., Primary structure of rat brain sodium channel III deduced from the cDNA sequence. *FEBS Lett.* 288, 187-194, 1988.

Leonard, J.P., Nargeot, J., Snutch, T.P., Davidson, N., and Lester, H.A., Ca channels induced in *Xenopus* oocytes by rat brain mRNA. *J Neurosci.* 7, 875-881, 1987.

Leung, H.T., Branton, W.D., Phillips, H.S., Jan, L., and Byerly, L., Spider toxins selectively block calcium currents in *Drosophila. Neuron* 3, 767-772, 1989.

Llinás, R., Sugimori, M., Lin, J.W., and Cherksey, B., Blocking and isolation of a calcium channel from neurons in mammals and cephalopods utilizing a toxin fraction (FTX) from funnel-web spider poison. *Proc. Natl. Acad. Sci. USA* 86, 1689-1693, 1989.

Loughney, K., Kreber, R., and Ganetzky, B., Molecular analysis of the para locus, a sodium channel gene in Drosophila. Cell 58, 1143-54, 1989.

Loret, E.P., Mansuelle, P., Rochat, H., and Granier, C., Neurotoxins active on insects: amino acid sequences, chemical modifications, and secondary structure estimation by circular dichroism of toxins from the scorpion Androctonus australis Hector. Biochemistry 29, 1492-1501, 1990.

Mann, E. and Enna, S.J., Phylogenetic distribution of bicuculline-sensitive γ–aminobutyric acid (GABA) receptor binding. *Brain Res.* 184, 367-373, 1980.

Mintz, I.M., Adams, M.E., and Bean, B.P., P-type calcium channels in rat central and peripheral neurons. *Neuron* 9:85-95, 1992.

Miranda F., Kupeyan C., Rochat H., Rochat C., and Lissitzky, S., Purification of animal neurotoxins. Isolation and characterization of four neurotoxins from two different sources of *Naja haje* venom. *Eur. J .Biochem.* 17, 477-84, 1970.

Nakanishi, N., Schneider, N.A., and Axel, R., A family of glutamate receptor genes: evidence for the formation of heteromultimeric receptors with distinct channel properties. *Neuron* 5, 569-81, 1990.

Noda, M., Ikeda, T., Kayano, T., Suzuki, H., Takshima, H., Kurasaki, M., Takahashi, H., and Numa, S., *Nature* 320, 188-192, 1986.

Nowycky, M.C., Fox, A.P., and Tsien, R.W., Three types of neuronal calcium channel with different calcium agonist sensitivity. *Nature* 316, 440-443, 1985.

Pelhate, M. and Zlotkin, E., Actions of insect toxin and other toxins derived from the venom of the scorpion *Androctonus australis* on isolated giant axons of the cockroach (*Periplaneta americana*). *J. Exp Biol.* 97, 67-77, 1982.

Poli, M.A., Mende, T.J., and Baden, D.G., Brevetoxins, unique activators of voltage-sensitive sodium channels, bind to specific sites in rat brain synaptosomes. *Mol. Pharmacol.* 30, 129-135, 1986.

Ragsdale, D., Gant, D.B., Anis, N.A., Eldefrawi, A.T., Eldefrawi, M.E., Konno, K., and Miledi, R., Inhibition of rat brain glutamate receptors by philanthotoxin. *J. Pharmacol. Exp. Ther.* 251, 156-163, 1989.

Ramaswami, M., and Tanouye, M.A., Two sodium-channel genes in Drosophila: implications for channel diversity.*Proc. Nat. Acad. Sci. USA*, 86, 2079-2082, 1989.

Rathmayer, M.L. and Miller, R.J., Excitatory amino acid receptors, second messengers and regulation of intracellular Ca2+ in mammalian neurons. *Trends Pharmacol. Sci.* 11, 254-260, 1990

Rathmayer, W., Walther, C., and Zlotkin, E., The effect of different toxins from scorpion venom on neuromuscular transmission and nerve action potential in the crayfish. *Comp Biochem. Physiol.* 56c, 35-39, 1977.

Regan. L.J., Voltage-dependent calcium currents in Purkinje cells from rat cerebellar vermis. *J. Neurosci.* 11, 2259-2269, 1991.

Regan, L.J., Sah, D.W.Y., and Bean, B.P., Ca^{2+} channels in rat central and peripheral neurons, high-threshold current resistant to dihydropyridine blockers and ω-conotoxin. *Neuron* 6, 269-280, 1991.

Ruhland, M., Zlotkin, E., and Rathmayer, W., The effect of toxins from the venom of the scorpion *Androctonus australis* on a spider nerve-muscle preparation. *Toxicon* 15, 157-160, 1977.

Salkoff, L., Butler, A.,Wei. A., Scavada, N., Griffin, K., Ifune, C., Goodman, R., and Mandell, G., Genomic organization and deduced amino acid sequence of a putative sodium channel gene in Drosophila. *Science* 237, 744-749, 1987.

Shank, R.P., Pong, S.F., Freeman, A.R., and Graham, L.T., Bicuculline and picrotoxin as antagonists of γ-amino-butyrate and neuromuscular inhibition in the lobster. *Brain Res.* 72, 71-78, 1974.

Simmonds, M.A., Classification of some GABA antagonists with regard to site of action and potency in slices of rat cuneate nucleus. *Eur. J. Pharmacol.* 80, 347-358, 1982.

Takeuchi, A. and Onodera, K., Effect of bicuculline on the GABA receptor of the crayfish neuromuscular junction. *Nature New Biol.* 236, 55-56, 1972.

Takeuchi, A. and Takeuchi, N., A study of the action of picrotoxin on the inhibitory neuromuscular junction of the crayfish. *J. Physiol. (London)* 205, 377-391, 1969.

Tazaki, K. and Cooke, I.M., Characterization of Ca currents underlying burst formation in lobster cardiac ganglion motoneurons. *J. Neurophysiol.* 63, 370-384, 1990.

Tintpulver, M., Zerachia, T., and Zlotkin, E., The action of the toxin derived from scorpion venom on the ileal smooth muscle preparation. *Toxicon* 15, 371-377, 1976.

Tsien, R.W., Lipscombe, D., Madison, D.V., Bley, K.R., and Fox, A.P., Multiple types of neuronal calcium channels and their selective modulation. *TINS* 11, 431-438, 1988.

Unwin, N., The structure of ion channels in membranes of excitable cells. *Neuron* 3, 665-676, 1989.

Usherwood, P.N.R., Glutamate synapses and receptors on insect muscle, in *Glutamate as a Neurotransmitter*, Di Chiara, G. and Gessa, G.L., Eds., Raven Press, New York, pp 183-193, 1981.

Wafford, K.A., Sattelle, D.B., Abalis, I., and Eldefrawi, A.T., γ-Aminobutyric acid-activated 36Cl- influx: a functional *in vitro* assay for CNS γ-aminobutyric acid receptors of insects. *J. Neurochem.* 48, 177-180, 1987.

Waldrop, B., Christensen, T.A., and Hildebrand, J.G., GABA-mediated synaptic inhibition of projection neurons in the antennal lobes of the sphinx moth, *Manduca sexta. J. Comp. Physiol. A* 161, 23-32, 1987.

Walker, R.J., Crossman, A.R., Woodruff, G.N., and Kerkut, G.A., The effect of bicuculline on the gamma-aminobutyric acid (GABA) receptors of neurones of *Periplaneta americana* and *Helix aspersa. Brain Res.* 33, 75-82, 1971.

Walther, C., Zlotkin, E., and mayer, W., Action of different toxins from the scorpion *Androctonus australis* of a locust nerve-muscle preparation. *J. Insect Physiol.* 22, 1187-1194, 1976.

Watkins, J.C., Krogsgaard-Larsen, P, and Honore, T., Structure-activity relationships in the development of excitatory amino acid receptor agonists and competitive antagonists. *Trends. Pharmacol. Sci.* 11, 25-33, 1990.

Young, A.B. and Fagg, G.E., Excitatory amino acid receptors in the brain: membrane binding and receptor autoradiographic approaches. *Trends Pharmacol. Sci.* 11, 126-133, 1990.

Zlotkin, E., Kadouri, D., Gordon, M. D., Pelhate, M., Martin, M.F.., and Rochat, H., An excitatory and a depressant insect toxin from scorpion venom both affect sodium conductance and possess a common binding site. Arch Biochem Biophys 240, 877-887, 1985.

Zlotkin, E., The interaction of insect-selective neurotoxins from scorpion venoms with insect neuronal membranes, in *Neuropharmacology and Pesticide Action*, Frod, M.G., Lunt, G.G., Reay, R.C., and Usherwood, P.N.R., Eds, Ellis Horwood, Chichester, pp 352-383, 1986.

Zlotkin, E., Rochat, H., Kupeyan, C., Miranda, F, and Lissitzky, S., Purification and properties of the insect toxin from the venom of the scorpion *Androctonus australis* Hector, *Biochimie*, 53, 1073-1078, 1971.

Zlotkin, E., Miranda, R., and Lissitzky, S., A factor toxic to crustacean in the venom of the scorpion *Androctonus australis* Hector. *Toxicon* 10, 211-216, 1972.

Zlotkin, E., Eitan, M., Bindokas, V.I., Adams, M.E., Moyer, M., Burkhart, W., and Fowler, E., Functional duality and structural uniqueness of depressant insect-selective neurotoxins. *Biochemistry* 14, 4814-4821, 1991.

Reports of Current Research

Neuroanatomy and Neural Function

ULTRASTRUCTURAL CHARACTERISTICS OF NEUROSECRETORY A CELLS IN IN CENTRAL NERVOUS SYSTEM OF SILKWORM, *BOMBYX MORI*

Hiromu Akai[1], Takayuki Nagashima[1], Shinji Aoyagi[1], and Emiko Kobayashi[2]

[1]Tokyo Univ. of Agriculture, Sakuragaoka, Setagaya, Tokyo 156, Japan, and [2]Topcon Corporation, Hasunuma, Itabashi, Tokyo 174, Japan

INTRODUCTION

The neurosecretory A cells of *Bombyx mori* were observed in the pars intercerebralis of the brain under a light microscope (Kobayashi, 1957). These neurosecretory cells (NSCs) are characterized by their content of large vacuoles in the cytoplasm. Except for a short description in a chapter of a book by Akai (1992), however, there have been no papers published on the ultrastructural characteristics of the NSC or details of these large vacuoles. We have been focussing on the function and significance of these vacuoles, and, in this short communication, will describe the ultrastructural characteristics of the NSCs and the vacuoles during larval stages of *Bombyx mori*.

MATERIALS AND METHODS

The test animals used were the Japanese commercial silkworm, *Bombyx mori*. These were reared on an artificial diet.

The larvae were dissected at several stages during the 4th and 5th instars, and the brains were immediately fixed in a prefixative solution of 2% paraformaldehyde and 2.5% glutaraldehyde for 2 h at room temperature (Nagashima *et al.*, 1991). After prefixation the materials were repeatedly rinsed with 0.1 M cacodylate buffer, and postfixed in cacodylate-buffered 1% OsO_4 solution for 2.5 h at room temperature. After dehydration with a graded series of ethanol, these materials were finally embedded in Epon 812.

For light microscope observation, semithin sections (about 0.75 um thick) were cut with a glass knife inserted into a Richert OMU2 ultramicrotome, mounted on a glass slide, and stained with Azur B staining solution prior to observation.

For the transmission electron microscope, silver and gold sections were cut with a diamond knife on the same ultramicrotome and were stained with saturated uranyl acetate in distilled water and then in solution containing lead acetate, lead citrate, and lead nitrate (Sato, 1968). The microscope with which sections were observed was an LEM-20000. The combined use of light and electron microscopes allowed viewing of both light and electron images of the same area of a semithin section (Sakai, 1982).

0-8493-4591-X/94/$0.00 + $.50
© 1994 by CRC Press, Inc.

RESULTS AND DISCUSSION

Under the light microscope we found three pairs of NSCs each containing one large vacuole in the cytoplasmic area in the pars intercerebralis of the brain (Fig. 1). These NSCs were also visible in the larval, pupal, and adult stages, except during molting periods.

The vacuoles were usually seen as round or elliptical in shape, and they increased gradually during larval development from 4th to 5th instar. Diameters of the vacuoles were about 10 to 15 μm in the 4th and 20 to 30 μm in the 5th larval instar (Figs. 1 and 2). Other types of NSCs were also detected in the median, the lateral and the central areas in the brain, but they did not contain vacuoles of comparable size in any stage.

The electron microscope revealed that each of these NSCs have a round nucleus in the central area and a vacuole which is even larger than the nucleus (Fig. 3). Rough endoplasmic reticulum (RER) is seen throughout the entire cytoplasmic area with developed RER conspicuous around the vacuole. The latter contains numerous cisternae including fine granular materials, thus forming a mesh-work system of RER around the vacuole. The vacuole also contains fine granular materials as do the cisternae, showing that these materials move from the cisternae to the vacuole (Fig. 4).

Masses of neurosecretory granules (NSGs) are found in the area of the RER around the vacuole and Golgi complexes are also seen here. Large numbers of mitochondria are also found throughout the cell.

We can only speculate on the function of the large vacuole in the NSC: perhaps it is formed as a pool to store the materials which are produced in the RER, without passing through the Golgi complex. NSGs are usually formed in the Golgi complex and are sent to the corpus cardiacum and the corpus allatum by the neurosecretory axons. Many questions remain, for example, (1) What kind of material is in the vacuole?, (2) How are these materials transferred to the outside?, and (3) Is there a hormonal function, as with other typical NSCs?

REFERENCES

Akai, H. Endocrine system of insects, in *Atlas of Endocrine Organs, Vertebrates and Invertebrates*, Matsumoto, A. and Ishii, S., Eds., Springer-Verlag, Tokyo, 1992, pp 187-206.

Kobayashi, M., Studies on the neurosecretion in the silkworm, *Bombyx mori* L., *Bull. Sericul. Exp. Stat.* 15, 181-273, 1957. (Japanese with English summary)

Nagashima, T., Niwa, N., Okajima, S., and Nonaka, S., Ultrastructure of silk gland of webspinners, *Oligotoma japonica* (Insecta, Embioptera), *Cytologia* 56, 679-685, 1991.

Sakai, T., Super wide-field electron microscopy for biological specimens, in *The Ultrastructure and Functioning of Insect Cells*, Akai, H., King, R. C., and Morohoshi, S., Eds., Soc. for Insect Cells Japan, Tokyo, 1982, pp 189-193.

Sato, T., A modified method for lead staining of thin sections, *J. Electron Micrsc.* 17, 158-159, 1968.

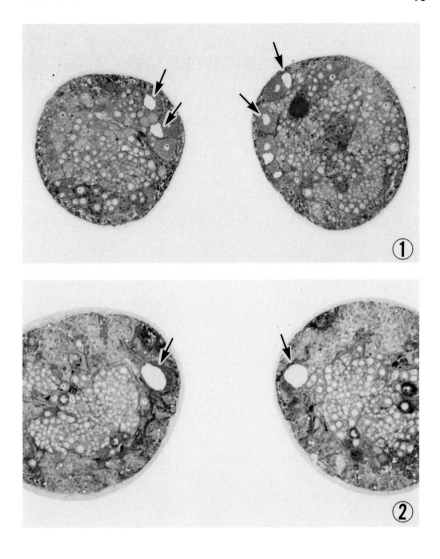

Fig. 1. Semithin section of brain of a 4th day 4th instar of *B. mori*. NSCs each containing a large vacuole are visible in pars intercerebralis (arrows). X 180

Fig. 2. Semithin section of a 9th day 5th instar. Large vacuole are visible (arrows). X 180

Fig. 3. Transmission electron micrograph of NSC with a large vacuole (v) in the cytoplasm. n : nucleus.

Fig. 4. Part of the enlarged vacuole containing fine granular materials. Developed RER containing numerous cisternae are visible around the vacuole. NSCs (arrows) and Golgi complex (G) are also seen. X 9,300

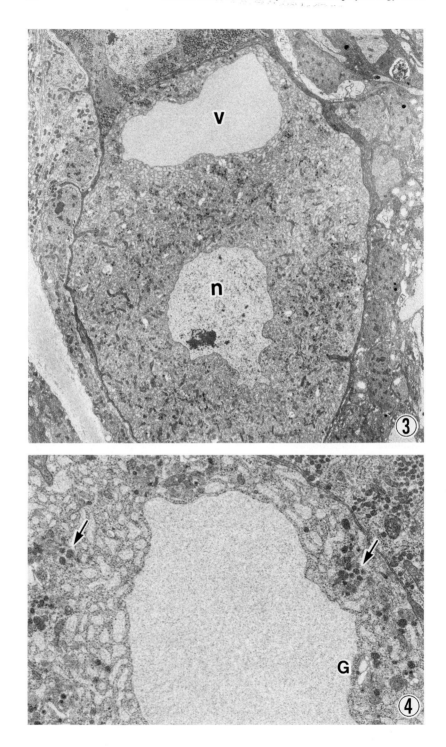

LOCALIZATION OF ^{14}C-LABELLED COMPOUNDS IN INSECT BRAIN

Richard Tykva[a], Blanka Bennettova[b] and Petr Jelinek[a]

[a]Institute of Organic Chemistry and Biochemistry, Academy of Sciences of the Czech Republic, 166 10, Praha 6, Czech Republic

[b]Institute of Entomology, Academy of Sciences of the Czech Republic, 370 05 Ceske Budejovice, Czech Republic

INTRODUCTION

Metabolic pathways in an insect organism may be precisely imaged by radiotracer methodology[1]. A decisive advantage of radiolabelling as compared to non-nuclear methods is a high detection efficiency.

This paper describes a newly developed PC-controlled rapid quantitative imaging of ^{14}C-labelled compounds in insect brain using a recently described method[2]. The values measured are discussed.

MATERIALS AND METHODS

Teneral females of the flesh fly, *Sarcophaga bullata* Parker, were used in the experiments. An acetone solution (5 µl) of 370 kBq of [benzen-U-^{14}C]ethyl 2-4-[(1,4-dioxaspiro[4,5]dec-6-yl)methyl]phenoxy]ethyl carbamate (2.294 GBq/mmol) was applied on the upper part of the insect thorax. After 48 hrs, the flies were sacrificed and briefly washed with 5 ml of methanol-chloroform solution (1:1). Then the head was separated from the body.

Standard histological sections were prepared by routine methodology and stained with Mallorey's staining technique. For radiometric analysis the head was embedded in fish filet and frozen in a cryostat at -20°C. Then 260 µm sections were prepared by coating the embedded head with a 1% solution of gelatine in water before each cutting. Sections were placed on a microscopic slide and dried.

Distribution of radioactivity within a section of the brain was measured by means of a PC-controlled, two-dimensional shift under a special silicon surface-barrier detector. The distribution of radioactivity was imaged by means of a rectangular PC constructed net the dimensions of which can be chosen according to the experimental conditions. In Fig. 1, the three possibilities (a, b, c) of imaging are shown. In comparison with autoradiography, the present

0-8493-4591-X/94/$0.00 + $.50
© 1994 by CRC Press, Inc.

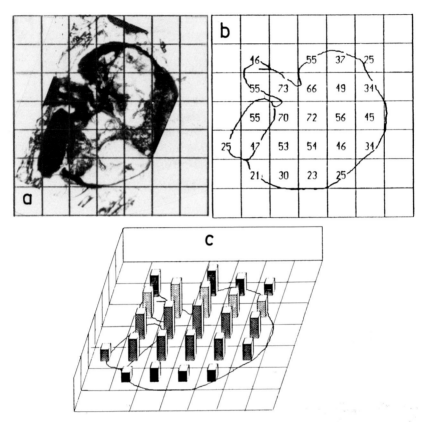

Figure 1. Imaging of radioactivity localization in the brain section (dimensions of 0.5x0.5 mm chosen in this case): a-photograph, b-net with the measured cpm and the sample outline on transparency, c-axonometric imaging of the equal measurement (in % relation of the highest value of counting rate measured).

method has advantages mainly in permitting a fully PC-controlled measurement and evaluation, quantitative values expressed in counting rate per each measured site, a relatively short time of exposition, no distortion of the measured sample, and higher long-term stability.

RESULTS AND DISCUSSION

Our previous analysis of *Scarcophaga bullata* and the same IGR[4] showed the presence of label in the head 48 hrs after the application the radioactivity as at a maximum. Surface washing removed most of the radioactivity outside the head cuticle.

When serial sections of the brain were compared, there were distinct differences in the distribution of radioactivity.

The pattern of distribution, which showed the highest radioactivity in the middle and lower parts of the brain, was identical in all experiments. Corresponding serial histological sections (Fig. 2) revealed the presence of neurosecretory cells in these areas. Therefore it is possible that the IGR or its metabolites interfered with those cells and that reproduction was partly affected in this way. Further investigations are under way.

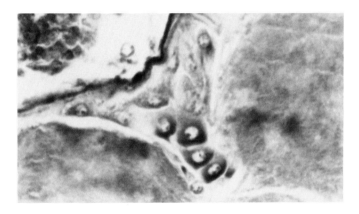

Figure 2. Histological section of the brain showing the area of neurosecretory cells.

The results obtained prove that the choice of the radiometric procedures determines the extent of possible radiotracer analyses of physiological processes in insects.

REFERENCES

Langley, P.A. et al., Formulation of pyriproxyfen, a juvenile hormone mimic, for tsetse control, *Med. Veter. Entomol.* 4, 127-133, 1990.

Tykva, R. et al., A topografix method for studying uptake, translocation and distribution of inorganic ions using two radiotracers simultaneously, *J. Exp. Bot.* 4, 1083-1087, 1992.

Rogers, W. Autoradiography, 3rd ed. North-Holland Publ., Amsterdam, 1974.

Tykva, R. and Bennettova, B., Quantitative analysis of the pesticide fate after its application to the insect. IAEA Rep. SM 327/50, International Atomic Energy Agency, Vienna, 1992.

DUM NEURONS IN THE COCKROACH TAG :
MORPHOLOGICAL ELECTROPHYSIOLOGICAL AND PHARMACOLOGICAL ASPECTS

Bruno LAPIED, Irina SINAKEVITCH,
Françoise GROLLEAU and Bernard HUE

Laboratoire de Neurophysiologie, URA CNRS 611, Université d'Angers,
F-49045 ANGERS Cedex, FRANCE

Dorsal Unpaired Median (DUM) neurons, corresponding to a distinctive group of neurosecretory cells have subsequently been reported in several insect orders. These neurons whose somata lie near the dorsal midline of each segmental ganglion, present a T-shaped morphology (Watson 1984). Unlike many insect neurons, their soma is electrically excitable with spontaneous action potentials (Crossman *et al.*, 1971). It has also been demonstrated that some of DUM neurons are octopaminergic and that their function is to modulate, directly or indirectly the muscle activities, via the release of octopamine (Evans 1985). Although, the anatomy and physiology of DUM neurons have extensively been studied in a number of insect species, very little information is available about DUM neurons of the terminal abdominal ganglion (TAG) of the male cockroach *Periplaneta americana*. The present study was undertaken to extend our knowledge in the morphology, electrophysiology and pharmacology of these TAG DUM neurons using complementary experimental approaches including cobalt staining, immunocytochemistry and patch-clamp technique.

I. COBALT STAINING AND IMMUNOCYTOCHEMISTRY

Intracellular cobalt stainings revealed that from each soma (40 μm to 60 μm), emerges a single primary neurite running anteriorly before dividing into two symmetrical lateral neurites projecting into the left and the right side of the ganglion. These neurons have a characteristic major branching pattern of secondary neurites, mainly located at the periphery of the ganglion, (Fig. 1a). After leaving this region, the lateral neurites run posteriorly towards the base of phallic nerves which ramify in a number of fine branches to innervate, for instance, the accessory glands. In fact, two groups of cells that have a typical DUM neuron morphology have been revealed along the dorsal median and posterior parts of the midline by retrograde cobalt filling through trunk 5C1 of

0-8493-4591-X/94/$0.00 + $.50
© 1994 by CRC Press, Inc.

phallic nerve (Fig. 1b). Using immunocytochemical procedures with polyclonal anticonjugated octopamine antibodies, it has been possible to reveal the presence of octopamine-like immunoreactive DUM neurons according to Eckert *et al.* (1992). In our case, both somata and corresponding lateral neurites of these two groups have been labelled. Some of the OA-ir neurites run into branch 5C1 of phallic nerve. This branch 5C1 laterally carries several gathered labelled neurites towards the mushrooms-shaped accessory gland (MSG) and ejaculatory duct (ED) where they ramify into different fine branches to supply some MSG tubules and EJ muscles. This suggest an octopaminergic innervation of the musculature of accessory gland by some of TAG DUM neurons.

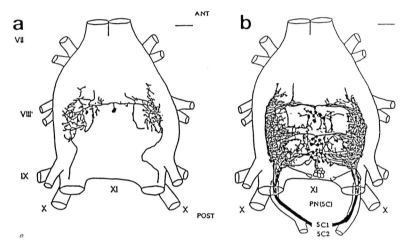

Figure 1. Dorsal view drawings of general morphology of TAG DUM neurons revealed by anterograde (a) and retrograde (b) cobalt chloride fillings. PN : phallic nerves, ANT : anterior, POST : posterior. Scale bar : 100 μm.

II. ELECTROPHYSIOLOGY

The complexity of the anatomy of TAG DUM neurons with the presence of a glial blood-brain barrier surrounding the neurons and restricting the penetration of ions and molecules, justified the possibility of using an alternative approach i.e., isolated DUM neurons maintained in short-term culture (Lapied *et al.*, 1989). Isolated DUM neurons showed spontaneous electrical activity (Fig. 2a), as recording with patch-clamp technique in the whole-cell recording configuration that was generally the same as that of the *in situ* DUM neurons (Lapied *et al.* 1989).

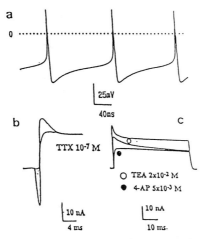

Figure 2. a) Spontaneous AP recorded with the patch-clamp technique in an isolated DUM neuron. b) Blocking effect of 10^{-7} M TTX on the inward current which produced a reduction of outward current. Depolarizing pulse to -10 mV from -80 mV HP. c) Blocking effects of TEA and 4-AP on TTX resistant transient potassium current elicited with a 100 ms depolarizing pulse to +40 mV from -80 HP.

The different ionic conductances mediating the spontaneous electrical activity were studied with the whole-cell clamp technique. The inward sodium current activated at -35 mV, was maximal at -10 mV and reversed at +48 mV very close to the equilibrium potential for sodium. The sodium channels were half-inactivated at -41 mV and half-activated at -26 mV (Lapied *et al.*, 1990). The characterization of the potassium current involved in electrical activity was more complicated. Pharmacological dissection of the potassium current using TTX, 4-aminopyridine, TEA and charybdotoxin suggest the presence of at least four types of potassium currents (Ik) including Ik/Na, Ik/Ca, transient outward current (IA) and delayed rectifier outward current (Ikdr) (Fig. 2b,c). Ik/Na activated under physiological conditions could play a role in the progressive slowing down of the discharge frequency of AP during a depolarization. Moreover the intracellular sodium increases, Ik/Na will promote an hyperpolarization of the neuron. This hyperpolarization also mediated by Ik/Ca and Ikdr may contribute to gradually remove inactivation of the IA (inactivated by depolarization), which shuts down other potassium channels and allow the membrane slowly to depolarize again, to reach the threshold of the spontaneous AP.

III. PHARMACOLOGY

Although the somata of TAG DUM neurons do not possess synaptic contacts, they possess receptors for a number of neurotransmitters.

The pharmacology of these DUM neurons, mainly in the field of cholinergic receptors, however, have not been well characterized. In fact, the nicotinic and/or musarinic nature of the cholinergic receptors have never been determined as the acetylcholine (ACh)-induced response was insensitive to both classical nicotinic and muscarinic antagonists, α-bungarotoxin (α-BGT) and atropine respectively (Lapied *et al.*, 1990). However, the use of selective verterate cholinergic antagonists (see table 1) have given evidence for the presence of a diversity of functional ACh receptors which have been classified in terms of nicotinic (resistant to α-BGT) muscarinic (M1-like, M2-like receptor subtypes) and "mixed" (sensitive to both nicotinic and muscarinic ligands) (Lapied *et al.*, 1990, 1992).

| | ACh | Nicotine | | McN-A-343 | | Arecoline | | |
	FD	FD	SD	H	SD	FD	SD1	SD2
α-bgt	-	-	+			-	-	+
d-TC	±	±	+			+	-	+
mecamylamine						+	-	+
gallamine		-	+					
scopolamine	±			±	±	-	+	+
atropine	±			±	±	-	±	±
QNB						-	+	+
PZP		-	+	-	+	-	+	+
4-DAMP						-	-	-
MET				+	-	-	-	-

ACh : acetylcholine ; α-bgt : α-bungarotoxin ; d-TC : d-tubocurarine ; QNB : quinuclidinyl benzilate ; PZP : pirenzepine ; 4-DAMP : diphenylacetoxy-4-methylpiperidine ; MET : methoctramine ; FD : fast depolarization ; SD : slow depolarization ; H : hyperpolarization

REFERENCES

Crossman, A.R., Kerkut, G.A., Pitman, R.M. and Walker, R., *Comp. Biochem. Physiol.*, 40A, 579-594, 1971.

Eckert, M., Rapus, J., Nürnberger, A. and Penzlin, H., *J. Comp. Neurol.*, 322, 1-15, 1992.

Evans, P.D., Octopamine, In *Comprehensive Insect Physiology, Biochemistry and Pharmacology*, Kerkut, G.A. and Gilbert, L., Eds, Pergamon Press, Oxford, 1985.

Lapied, B. Malécot, C.O. and Pelhate, M., *J. exp. Biol.*, 144, 535-549, 1989.

Lapied, B., Malécot, C.O. and Pelhate, M., *J. exp. Biol.*, 151, 387-403, 1990.

Lapied, B., Le Corronc, H. and Hue, B., *Brain Res.*, 533, 132-136, 1990.

Lapied, B., Tribut, F. and Hue, B., *Neurosci. Lett.*, 139, 165-168, 1992.

Watson, A.H.D., *J. Neurocytol.*, 13, 303-327, 1984.

OCCURRENCE AND FUNCTION OF AN INSULIN-LIKE PEPTIDE IN THE FIFTH INSTAR MALE LARVAE OF *RHODNIUS PROLIXUS*

V.L. Sevala, V.M. Sevala, K.G. Davey and B.G. Loughton
Department of Biology, York University
North York, Ontario, Canada M3J 1P3

Insulin plays a key role in growth, development, metabolism and reproduction in vertebrates (Froesh et al., 1985). Insulin-like molecules with physiological effects similar to those of vertebrates also have been described in invertebrates (Loughton, 1987; Smit et al., 1988). Of particular interest are the bombyxins, insulin-like peptides (ILP) which regulate molting in some insects, particularly silk moths, and which originate from insulin immunoreactive neurons in the brain (Ishizaki and Suzuki, 1988). In the present investigation we have used immunocytochemistry, radioimmunoassay (RIA) and Western blotting to investigate the role of an insulin-like peptide in molting in the hemipterous bug *Rhodnius prolixus*. Insulin-like immunoreactivity was detected in the brain, suboesophageal ganglion and ventral nerve cord of fifth instar larvae of *R. prolixus*. In the brain, two prominent cells were stained in the protocerebrum with bovine insulin antiserum (Fig. 1A), and, significantly, the staining intensity varied during the instar. The cells were stained intensely before feeding and the staining intensity was lost within a few minutes after feeding in both males and females, indicating the release of the insulin-like material. The staining intensity gradually increased up to day 3 followed by a decrease once again on days 5 and 6 in female larvae or 6 and 7 in male larvae, which have been described earlier as defining the end of head critical period (Knobloch and Steel, 1987). These cells gradually regained immunoreactivity reaching a maximum intensity on the day of ecdysis and they retained immunoreactivity even after adult emergence. The prothoracic ganglion and mesothoracic ganglionic mass contained 3 and 9 pairs of immunoreactive cells respectively (25-30 um)(Fig. 1B and C).

The titre of insulin-like peptide in the hemolymph as measured by RIA revealed 3 peaks of activity in the male fifth instar larvae. A peak of insulin immunoreactivity appeared at 2 hours after feeding and again five days after feeding. The titre increased once more on day 13 and remained high until eclosion at about day 21. At the time of the appearance of the last peak of immunoreactivity in the hemolymph, the staining intensity of

0-8493-4591-X/94/$0.00 + $.50
© 1994 by CRC Press, Inc.

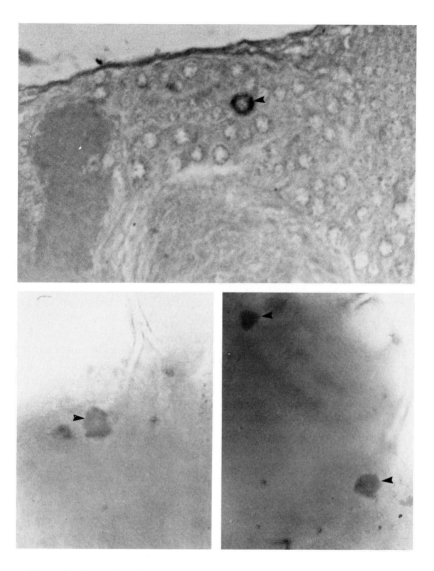

Figure 1: A. Section of the brain of a male 5th instar larva showing an insulin immunoreactive cell body in the protocerebrum (x600).

B and C. Whole mount staining of insulin immunoreactive cell bodies in the prothoracic ganglion (x300) and mesothoracic ganglionic mass (x350). Arrowheads indicate insulin immunoreactive cell bodies.

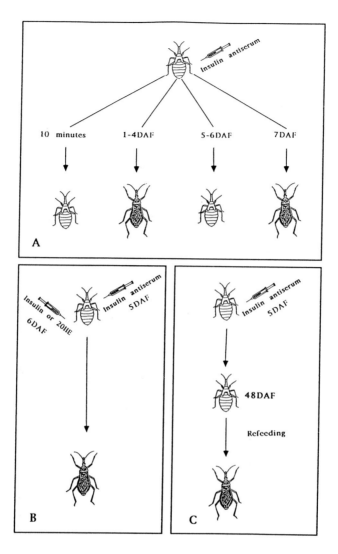

Figure 2: A. Effect of injection of bovine insulin antiserum (5 ul) on molting. Antiserum was injected immediately after feeding (10 min) and 1-7 DAF during the last larval instar. Control animals received normal rabbit serum.
B. Effect of injection of bovine insulin (1 ug) or 20HE (0.5 ug) on molting.
C. Effect of refeeding on molting. The animals remained as larvae after injection of antiserum were refed on 48DAF and observed for adult emergence. (DAF-days after feeding).

insulin immunoreactivity in brain increased, indicating that the brain is not the source of insulin-like material at this time. The thoracic ganglion may thus be the source of the insulin-like peptide late in the instar.

The significance of the insulin-like material was investigated by injecting bovine insulin antiserum at different times during development. The administration of antiserum immediately after feeding and on days 4 and 5 in female larvae, or 5 and 6 in male larvae prevented molting (Fig. 2A). Injection of antiserum at other times during the instar had no effect on adult development (Sevala et al., 1992). The injection of 20E or bovine insulin into such insects relieved the inhibition imposed by anti-insulin and such insects develop normally (Fig. 2B). Furthermore, the larvae in which development was suppressed due to the administration of antiserum were also rescued by later refeeding; all such larvae completed adult development within 21-25 days after the second feed (Fig. 2C). An ILP of approximately 6 kDa was identified in the hemolymph by immunoblotting. These data indicate that the ILP which appears during the head critical period is involved in the control of molting in *R. prolixus*.

REFERENCES

Froesh, E.R., Schmid Chr., Schwander, J. and Zapf, J., Actions of insulin like growth factors, *Ann. Rev. Physiol.*, 47, 443, 1985.

Ishizaki, H. and Suzuki, A., An insect brain peptide as a member of insulin family, *Horm. Metabol. Res.*, 20, 426, 1988.

Knobloch, C.A. and Steel, C.G.H., Effects of decapitation at the head critical period for moulting on hemolymph ecdysteroid titres in final instar male female *Rhodnius prolixus* (Hemiptera), *J. Insect Physiol.*, 33, 967, 1987.

Loughton, B.G., Studies on locust hypolipaemic hormone, *J. Insect Physiol.*, 33, 569, 1987.

Smit, A.B., Vreugdenhil, E., Ebberink, R.H.M., Geraerts, W.P.M., Klootwijk, J. and Joosse, J., Growth controlling molluscan neurons produce the precursor of an insulin related peptide, *Nature*, 331, 535, 1988.

Sevala, V.M., Sevala, V.L., Loughton, B.G. and Davey, K.G., Insulin immunoreactivity and molting in *Rhodnius prolixus*, *Gen. Comp. Endocrinol.*, 86, 231, 1992.

METAMORPHIC CHANGES IN THE LOCALIZATION OF LEUCOKININ LIKE IMMUNOREACTIVE NEURONS IN THE VENTRAL GANGLION OF THE FLY *SARCOPHAGA BULLATA*

P. Sivasubramanian

Department of Biology, University of New Brunswick, Fredericton, New Brunswick, Canada E3B 6E1

Leucokinins I - VIII are a family of octapeptides isolated from the head extracts of the cockroach *Leucophaea maderae* that stimulate contraction of the cockroach hindgut and fluid secretion in the Malpighian tubules of the yellow fever mosquito (see Cook and Wagner, 1991). With an antiserum raised against leucokinin I, Nässel and Lundquist (1991) were able to locate the cellular distribution of this neuropeptide within the the the brains of the cockroach *L. maderae* and blowfly, *Phormia terraenovae*. Subsequently, Cantera and Nässel (1992) demonstrated the localization of leucokinin I immunoreactive (LKIR) neurons in the ventral ganglion (VG) of three species of higher Diptera namely, *Drosophila melanogaster*, *Calliphora vomitoria* and *P. terraenovae*. Interestingly, there were only seven pairs of LKIR neurons (one pair in each of the abdominal neuromeres A1 - A7) in the larval VG. The adult abdominal ganglion had three additional pairs of LKIR neurons. The present study examines the distribution and metamorphic changes in the localization of LKIR neurons within the VG of the fleshfly, *Sarcophaga bullata*.

The postembryonic development of the fleshfly lasts about 18 days at $25^{\circ}C$ (6 days of larval and 12 days of pupal period). At regular intervals during the course of this period the VG were dissected out and the whole ganglia were examined for LKI immunoreactivity by indirect immunofluorescent method. The LKI antiserum was raised by injecting rabbits with synthetic LKI coupled to bovine serum albumin (Nässel and Lundquist, 1991).

The larval VG is a fused mass of three thoracic and eight abdominal neuromeres. There are only seven pairs of LKIR neurons, all located exclusively within the abdominal ganglion and distributed ventrolaterally, one pair in each of the first seven abdominal neuromeres. Immunofluorescent axons are seen in the lateral abdominal nerves that exit the abdominal neuromeres 1 - 7. There are no immunoreactive neurons in the last abdominal neuromere as well as in the three thoracic neuromeres. This pattern is consistent throughout the larval period, from the time of hatching till pupariation.

0-8493-4591-X/94/$0.00 + $.50

© 1994 by CRC Press, Inc.

During metamorphosis the VG undergoes dramatic reorganization including further fusion of abdominal neuromeres. Throughout this period the seven pairs of LKIR neurons retain their immunoreactivity. However, consistent with the metamorphic changes the LKIR neurons also aggregate into ventrolateral clusters. This shift in the localization of the seven pairs of larval LKIR neurons occurs around 5th day after pupariation. New LKIR neurons, a median dorsal pair and two ventrolateral pairs appear on day 8 and 9 respectively, after pupariation. The axons within the lateral abdominal nerves become immunoreactive only after the completion of metamorphosis, i.e. on day 11 after pupariation.

The distribution of LKIR neurons in the fleshfly, *S. Bullata* resembles very closely that of other Diptera (Cantera and Nässel, 1992). The continued presence of LKI immunoreactivity within the neurons suggests some basic function for this neuropeptide, particularly during the larval and adult life, since it is only at these stages the efferent axons also exhibit LKI immunoreactivity. The detailed examination of LKI immunoreactivity in other flies by Cantera and Nässel (1992) revealed several targets for these neurons such as, the body wall muscles of the larva and the spiracles, heart and neurohemal areas of the abdominal wall of the adult flies. Surprisingly, the hindgut whose contractility was used as a bioassay for this neuropeptide by Holman et al. (1986) was not directly innervated by LKIR neurons even in *Leucophaea maderae* (Nässel et al., 1992). However, it is possible the peptide is released into the system from the other neurohemal areas and thus may have an indirect action on the hindgut.

REFERENCES

Cantera, R. and Nässel, D. R., Segmental peptidergic innervation of abdominal targets in larval and adult Dipteran insects revealed with an antiserum against Leucokinin I. *Cell Tiss. Res.* 269, 459–471, 1992.

Cook, B. J. and Wagner, R. M., Myotropic neuropeptides. Physiological and pharmacological actions, in *Insect neuropeptides. Chemistry, biology and action.* Menn, J. J., Kelly, T. J. and Masler, E. P., Eds. American Chemical Society, Washington, D. C. 1991. Chap. 6.

Holman, G. M., Cook, B. J. and Nachman, R. J., Isolation, primary structure and synthesis of two neuropeptides from *Leucophaea maderae. Comp. Biochem. Physiol.* 84C, 204–211, 1986.

Nässel, D. R. and Lundquist, C. T. , Insect tachykinin-like peptide: Distribution of leucokinin immunoreactive neurons in the cockroach and blowfly brains. *Neurosci. Letters* 130, 225–228, 1991.

Nässel, D. R., Cantera, R. and Karlsson, A., Neurons in the cockroach nervous system reacting with antisera to the neuropeptide Leucokinin I. *Jour. Comp. Neurol.* 322, 45 – 67, 1992.

STRUCTURE OF THE ENTERIC NERVOUS SYSTEM IN *PERIPLANETA AMERICANA*, WITH SPECIAL REFERENCE TO THE MUSCLE INNERVATION PATTERN COMPARED WITH THAT OF THE VERTEBRATE SYSTEM

Yasuhisa Endo

Department of Applied Biology, Kyto Institute of Technology, Matsugasaki, Sakyo, Kyoto 606, JAPAN

INTRODUCTION

In most of animals, digestive organs have a well-developed nervous system, which mainly innervates the musculature and plays roles of sensory and motor. In mammals, visceral muscles are composed of smooth muscle cells, except for heart and diaphragm. In insects, on the other hand muscle cells of intestine are a striated type like skeletal muscles. Although several studies have been described on the structure of stomatogastric nervous system in several species of insects (cf. Huddart 1985; Penzlin 1985), precise pattern of muscle innervation remains to be fully elucidated.

The aim of this study is to clarify whether or not each of muscle cells are innervated by the nerve fibers like the case of skeletal muscles, and whether or not the neurotransmitters are released at the classical synaptic sites.

MATERIALS AND METHODS

Embryos, first and second instar nymphs, and both sexes of adults of American cockroaches, *Periplaneta americana*, were used. The digestive tracts were removed, and fixed in Bouin's fixative or 4% paraformaldehyde plus 0.2% picric acid in 0.1M phosphate buffer, pH 7. 4, for 2 to 3 hr. For immunocytochemistry using an antiserum to FMRF-amide (Cambridge Res. Biochem., diluted at 1:1000 with PBS), paraffin sections or whole-mount preparations were stained with PAP method or avidine-biotine method (Vectastain).

For transmission electron microscopy, the tissues were fixed with 2.5% glutaraldehyde and 1% OsO_4, dehydrated in ethanol series and embedded in epoxy resin. The stimulation of high K^+ and Ca^{2+} was applied for 10 to 30 sec just before the fixation to induce an artificial depolarization and exocytotic release of neurotransmitters (Endo 1988a,b, 1989, 1991). For scanning electron microscopy, HC1 treatment was applied to remove a collagenous connective tissue (Endo and Kobayashi, 1987).

0-8493-4591-X/94/$0.00 + $.50
© 1994 by CRC Press, Inc.

RESULTS AND DISCUSSION

1. Immunocytochemistry of FMRF-amide and scanning electron microscopy

The several cell bodies with FMRF-amide immunoreactivity are found in one ingluvial and two proventricular ganglia (Fig. 1), similar to those of *Galleria* (Zitnan et al. 1989), which locate on the dorsal and lateral surface of foregut. A few positive perikarya are present in the nerve cord outside the ganglia (Fig. 1. arrow). In the foregut, varicose nerve fibers ramify from the nerve bundles and traverse across the outer circular muscles (Fig. 3). In the midgut, relatively thick nerve bundles run along the longitudinal muscles and then ramify into fine branches which enter into the inner circular muscle layer (Figs. 2, 4).

In the mammalian enteric nervous system, the terminal portion of innervation is a continuous network of unmyelinated fibers which consist of a network of Schwann cells and nerve fibers (Endo and Kobayashi, 1987). There are no 'true' naked nerve endings found like the neuromuscular junction of skeletal muscles. In the cockroach midgut system, however, I could not detect a network formation of Schwann cells at least in the specimens observed after the HC1 treatment.

2. Exocytotic release of neurotransmitter substances

By transmission electron microscopy of the midgut, the synapse-like contacts are sometimes found in both of the longitudinal and circular muscles. After the stimulation of high K^+ and Ca^{2+}, omega figures indicative of exocytosis are seen not only in the synaptic contact but also in the free surface of nerve fibers (Figs. 5, 6). These non-synaptic release of neurotransmitters has been described in various peripheral nervous system (Endo 1988a,b, 1991; Golding and Pow 1991). The non-synaptic innervation of visceral muscles is consistent with the ultrastructural and electrophysiological data in locust by Anderson and Cochrane (1977, 1978) who described that the muscle fibers appear to be multiterminally and polyneuronally innervated.

3. Conclusion

The patterns of nerve fibers are completely different among the foregut, midgut and hindgut. This is mainly owing to the difference of muscle cells pattern (cf Huddart, 1985). In the midgut, where there are no neuronal cell bodies, thick nerve bundles run along the longitudinal muscles and then the ramified fibers run laterally into the circular muscles. The nerve fibers contain many cored vesicles, which correspond to the various neuropeptide immunoreactivities. Some of the vesicles are released non-synaptically in the muscle layer. The enteric nerves may innervate not each of "target" muscle cells but a relatively broad area, like "paracrine" or local neurosecretory fashion.

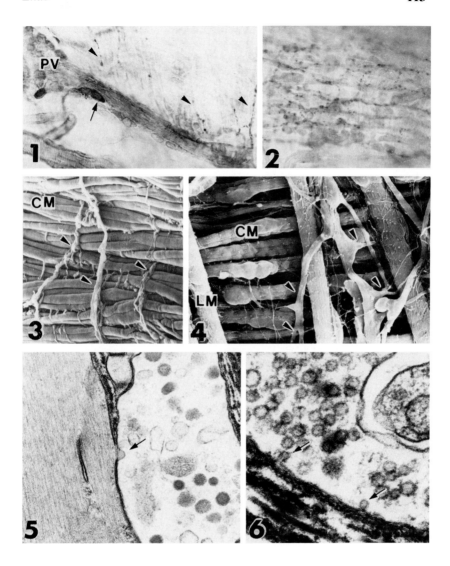

Fig. 1. FMRFamide neuron (arrow) and fiber (arrowheads) in the foregut (2nd instar nymph). whole-mount preparation. PV: proventricular ganglion. x 300

Fig. 2. FMRFamide nerve fibers in the midgut (2nd instar nymph). whole-mount preparation. x 600

Fig. 3. Scanning electron micrograph of nerve fibers (arrowheads) in the foregut (adult). HCl treatment. Cm: circular muscles. x 100

Fig. 4. Scanning electron micrograph of nerve fibers (arrowheads) in the midgut (adult). HCl treatment. CM: circular muscles. LM: longitudinal muscle. x 500

Fig. 5. Exocytotic release of neurotransmitters (arrow) at the synapse-like neuro-muscular junction site in the circular muscle of the midgut. x 25,000

Fig. 6. Non-synaptic release of neurotransmitters (arrows) in the connective tissue around the circular muscle. x 67,000

REFERENCES

Anderson, M. and Cochrane, D.G., Studies on the midgut of the desert locust, *Schistocerca gregaria.* I. Morphology and electrophysiology of the muscle coat. *Physiol. Ent.,* 2, 247-243, 1977.

Anderson, M. and Cochrane, D. G., Studies on the midgut of the desert locust, *Schistocerca gregaria.* II. Ultrastructure of the muscle coat and its innervation. *J. Morphol.,* 156, 257-278, 1978.

Endo, Y., Non-synaptic release of transmitter-containing vesicles from the enteric neurons of the small intestine. *Zoological Science,* 5, 965-971, 1988a.

Endo, Y., Exocytotic release of neurotransmitter substances from nerve endings in the taste buds of rat circumvallate papillae. *Arch. Histol. Cytol.,* 51, 489-494, 1988b.

Endo, Y., Exocytotic release of neurotransmitters form nerve fibers in the endocrine pancreas (Brockmann body) of the teleost, *Takifugu niphobles. Neurosci. Lett.,* 126, 60-622. 1991.

Endo, Y. and Kobayashi, S. A scanning electron microscope study on the autonomic groundplexus in the lamina propria mucosae of the guinea-pig small intestine. *Arch. Histol. Jpn.,* 50, 243-250, 1987.

Golding, D. W. and Pow, D. V., 'Neurosecretion' by synaptic terminals in the locust corpus cardiacum: is non-synaptic exocytosis part of the regulated or the constitutive pathways?, *Biol. Cell,* 73, 157-162, 1991.

Huddart H., Visceral muscle. In *Comprehensive Insect Physiology,* Kerkut, G. A. and Gilbert, L. I., Eds., vol. 11, pp. 131-194, Pergamon Press, Oxford, 1985.

Penzlin, H., Stomatogastric nervous system, In *Comprehensive Insect Physiology,* Kerkut, G. A. and Gilbert, L. I., Eds., vol. 5, pp. 371-406, Pergamon Press, Oxford, 1985.

Zitnan, D., Endo, Y. and Sehnal, F., Stomatogastric nervous system of *Galleria mellonella* L. (Lepidoptera: Pyralidae): Changes during metamorphosis with special reference to FMRFamide neurons. *Int. J. Insect Morphol. Embryol.,* 18, 227-237, 1989.

MIDGUT ENDOCRINE CELLS IN *THYSANURA*

Dušan Žitňan[1], František Weyda[2], and František Sehnal[2]

[1]Department of Entomology, University of California, Riverside,
CA 92521, USA and [2]Institute of Entomology, Academy of Sciences,
370 05 České Budějovice, Czech Republic

I. INTRODUCTION

Insect midgut contains endocrine cells (EC) which harbour a variety of regulatory peptides related to those known from the gastroenteropancreatic system of vertebrates (Žitňan et al. 1993). EC were first recognized by the presence of secretory granules. One of the very first ultrastructural studies concerned *Petrobius maritimus* which was referred to as apterygote insect (Cassier et al. 1972). EC of this species are clustered and present only in the middle of clusters of regenerative cells (nidi).

The studies on pterygote insects revealed that differentiated EC were singly scattered among epithelial cells and only the smallest EC were found in the periphery, or adjacent to the nidi indicating that they are derived from regenerative midgut cells (Endo et al. 1983). Since *Petrobius* is actually not an insect, but a representative of their sister group, *Archaeognatha*, it is possible that the clustering of endocrine cells represents an apomorphic feature of this group. This point is examined in the present study which compares endocrine cells of *Archeognatha* with those of *Zygentoma*,which are true apterygote insects. *Archeognatha* and *Zygentoma* are often referred to as *Thysanura* in sensu lato.

II. MATERIALS AND METHODS

Adults of *Machilis helleri* (Verh.), *Archeognatha*, were collected on lime rocks near Prague or in the vicinity of Bratislava, and *Lepisma saccharina* L., *Zygentoma*, in the living houses of the mentioned towns. For immunohistochemistry, the midguts were processed as described by Žitňan et al. (1993).

Rabbit antiserum to FMRFamide (kindly provided by Dr. C.J.P.Grimmer-likhuijzen) and biotin-streptavidin-peroxidase system (Amersham) were used to identify the endocrine cells.

For the ultrastructural examinations, midguts were fixed in 2.5% paraformaldehyde/2% glutaraldehyde in 0.1 M PIPES buffer, pH 7.2, for 2 hr at room temperature, and postfixed in 1% osmium tetroxide for 30 min. Sections in Epon-araldit were contrasted with uranyl acetate and lead citrate.

Immunocytochemistry was performed on midguts fixed in 4% paraformaldehyde in 0.1 M PIPES for 1 hr, partially dehydrated in 50% and 75% ethanol (10 min each), soaked in LR White/70% ethanol (20 min), and saturated (20 min) and embedded (24 hr polymerization at 70xC) in LR White.

0-8493-4591-X/94/$0.00 + $.50
© 1994 by CRC Press, Inc.

Sections on nickel mesh were incubated for 45-60 min in primary antiserum (1:500 in PBS/BSA, room temperature). After 3 washings in PBS/BSA and three in TRIS/BSA, they were incubated for 30-60 min in protein A labelled with gold particles 1:20 in TRIS/BSA, room temperature), rinsed in water, and contrasted with lead citrate.

III. RESULTS

Most of the FMRFamide-positive cells in the midgut of *Machilis* were single.

All of them were located in the middle of regenerative nidi. In rare cases we found nidi with two adjacent endocrine cells, of which one was always of the closed type.

FMRFamide-positive cells in the midgut of *Lepisma* were also found in the regenerative nidi only (Fig. 1). A nidus contained either a single, centrally localized endocrine cell, or a cluster of endocrine cells. In the latter case, 1-5 small closed-type cells adhered to a large central cell of the open type.

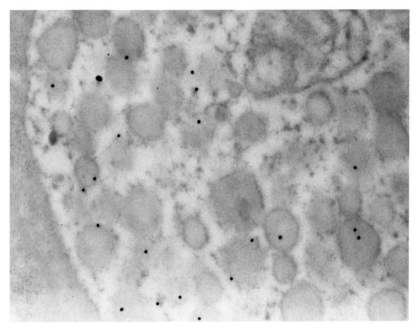

Figure 1. Immunolocalization of FMRFamide in *Lepisma.*
Magnification: x41,000

Ultrastructural examinations revealed presence of apparently endocrine cells with secretory granules in the nidi of both *Machilis* (Fig. 2) and *Lepisma.* The cells contained electron-dense granules and electron-lucent vesicles. Material labelled with immunogold was mostly localized in the vesicles.

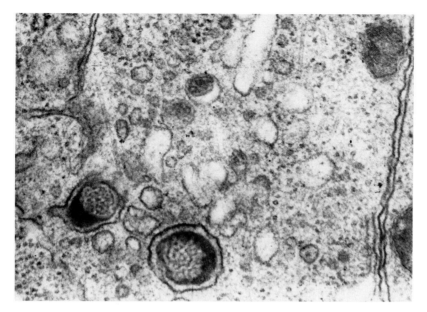

Figure 2. Ultrastructure of the endocrine cell of *Machilis.*
Magnification: x63,000

IV. DISCUSSION

The arrangement of midgut endocrine cells in the examined representatives of *Archeognatha* and *Zygentoma* is similar. Both these groups differ from the pterygote insects by having EC in the regenerative nidi and nowhere else, and by the occurrence of some endocrine cells in pairs or larger groups. Such clustering of EC is quite exceptional in *Pterygota* (Žitňan et al. 1993). Presence of fully differentiated EC in the center of nidi is characteristic exclusively for representatives of *Archaeognatha* and *Zygentoma*. Examination of numerous species representing 17 pterygote orders clearly showed that differentiated EC are never present in the nidi.

V. REFERENCES

Cassier P., Albert J., Fain-Maurel M.A., *C.R. Acad Sc. Paris* 275, 2691-2693, 1972.
Endo Y., Sugihara H., Fujita S., Nishiitsui-Uwo J., *Biomed. Res.* 4, 51-60, 1983.
Žitňan D., Šauman I., Sehnal F., *Arch. Insect Biochem. Physiol.* 22, 113-132, 1993.

THE NERVOUS SYSTEM CONTROLS CUTICULAR
TANNING IN TSETSE LARVAE

J. Žďárek[1] and D.L. Denlinger[2]

[1]Insect Chemical Ecology Unit, Institute of Organic Chemistry and Biochemistry, AS CR, Flemingovo nám. 2, 166 10 Praha 6, Czech Republic; [2]Department of Entomology, The Ohio State University, Columbus, Ohio 43210, U.S.A.

The larval integument was deprived of nervous connections with the CNS by a new technique of non-invasive nerve severance. This enabled us to separate nervous from hormonal action on metamorphosis of larval integument and to obtain the first evidence that the nervous system is directly involved in regulation of puparial sclerotization. Formation of the new cuticle after ecdysis is completed by sclerotization, a process of stiffening and hardening associated with dehydration, loss of plasticity and phenolic tanning. While the ecdysial behavior is initiated by eclosion hormone, the cuticular transformation is regulated by bursicon (see Reynolds, 1986 for review). Formation of a fly puparium is a developmental event different from the moult. No new cuticle is secreted and the old larval cuticle sclerotizes. During this process another set of neurohormones is engaged. These are called pupariation factors (see Žďárek, 1985 for review). Both ecdysis and pupariation are potentiated by the secretion of the steroid moulting hormone ecdysone. Thus it is generally accepted that both neural and hormonal mechanisms are involved in the regulation of ecdysis and pupariation. The nerves control the muscular activity underlying the behavior while the control of the cuticular changes is the exclusive domain of blood-borne hormones.

While working with the larviparous tsetse fly we found that the regulatory mechanisms of pupariation in this insect do not fit the above scheme. As in other cyclorrhaphous Diptera, ecdysteroids play a role in preparation of epidermal cells for pupariation (Žďárek and Denlinger 1991). However, regulation of pupariation cannot be readily explained by the classic pattern of hormonal control. Tsetse larvae ligated shortly before pupariation tan their cuticle in the hind part of the body but fail to do so in the front section containing the CNS and the ring gland. A similar phenomenon of anterior inhibition has been observed in several other fly species (Price, 1970; Chang, 1972; Ratnasiri and Fraenkel, 1973). In an attempt to explain these curious

0-8493-4591-X/94/$0.00 + $.50
© 1994 by CRC Press, Inc.

results, Chang (1972) and Ratnasiri and Fraenkel (1974b) proposed a hypothesis that lack of oxygen in the ligated front part causes inhibition. This hypothesis was challenged and rejected by Whitehead (1974), who, on the basis of his own experiments on tsetse larvae, supported the idea of Langley (1967) that the epidermis is prevented from responding to hormones by a nervous inhibition. However, it has been difficult to produce a direct proof for the nervous inhibition hypothesis because a ligature separating the two portions of the body cuts both the hormonal and neural connections. Using a new non-invasive nerve severance technique we have separated nervous and hormonal effects on larval integument and obtained the first direct proof that the nervous system exerts a direct action on individual epidermal cells (Žďárek and Denlinger, 1992).

When aborted tsetse larvae are ligated 12 to 24 h before pupariation only the front parts tan in most cases, whereas larvae ligated after deposition during the mobile wandering state (1 to 2 h before pupariation) seldom pupariate in either part of the body. However, when the larvae were ligated still later, i.e. after immobilization and during retraction of the anterior segments, they consistently tanned in the hind part only. The absence of pupariation in the front part of the immobile larvae suggests that some kind of inhibition is established in larvae around the time of parturition, and this inhibition is removed when pupariation has proceeded to the point where the body is contracting to its final puparial shape. The sensory input responsible for the inhibition removal originates from the region of the hind spiracles (Žďárek and Denlinger, 1992).

To separate the effect of CNS innervation from the effects of hormonal factors, nerves behind the CNS were non-invasively severed. The operated larvae had their hormonal channels intact but nervous connections destroyed. When the operation was done on the retracting larvae, nerve severance produced individuals that tanned only in the denervated zone. This effect was demonstrated even more dramatically by destroying nerve connections in one side of the body. In such cases only the quadrant directly anterior to the site of nerve severance was inhibited and the rest of the body pupariated normally. The transition between inhibited and uninhibited epidermal cells in the sagittal plane is abrupt. The tanned cells are clearly segregated from the inhibited ones, thus forming a straight border that bisects the larva longitudinally. Obviously, such a response cannot be explained by hormonal factors Žďárek and Denlinger, 1992).

A curious effect, as yet not fully explained, was obtained when the integument of wandering (mobile) larvae was cauterized or mechanically destroyed around the "waist" region. The operated larvae pupariated only in the anterior part, and tanning was greatly accelerated in comparison to intact controls (Žďárek and Denlinger, 1992). Results of this experiment further support our conclusion that nerves rather than blood borne hormones are either directly involved in the control of puparial sclerotization in tsetse larvae or possibly regulate local release of a humoral factor from their endings. Anatomical evidence well documents the extensive innervation of the epidermal cells in dipteran larvae, and neurosecretory release sites are abundant in this area (Finlayson, 1972).

We may only speculate what evolutionary advantage this modification of regulation of metamorphosis offers the larviparous tsetse. Direct involvement of nerves in the control of both pupariation behavior and tanning permits a fast initiation of the response and a more precise coordination of these two events. While a newly deposited larva normally initiates pupariation 1 to 2 h after parturition, it will pupariate within 10 to 15 minutes if it is stimulated mechanically or thermally (Žďárek and Denlinger, 1991). This capacity for rapid response to external stimuli suggests that the epidermal cells are competent to pupariate at any time during the wandering stage, but only when the nervous system releases its inhibition will pupariation begin. If the larva gets into an unsuitable site, such as an excessively wet environment, it can delay the onset of pupariation. Thus the final "go-ahead" command for pupariation can be advanced or delayed, and this decision is relayed to the epidermal cells by nervous conduits. For a species with adenotrophic viviparity this may be a beneficial adaptation. Larvae spend only a brief period as vulnerable free living individuals and, consequently, they need an effective decision-making mechanism with a capacity for rapid response during this time.

How widespread this reliance of the integument on the nervous system may be remains unknown, but several reports of anterior inhibition in ligated larvae of phylogenetically advanced flies (Price, 1970; Fraenkel and Žďárek, 1971; Chang, 1972; Ratnasiri and Fraenkel, 1974a, b) suggest that similar mechanisms may be operating in other species. Our results of non-invasive nerve severance performed on larvae of *Sarcophaga* (Žďárek and Denlinger, unpublished observations) suggest that this kind of cuticular control may also be operating in this genus but as a less significant component of the regulatory mechanism controlling puparial sclerotization. Direct neural control of cuticular plasticization was also demonstrated in the blood sucking bug, *Rhodnius prolixus* (Maddrell, 1965).

REFERENCES

Chang, F., The effect of ligation on tanning in the larva of tachinid parasite, *Lespesia archippivora. J. Insect Physiol.*, 18, 729-735, 1972.

Denlinger, D.L. and Žďárek, J., Commitment to metamorphosis in tsetse (*Glossina morsitans centralis*): Temporal, nutritional and hormonal aspects of the decision. *J. Insect Physiol.*, 37, 333-338, 1991.

Finlayson, L.H., Chemoreceptors, cuticular mechanoreceptors and peripheral multiterminal neurons in the larva of the tsetse fly (*Glossina*). *J. Insect Physiol.*, 18, 2265-2276, 1972.

Fraenkel, G. and Žďárek, J., The evaluation of the *"Calliphora-test"* as an assay for ecdysone. *Biol. Bull. Woods Hole*, 139, 138-150, 1970.

Langley, P.A., Effect of ligaturing on puparium formation in the larva of the tsetse fly *Glossina morsitans* Westwood. *Nature Lond.*, 180, 389-390, 1967.

Maddrell, S.H.P., Neurosecretory supply to the epidermis of an insect. *Science*, 150, 1033, 1965.

Ratnasiri, N.P. and Fraenkel, G., Inhibition of pupariation in *Sarcophaga bullata. Nature, Lond.*, 243, 91-93, 1973.

Ratnasiri, N.P. and Fraenkel, G., Anterior inhibition of pupariation in ligated larvae of *Sarcophaga bullata* and other fly species: Incidence and expression. *Ann. Ent. Soc. Am.*, 67, 195-203, 1974a.

Ratnasiri, N.P. and Fraenkel, G., The physiological basis of anterior inhibition of puparium formation in ligated fly larvae. *J. Insect Physiol.*, 20, 105-111, 1974b.

Reynolds, S.E., Endocrine timing signals that direct ecdysial physiology and behavior, in *Insect Neurochemistry* and *Neurophysiology* 1986. Bořkovec, A.B. and Gelman, D.B., Eds. The Humana Press, Clifton, 53-77, 1986.

Whitehead, D.L., The retardation of puparium formation in Diptera: could factors other than ecdysone control cuticle stabilisation in *Glossina Sarcophaga* species? *Bull. Ent. Res.*, 64, 223-240, 1974.

Žďárek, J., Regulation of pupariation in flies, in *Comprehensive Insect Physiology, Biochemistry and Pharmacology*, Vol. 8, Kerkut, G.A. and Gilbert, L.I., Eds., Pergamon Press, Oxford, 1985, chap. 9.

Žďárek, J. and Denlinger, D.L., Wandering behavior and pupariation in tsetse larvae. *Physiol. Ent.*, 16, 523-529, 1991.

Žďárek, J. and Denlinger, D.L., Neural regulation of pupariation in tsetse larvae. *J. Exp. Biol.*, 173, 11-24, 1992.

ENDOCRINE AND NEUROENDOCRINE MECHANISMS OF ARRESTED HOST DEVELOPMENT IN PARASITIZED INSECTS: LESSONS FROM THE TOBACCO HORNWORM?

Nancy E. Beckage[1], Dusan Zitnan[1], and Frantisek Sehnal[2]

[1]Department of Entomology, Boyce Hall, University of California-Riverside, Riverside, CA 92521-0314, USA, and [2]Institute of Entomology, Academy of Sciences, 370 05 Ceské Budejovice, Czech Republic

I. INTRODUCTION

Parasites exploit myriad physiological strategies to achieve successful development in insect hosts, several of which may act in concert to enhance the probability of the parasites' success. Many braconid and ichneumonid wasp parasites induce an irreversible developmental arrest in larval stage lepidopteran hosts, and the induction of host arrest at this time (usually at the pre- or post-wandering stage) assures that the wasps have sufficient time to mature and emerge before the host transforms itself into a pupa shielded by highly sclerotized (i.e. impenetrable) cuticle. While many solitary parasites eviscerate the host carcass, leaving little but the cuticle intact, gregarious parasitoids draw nutrients solely from the host's hemolymph. Following emergence of the wasps, the host larva often lingers for several days or weeks without feeding or molting again. The physiological mechanisms inducing arrest are the focus of this paper, with emphasis on *Manduca sexta*.

Endocrine and neuroendocrine mechanisms are of paramount importance in causing arrest although with solitary parasites, other factors such as degeneration (or actual ingestion) of the host's prothoracic glands may be more important. In contrast, with gregarious parasites, the glands often persist; for example, in parasitized tobacco hornworm larvae, the host's prothoracic glands remain intact following emergence of *Cotesia congregata* (Beckage and de Buron 1993). In this system, the absence of the appropriate molting cues, rather than the lack of the physiological capacity to respond to molt-inducing signals, likely constitutes the primary arrest mechanism; hence, effects are mediated at the level of hormone synthesis or release, rather than interference with the host's response.

This paper presents a survey of the various mechanisms of host arrest thus far identified in parasitized insects, focusing sequentially on the potential roles of juvenile hormone (JH), ecdysteroids, and neuropeptides, and their interactions in causing host arrest. Multiple molecules appear involved in causing arrest in most species, and a simple universal mechanism does not appear to exist.

0-8493-4591-X/94/$0.00 + $.50
© 1994 by CRC Press, Inc.

123

II.   CRITICAL ARREST-INDUCING FACTORS

A cross-species comparison of parasitic effects on host JH titers reveals three distinct trends: parasitism has no discernable effect, causes an increase in hemolymph JH concentration, or induces a premature decline in the JH titer (Hurd 1993; Lawrence and Lanzrein 1993; Beckage 1985). In hosts induced to metamorphose early, arrest often occurs in the wandering or prepupal phases, whereupon the parasite completes its development and emerges from the host before its pupation is completed.

In instances where parasitism induces a JH increase in the host hemolymph, several factors may synergistically elevate the host's JH titer. In tobacco hornworm larvae parasitized by *C. congregata*, the level of hemolymph JH esterase activity drops off dramatically following ecdysis to the fifth instar, and the JH titer simultaneously increases (Beckage and Riddiford 1982; Beckage 1985). Moreover, JH III is detectable in the host's hemolymph, which is the JH homolog produced by immature stages of Hymenoptera and Diptera. Its mere presence is suggestive of the parasites' production and secretion of JH III into the hemocoel of the host larva since the host itself produces only JH I and JH II, and the parasites have high titers of JH III, particularly during the first and second larval instars (Lawrence *et al.* 1990). The polar metabolite JH III acid is detectable in the hemolymph of cabbage looper larvae parasitized by *Chelonus*, and presumably likewise is of wasp origin   (Jones *et al.* 1990). Other parasitized insects showing a decline in hemolymph JH esterase activity include hosts of *Microplitis* (Zhang *et al.* 1992) and other parasitoids closely related to *C. congregata* (Tanaka *et al.* 1987). How JH esterase is reduced is not known. Hayakawa (1991) isolated a JH esterase inhibiting peptide from parasitized insects but its action remains enigmatic. A survey of JH-elevating effects suggests a combination of a reduced level of hemolymph JH esterase and parasite secretion of JH III may be common during parasitism.

Ecdysteroid deficiency may constitute one of the most common arrest-inducing mechanisms, and the species in which ecdysteroid deficiency has been documented include hosts of *Chelonus* (Jones *et al.* 1992), *Cardiochiles* (Tanaka and Vinson 1991), *Campoletis* (Dover *et al.* 1988), *Cotesia* (Beckage and Riddiford 1982; Beckage *et al.* 1992), and *Microplitis* (Dahlman and Vinson 1993). In exceptional cases, host ecdysteroid titers appear normal or even elevated compared to unparasitized insects, yet the host fails to respond; such a result is suggestive of refractoriness of host target tissues to ecdysteroids normally produced during metamorphosis, as occurs in hosts of *Euplectrus* (Kelly and Coudron 1990).

Few studies have ascertained how parasitism affects the neuroendocrine system. In parasitized *M. sexta*, immunohistochemical and Western blot data show the presence of large accumulations of both "small" (bombyxin) and "large" PTTHs in the host's cerebral neurosecretory cells and retrocerebral complex in arrested hosts with emerged wasps Several other neuropeptides (allatotropin, allatostatin, diuretic hormone, eclosion hormone, adipokinetic hormone, proctolin, and FMRFamide-like peptides) similarly accumulate to high levels far exceeding the neuropeptide contents in unparasitized fed or starved fifth instar larvae (Zitnan *et al.* 1993). Accumulation occurs in the enteric neurons and midgut endocrine cells also. The mechanisms causing accumulation are now under study. In a related system, an earlier paper suggested that either the synthesis of PTTH or its secretion was prevented in *Pseudaletia separata* larvae followingemergence of *Apanteles kariyai* (Tanaka *et al.* 1987).

Unparasitized larvae injected with parasitoid polydnaviruses often display symptoms akin to developmental arrest, as occurs in *M. sexta* (Beckage *et al.* 1993). Effects of polydnaviruses on host insect development have been described elsewhere (Stoltz 1993; Lawrence and Lanzrein 1993), so their regulatory function will not be explored further here. Often the injection of polydnaviruses slows growth, which complicates an endocrine analysis, and the precise nature of their action on the host's endocrine and nervous systems yet remains to be discovered.

Parasitoid teratocytes (derivatives of serosal cells) have been implicated as secreting JH into the hemolymph of the host (Lawrence *et al.* 1990; Grossniklaus-Bürgin and Lanzrein 1990). Moreover, injection of parasitoid teratocytes induces partial or complete arrest of unparasitized insects. In *Heliothis virescens* larvae injected with *Microplitis demolitor* teratocytes, the level of hemolymph JH esterase activity is reduced, which then may act to prevent PTTH release by the brain; as a consequence, the ecdysteroid titer of the host is depressed (Zhang *et al.* 1992). A different scenario ensues following injection of *Cardiochiles nigriceps* teratocytes into surrogate "host" *H. virescens* larvae, which develop extremely high ecdysteroid titers (Pennacchio *et al.* 1992). Wani *et al.* (1990) demonstrated that a potent cocktail of *A. kariyai* polydnavirus, venom, and teratocytes is required to induce arrest in *P. separata* larvae. Obviously no clear trend is evident yet as to the teratocytes' physiological role (Dahlman and Vinson 1993). Perhaps an endocrine function is not implausible given that in several free-living species the serosa transiently secretes hormones (Lawrence and Lanzrein 1993).

In *M. sexta*, arrest likely arises from multiple effects acting on neuropeptides, ecdysteroids, JH, and possibly other regulatory pathways. Our studies emphasize the value of this host-parasite system as a model for unraveling intricacies of the hormonal regulation of normal development and metamorphosis of unparasitized insects, as well as effects of parasitism.

Research on the tobacco hornworm system has been supported by grants from the USDA (92-37302-7470) and NSF (IBN-9006003-04).

REFERENCES

Beckage, N.E.   *Ann. Rev. Ent.* 61, 103-106, 1985.

Beckage, N.E.   and de Buron, I.   *J. Inv. Pathol.* 61, 103-106, 1993.

Beckage, N.E.   and Riddiford, L.M.   *Gen. Comp. Endo.* 47, 308-322, 1982.

Beckage, N.E., Tan, F.F., Schleifer, K.W., Lane, R.D. and Cerubin, L.L. *Arch. Insect Biochem. Physiol.*, in press, 1993.

Beckage, N.E., Zitnan, D., and Gelman, G.D.   *Am. Zool.* 32, 130A, 1992.

Dahlman, D.   and Vinson, S.B.   In *Parasites and Pathogens of Insects* Vol. 1, *Parasites,* Beckage, N.E., Thompson, S.N., and Federici, B.A., Eds., Academic Press, San Diego, 1993, pp. 145-165.

Dover, B.A., Davies, D.H., and Vinson, S.B.   *Arch. Insect Biochem. Physiol.* 8, 113-126, 1988.

Grossniklaus-Bürgin, C. and Lanzrein, B. *Arch.Insect Biochem. Physiol.* 14, 14-30, 1990.

Hayakawa, Y.   *J. Biol. Chem.* 266, 7982-7984, 1991.

Hurd, H.   In *Parasites and Pathogens of Insects* Vol. 1, *Parasites,* Beckage, N.E., Thompson, S.N., and Federici, B.A., Eds., Academic Press, San Diego, 1993, pp. 87-105.

Jones D., Gelman, D.B., and Loeb, M.   *Arch. Insect Biochem. Physiol.* 21, 155-165, 1992.

Jones G., Hanzlik, T., Hammock, B.D., Schooley, D.A., Miller, A., Tsai, L.W. and Baker, F.C.   *J. Insect Physiol.* 36, 77-83, 1990.

Kelly T.J.   and Coudron, T.A.   *J. Insect Physiol.* 36, 463-470, 1990.

Lawrence, P.O., and Lanzrein, B.   In *Parasites and Pathogens of Insects* Vol. 1, *Parasites,* Beckage, N.E., Thompson, S.N., and Federici, B.A., Eds., Academic Press, San Diego, 1993, pp. 59-86.

Lawrence, P.O., Baker, F.C., Tsai, L.W., Miller, C.A., Schooley, D.A., and Geddes, L.G.   *Arch. Insect Biochem. Physiol.* 13, 53-62, 1990.

Pennachio, F., Vinson, S.B., and Tremblay, E.   *Arch. Insect Biochem. Physiol.* 19, 177-192, 1992.

Stoltz, D.B.   In *Parasites and Pathogens of Insects* Vol. 1, *Parasites,* Beckage, N.E., Thompson, S.N., and Federici, B.A., Eds., Academic Press, San Diego, 1993, pp. 167-187.

Strand, M.R., Dover, B.A., and Johnson, J.A.   *Arch. Insect Biochem. Physiol.* 13, 41-52, 1990.

Tanaka, T.   and Vinson, S.B.   *J. Insect Physiol.* 37, 139-144, 1991.

Tanaka, T., Agui, N., and Hiruma, K.   *Gen. Comp. Endo.* 67, 364-374, 1987.

Wani, M., Yagi, S., and Tanaka, T.   *Ent. Exp. Appl.* 101-104, 1990.

Zhang, D., Dahlman, D.L., and Gelman, D.B.   *Arch. Insect Biochem. Physiol.* 20, 231-242, 1992.

Zitnan, D., Kramer, S.J., and Beckage, N.E.   *Develop. Biol.* submitted, 1993.

Signalling Within the Nervous System

NITRIC OXIDE SIGNALLING IN THE
INSECT NERVOUS SYSTEM

Maurice R. Elphick, Irene C. Green* and Michael O'Shea

Sussex Centre for Neuroscience and *Biochemistry Laboratory,
School of Biological Sciences, University of Sussex, Brighton, U.K.

The gas nitric oxide (NO) is now recognized as an important signalling molecule in mammalian nervous systems (Moncada *et al.*, 1991; Bredt and Snyder, 1992). It functions as a signal by diffusing through cell membranes from neuronal sites of synthesis and activating soluble guanylyl cyclase in target cells. NO is involved in the regulation of a variety of physiological processes in mammals and is thought to function as an anterograde transmitter (e.g. in the regulation of smooth muscle tone) and a retrograde transmitter (e.g. in long-term potentiation in the hippocampus). We have recently obtained evidence that the NO - cyclic GMP signalling pathway also operates in the nervous systems of invertebrates such as insects (Elphick *et al.*, 1993).

NO release is stimulated by Ca^{2+} influx into neurons containing Ca^{2+}/calmodulin-activated NO synthase, an NADPH requiring enzyme which generates NO from molecular oxygen and L-arginine. Equimolar citrulline is co-synthesized with NO, so NO synthase activity can be assayed by measuring conversion of [^{14}C]-arginine to [^{14}C]-citrulline (Knowles *et al.*, 1989). We have used this assay to test for NO synthase activity in extracts of cerebral ganglia from the locust *Schistocerca gregaria*. An enzyme with similar properties to mammalian neuronal NO synthase is present in the locust nervous system; it requires NADPH, it is activated by Ca^{2+}/calmodulin and it is inhibited by N^{ω}-Nitro and N^{ω}-Methyl analogs of L-arginine (Elphick *et al.*, 1993). We are now interested in purifying and characterizing this enzyme and identifying neurons in the locust nervous system that contain it.

Mammalian NO synthases have been purified using a variety of chromatographic protocols but by far the most effective involve the use of 2',5'-ADP agarose (Bredt and Snyder, 1990). NADPH requiring enzymes, such as NO synthase, are retained by 2',5'-ADP agarose and can then be eluted from this substrate with NADPH. We have applied this technique to purify NO synthase from extracts of cerebral ganglia from *Schistocerca gregaria* using a method based on that described by Schmidt *et al.* (1991).

Sixty cerebral ganglia (without optic lobes) were homogenized in 2ml ice-cold buffer A (50mM Tris-HCl, pH 7.6/0.5mM EDTA/0.5mM EGTA/10μM leupeptin/1μM pepstatin A/0.1mM phenylmethylsulfonyl fluoride (PMSF)/1mM dithiothreitol). The homogenate was centrifuged at 100,000g at 4°C for 60 minutes and the supernatant mixed with 1ml of

0-8493-4591-X/94/$0.00 + $.50
© 1994 by CRC Press, Inc.

2',5'-ADP agarose equilibrated in buffer B (10mM Tris-HCl, pH 7.6/0.1mM EDTA/0.1mM EGTA/10μM leupeptin/1μM pepstatin A/0.1mM PMSF/1mM dithiothreitol) for 30 minutes at 4°C. The supernatant/ADP agarose mixture was then poured into a Poly-Prep column (Bio-Rad) and washed at 4°C with 5 x 1ml buffer B, 5 x 1ml buffer C (0.5M NaCl in buffer B), 10 x 1ml buffer B, 5 x 1ml buffer D (10 mM NADPH in buffer B) and 8 x 1ml buffer B.

50μl samples of the 1ml fractions collected were assayed for NO synthase activity as described above and by Elphick *et al.* (1993), but with the addition 50 units of bovine brain calmodulin to replace the locust neural calmodulin lost through purification. The column eluates were also tested for NADPH diaphorase activity since it has been demonstrated that NO synthase is responsible for this enzyme activity in mammalian nervous systems (Hope *et al*, 1991). NADPH diaphorase activity was assayed by measuring the reduction of 1mM nitro blue tetrazolium in 50μl 50mM Tris-HCl (pH 7.6) in the presence of 1mM NADPH at room temperature for at least 60 minutes in microtiter plates. The purple formazan reaction product that is indicative of diaphorase activity was measured by monitoring the absorbance of the incubation medium at 570nm. Incubations in which enzyme extract was replaced with 50μl Tris-HCl were conducted as controls.

Material exhibiting <u>both</u> NO synthase-like and NADPH diaphorase activity was retained by the 2',5'-ADP agarose and eluted with 10mM NADPH (fractions 24-28; Fig. 1). Some material that exhibited both of these activities was not retained by the column (fractions 1-3; Fig. 1), possibly due to overloading of the column. These results demonstrate that an enzyme which exhibits both NO synthase and NADPH diaphorase activity is present in the locust nervous system and can be purified using 2',5'-ADP agarose affinity chromatography. Moreover, these results indicate that NADPH diaphorase histochemistry could be used to localize NO synthase in the insect nervous system, as in mammals. Therefore, we have applied this method to the cerebral ganglion of *Schistocerca gregaria* in order to identify the neurons which contain the NO synthase/NADPH diaphorase activity we have detected in extracts.

Cerebral ganglia (with optic lobes intact) dissected from paraformaldehyde (4% in PBS) fixed locust heads were incubated sequentially in PBS (1hr), 2% Triton X-100 in 50mM Tris-HCl (2 x 1hr) and Tris-HCl (2 x 30min). NADPH diaphorase activity was revealed by incubating the ganglia in 1mM NADPH/0.25 mM nitro blue tetrazolium in Tris-HCl at room temperature for about 30 minutes. Staining was observed in several parts of the locust brain, including the optic lobes, the tritocerebrum and the antennal lobes of the deutocerebrum. In the antennal lobes the staining was clearly associated with a group of neuronal cell bodies (Fig. 2). Particularly striking in most of the preparations examined was a pair of intensely stained cell bodies situated at the periphery of the antero-medial quadrant of each antennal lobe (Fig. 2). These are the first putative nitrergic neurons to be identified in an insect nervous system.

We know very little about the functions of NO in the insect nervous system but we have demonstrated that NO causes activation of guanylyl

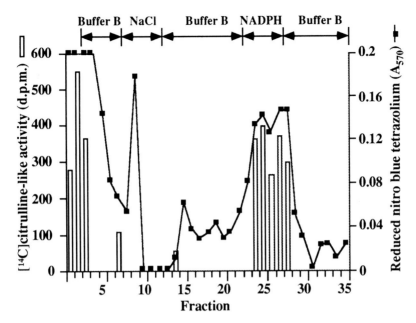

Figure 1. Fractionation of an extract of locust cerebral ganglia by 2',5'-ADP agarose affinity chromatography. NO synthase activity (assayed by monitoring generation of [14C]-citrulline from [14C]-arginine) and NADPH diaphorase activity (assayed by measuring reduction of nitro blue tetrazolium) are co-eluted from the column by 10 mM NADPH.

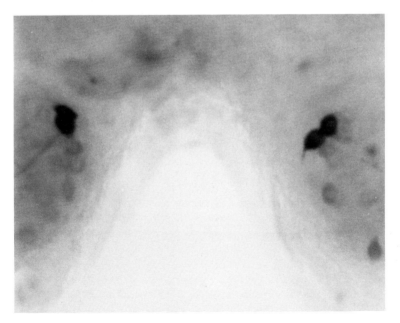

Figure 2. NADPH diaphorase-positive neurons in the antennal lobes of the cerebral ganglion of the locust *Schistocerca gregaria*. Magnification = X 100.

cyclase in the locust brain (Elphick *et al.*, 1993). Thus, at the cellular level NO appears to have the same action in the insect nervous system as in mammals. We can only speculate about the physiological roles of NO in insects at present. Identification of NADPH diaphorase-positive neurons in the antennal lobes of the locust brain (Fig. 2) is indicative of an involvement of the NO-cGMP pathway in insect olfaction. Interestingly, the olfactory bulb in the rat brain is also rich in NO synthase (Bredt and Snyder, 1992) and the NO-cGMP pathway is thought to participate in neural processing of olfactory sensory input in mammals (Breer and Shepherd, 1993). There is also evidence that NO plays an important role in the cellular processes associated with learning and memory in the mammalian brain (Bredt and Snyder, 1992). Intriguingly, D'Alessio *et al.* (1982) have demonstrated that the biological precursor to NO, L-arginine, is required for memory consolidation in the brain of the praying mantis *Stagmatoptera biocelta*. These observations suggest similar roles for NO in insects and mammals.

Acknowledgments: This work was supported by the Science and Engineering Research Council, U.K. (grant GR/G52524 to M.O. and M.R.E.) and the Medical Research Council , U.K. (grant G910 583 9SB to I.C.G.). I.C.G is holder of the B.D.A.-Ames Senior Research Fellowship.

REFERENCES

1. Bredt, D.S. and Snyder, S.H., Isolation of nitric oxide synthetase, a calmodulin requiring enzyme, *Proc. Natl. Acad. Sci. U.S.A.* 87, 682-685, 1990.
2. Bredt, D.S. and Snyder, S.H., Nitric oxide, a novel neuronal messenger, *Neuron* 8, 3-11, 1992.
3. Breer, H. and Shepherd, G.M., Implications of the NO/cGMP system for olfaction, *TINS* 16, 5-9, 1993.
4. D'Alessio, G., Di Donato, A., Jaffé, K., Maldonado, H. and Zabala, N.A., Arginine and memory consolidation in praying mantis, *J. Comp. Physiol. A* 47, 231-235, 1982.
5. Elphick, M.R., Green, I.C. and O'Shea, M. Nitric oxide synthesis and action in an invertebrate brain, *Brain Res.* (in press), 1993.
6. Hope, B.T., Michael, G.J., Knigge, K.M. and Vincent, S.R. Neuronal NADPH diaphorase is a nitric oxide synthase, *Proc. Natl. Acad. Sci. U.S.A.* 88, 2811-2814, 1991.
7. Knowles, R.G., Palacios, M., Palmer, R.M.J. and Moncada, S., Formation of nitric oxide from L-arginine in the central nervous system: a transduction mechanism for stimulation of the soluble guanylyl cyclase, *Proc. Natl. Acad. Sci. U.S.A.* 86, 5159, 1989.
8. Moncada, S., Palmer, R.M.J. and Higgs, E.A. Nitric oxide: physiology, pathophysiology, and pharmacology, *Pharmacol. Rev.* 43, 109-142, 1991.
9. Schmidt, H.H.H.W., Pollack, J.S., Nakane, M., Gorsky, L.D. Förstermann, U. and Murad, F. Purification of a soluble isoform of guanylyl cyclase-activating-factor synthase, *Proc. Natl. Acad. Sci. U.S.A.* 88, 365-369, 1991.

SEROTONIN IN THE ANTENNAL LOBE OF *MANDUCA SEXTA*: POSSIBLE FUNCTION AS A NEUROMODULATOR

John G. Hildebrand[1], Jon H. Hayashi[2], Peter Kloppenburg[1], Alison Mercer[3], and Xue Jun Sun[1]

[1]Division of Neurobiology, Univ. of Arizona, Tucson, AZ 85721, USA
[2]Insecticide Discovery, American Cyanamid, Princeton, NJ 08543, USA
[3]Dept. of Zoology, University of Otago, Dunedin, New Zealand

I. INTRODUCTION

A principal long-term goal of research in our laboratory is to understand the neurobiological mechanisms through which information about particular olfactory stimuli is detected, processed, and integrated with inputs of other modalities in the moth's brain and how the olfactory information ultimately controls behavioral responses (for reviews, see Christensen & Hildebrand, 1987; Homberg et al., 1989; Hildebrand et al., 1992).

In the giant sphinx moth *Manduca sexta*, as in other insects, olfactory information in the environment is detected by antennal receptor cells, which send axons into the antennal lobes (ALs) in the deutocerebrum of the brain. The neuropil of the AL is characterized by an array of sexually isomorphic, spheroidal glomeruli surrounding a central region of coarse neuropil and bordered by lateral, medial, and anterior groups of neuronal somata. The glomeruli are condensed neuropil structures, which contain terminals of primary-afferent axons, dendritic arborizations of AL neurons, and apparently all of the synapses between sensory axons and AL neurons and among AL neurons (Matsumoto & Hildebrand, 1981; Tolbert & Hildebrand, 1981). Sex-pheromonal information is processed in a prominent and distinct male-specific neuropil structure in each AL called the macroglomerular complex (Matsumoto & Hildebrand, 1981).

Most of the neurons in the AL fall into two main classes: local interneurons (LNs), which are confined to the AL, lack axons, and have multiglomerular arborizations, and projection neurons (PNs), which have dendritic arborizations in the AL neuropil and axons that project out of the AL (Homberg et al., 1989). Each antennal-receptor axon provides excitatory synaptic input to neurites of AL neurons -- usually LNs -- innervating the glomerulus to which it projects (Matsumoto & Hildebrand, 1981; Christensen et al., 1993). Thus LNs commonly are interposed between

0-8493-4591-X/94/$0.00 + $.50
© 1994 by CRC Press, Inc.

primary-afferent inputs to and PN outputs from the glomeruli. Both LNs and PNs are spiking interneurons that respond to olfactory stimulation of the antenna with monosynaptic excitation and/or polysynaptic excitation or inhibition (Matsumoto & Hildebrand, 1981; Christensen & Hildebrand, 1987; Christensen et al., 1993).

II. SEROTONIN-IMMUNOREACTIVE NEURON IN THE AL

Among the more than 1000 neurons in each of the ALs, one cell has been identified uniquely: the single serotonin (5-hydroxytryptamine or 5HT)-immunoreactive (5HT-ir) neuron (Kent et al., 1987; Homberg & Hildebrand, 1989). This extraordinary neuron has extensive projections and ramifications in the ipsi- and contralateral protocerebrum and the contralateral AL. Its anatomy, as revealed by intracellular staining, and immunocytochemical characteristics have led to speculations that this neuron may be a centrifugal cell that provides modulatory feedback from the protocerebrum to the AL (Homberg et al., 1989). Because it is the only 5HT-ir neuron in the moth's AL, and in view of the importance of 5HT for central and peripheral neuro-modulatory mechanisms in arthropods (e.g. Kravitz, 1988), we have begun to study the physiological role(s) of 5HT and the 5HT-ir neuron in the AL.

A. MORPHOLOGY OF THE 5HT-IMMUNOREACTIVE NEURON

The structure of the 5HT-ir neuron, in the AL of the adult female moth, has been revealed by means of laser-scanning confocal fluorescence microscopic studies of immunocytochemically stained preparations and electron-microscopic studies of thin-sectioned material prepared by peroxidase-antiperoxidase and immunogold labeling methods (Sun et al., 1993). The fine processes of the 5HT-ir neuron within individual glomeruli are largely confined to basal and core regions of the glomerulus and do not overlap with the region of termination of primary-afferent fibers. The membrane of the cell body is highly indented, and the cell body is very rich in endoplasmic reticulum, Golgi apparati, and clusters of dense-cored vesicles. No synapses have been found on the unbranched neurite in the ipsilateral AL, although large, dense-cored and small, clear vesicles are present. The neurites of the 5HT-ir neuron in the contralateral AL, however, do participate in synapses. The branches within the AL glomeruli, which are among the finest in those neuropil modules, participate in synapses at very low frequency: about 66% of the documented synapses were output synapses and 34% were input synapses, and among them, most appeared to be dyads in single sections.

B. PHYSIOLOGICAL EFFECTS OF 5HT ON AL NEURONS

Although we have not yet investigated the physiology of the 5HT-ir neuron, we assume that 5HT may be released by the 5HT-ir neuron and act on elements in the AL. We are studying the actions of 5HT on AL neurons by intracellular recording *in situ* and whole-cell patch recording *in vitro*.

Excitatory responses of AL LNs and PNs evoked by electrical stimulation of the ipsilateral antennal nerve are enhanced by superfusion of the AL with 10^{-4} M 5HT (Kloppenburg & Hildebrand, 1992). Experiments have shown that 5HT (10^{-4} M in the superfusing saline solution) leads to spike broadening and enhancement of responses of AL interneurons to primary-afferent synaptic input. This enhancement is accompanied by, and presumably at least partly due to, a 5HT-induced increase in the input resistance of the AL neurons. These effects of 5HT can be reversed by washing with saline for ca. 10 min. Thus, *in situ*, 5HT leads to spike broadening and increased excitability (anti-accommodation) of AL interneurons.

To probe for mechanisms underlying the actions of 5HT on AL neurons, we have begun to investigate whether 5HT produces similar effects in AL neurons in primary cell culture (Mercer et al., 1992, 1993). In this work we have used whole-cell patch-clamp recording techniques to analyze the effects of 5HT on electrophysiological properties of AL neurons grown in culture. AL neurons can be dissociated and cultured from animals at all 18 stages of adult development. A subset of neurons, recognized on the basis of morphological characteristics and studied under current clamp, respond to 5HT in a constant and predictable way that is essentially identical to the effects observed *in situ*. Anti-accommodation occurs rapidly (within msec) and is readily reversible, whereas spike-broadening develops more slowly (within minutes) and takes longer to reverse. These findings suggest that different cellular mechanisms may mediate these two 5HT-induced effects. In cells examined under voltage clamp, 5HT reduces the magnitude of voltage-activated outward current. Voltage-activated Na^+ currents are not affected by 5HT. This result supports our earlier finding that 5HT modulates voltage-activated K^+ currents in these cells. 5HT causes reversible reduction of both an A-type K^+ current and a more slowly activating K^+ current resembling the delayed rectifier. Modulation of K^+ currents may underlie the 5HT-induced increases in neuronal excitability and broadening of action potentials observed *in vivo*. Our observations suggest that modulation of K^+ channels is responsible for 5HT-induced increases in cell excitability and spike broadening observed in AL neurons both *in vivo* and *in vitro*.

REFERENCES

Christensen, T. A. and Hildebrand, J. G., Pheromonal information coding by projection neurons in the antennal lobes of the sphinx moth *Manduca sexta*, *Ann. N. Y. Acad. Sci.* 510, 224-228, 1987.

Christensen, T. A., Waldrop, B. R., Harrow, I.D. and Hildebrand, J. G., Local interneurons and information processing in the olfactory glomeruli of the moth *Manduca sexta, J. Comp. Physiol. A* 1993, in press.

Hildebrand, J. G., Christensen, T. A., Arbas, E. A., Hayashi, J. H., Homberg, U., Kanzaki, R. and Stengl, M., Olfaction in *Manduca sexta*: Cellular mechanisms of responses to sex pheromone, in *Proceedings of NEUROTOX 91 - Molecular Basis of Drug & Pesticide Action*, Duce, I.R., Ed., Elsevier Applied Science, London, 1992, chap. 23, pp. 323-338.

Homberg, U., Christensen, T. A. and Hildebrand, J. G., Structure and function of the deutocerebrum in insects, *Annu. Rev. Entomol.* 34, 477-501, 1989.

Homberg, U. and Hildebrand, J. G., Serotonin-immunoreactive neurons in the median protocerebrum and suboesophageal ganglion of the sphinx moth *Manduca sexta, Cell Tiss. Res.* 258, 1-24, 1989.

Kent, K. S., Hoskins, S. G. and Hildebrand, J. G., A novel serotonin-immunoreactive neuron in the antennal lobe of the sphinx moth *Manduca sexta* persists throughout postembryonic life, *J. Neurobiol.* 18, 451-465, 1987.

Kloppenburg, P. and Hildebrand, J.G., Modulatory effects of 5-hydroxytryptamine on interneurons in the antennal lobe of the sphinx moth *Manduca sexta, Soc. Neurosci. Abstr.* 18, 303, 1992.

Kravitz, E. A., Hormonal control of behavior: Amines and the biasing of behavioral output in lobsters, *Science* 241, 1775-1781, 1988.

Matsumoto, S. G. and Hildebrand, J. G., Olfactory mechanisms in the moth *Manduca sexta*: Response characteristics and morphology of central neurons in the antennal lobes, *Proc. Roy. Soc. Lond. B* 213, 249-277, 1981.

Mercer, A. R., Hayashi, J. H. and Hildebrand, J. G., Modulatory effects of 5-hydroxytryptamine on voltage-gated currents in cultured insect olfactory neurons, *Soc. Neurosci. Abstr.* 18, 303, 1992.

Mercer, A. R., Kloppenburg, P. and Hildebrand, J. G., Serotonin-induced changes in excitability of antennal-lobe neurons in *Manduca sexta, Soc. Neurosci. Abstr.* 19, 1993, in press.

Sun, X. J., Tolbert, L. P. and Hildebrand, J. G., Ramification pattern and ultrastructural characteristics of the serotonin immunoreactive neuron in the antennal lobe of the moth *Manduca sexta*: A laser scanning confocal and electron microscopic study, *J. Comp. Neurol.* 1993, in press.

Tolbert, L. P. and Hildebrand, J. G., Organization and synaptic ultrastructure of glomeruli in the antennal lobes of the moth *Manduca sexta*: A study using thin sections and freeze-fracture, *Proc. Roy. Soc. Lond. B* 213, 279-301, 1981.

AMINERGIC CONTROL OF LOCUST SALIVARY GLANDS

Declan W. Ali, Ian Orchard and Angela B. Lange.

Department of Zoology, University of Toronto,
Toronto, Ontario, Canada, M5S 1A1.

INTRODUCTION

The salivary glands of the locust, *Locusta migratoria*, are innervated by two pairs of motoneurons, SN1 and SN2, whose cell bodies are found in the sub-oesophageal ganglion (Altman and Kien, 1979). Immunohistochemistry reveals that the SN1 stain for tyrosine hydroxylase-like immunoreactivity (indicative of the presence of a catecholamine), while SN2 stain for serotonin-like immunoreactivity (Ali *et al.*, 1993). Axons from each of these neurons exit through nerve 7 and follow the salivary duct from where they branch over the salivary gland acini. HPLC-EC reveals that dopamine and serotonin are associated with the salivary glands, with values of approximately 40 pmol dopamine / mg of protein and 44 pmol serotonin / mg of protein (Ali *et al.*, 1993). In addition, it has been shown by radioenzymatic assay that dopamine is present in the SN1 cell body and serotonin is present in the SN2 cell body (Gifford *et al.*, 1991). Both dopamine and serotonin increase fluid secretion rate in isolated preparations of salivary glands (Baines *et al.*, 1989), and both dopamine and serotonin induce an elevation in cyclic AMP content of salivary glands (Ali *et al.*, 1993). Thus it would appear that locust salivary glands are innervated by both dopaminergic and serotonergic neurons and that these neurons may mediate their physiological actions by acting on receptors coupled to adenylate cyclase. In the present study we investigate the pharmacological properties of these receptors and examine the effects of various agonists and antagonists upon both neurally-induced and amine-induced elevations in cyclic AMP.

RESULTS AND DISCUSSION

Incubation of salivary glands with serotonin or dopamine in the presence of the phosphodiesterase inhibitor, IBMX, resulted in dose-dependent increases in cyclic AMP content with a maximum 7-fold elevation induced by 5 μM serotonin and 4-fold elevation induced by 10μM dopamine (Figure 1A). Salivary glands incubated in both dopamine and serotonin in the

0-8493-4591-X/94/$0.00 + $.50
© 1994 by CRC Press, Inc.

presence of IBMX resulted in an accumulation of cyclic AMP equivalent to the addition of the increases induced by each amine alone, thereby suggesting the presence of distinct receptors for dopamine and for serotonin.

Extracellular stimulation of nerve 7b resulted in the initiation of action potentials in each of SN1 and SN2 axons, and, at a stimulation frequency of 15 Hz for 5 minutes, resulted in an approximately 3-fold increase in cyclic AMP content of salivary glands. The increase in cyclic AMP was dependent upon both frequency and duration of stimulation. This result lends further support to the notion that the neurotransmitters released by SN1 and SN2 act via receptors coupled to adenylate cyclase.

The pharmacological properties of the receptors mediating an increase in cyclic AMP content were examined using vertebrate serotonergic and dopaminergic agonists and antagonists. The vertebrate 5-HT_2 receptor agonist, α-methylserotonin, induced a significant elevation in cyclic AMP content of 39% that induced by serotonin. 5-HT_2 receptor antagonists were capable of inhibiting the serotonin-induced elevation of cyclic AMP content. Of the antagonists tested, spiperone had the lowest IC_{50} value of 4.4 μM. The rank order of potency for inhibition was spiperone $>$ cyproheptadine $>$ mianserin $>$ methysergide $>$ ketanserin. The vertebrate dopamine D_1 receptor agonist, SKF-82958, induced a statistically significant elevation in cyclic AMP content. Furthermore, SCH-23390 (5 μM), a potent dopamine D_1 receptor antagonist, inhibited the response to 0.5 μM dopamine by 100% and had an IC_{50} value of 0.25 μM. The rank order of potency for D_1 receptor antagonists was found to be SCH-23390 $>$ butaclamol $>$ flupenthixol. The dopamine D_2 receptor antagonists, spiperone, sulpiride, and haloperidol were ineffective.

Using SCH-23390 and spiperone as the most effective antagonists of dopamine and serotonin respectively, we looked at the neurally-evoked increases in cyclic AMP content. Either of these antagonists were capable of reducing the neurally-evoked increases in cyclic AMP content, so confirming the probable release of both dopamine and serotonin by electrical stimulation (Figure 1B).

The results of this and earlier investigations point to the control of locust salivary glands via dopaminergic and serotonergic neurons. The biogenic amines, dopamine and serotonin, are associated with the salivary neurons SN1 and SN2 and with the salivary glands, where they have been shown to elevate cyclic AMP levels in a dose-dependent manner. Furthermore, these amines appear to bind to different receptors, both of which are probably linked to adenylate cyclase. A pharmacological profile of receptor sub-types on the salivary glands indicates that the receptors coupled to adenylate cyclase appear to be similar to vertebrate 5-HT_2 and vertebrate D_1 receptors. The description of identified dopaminergic and serotonergic neurons along with their target sites makes this an ideal preparation for studying the neurobiology of aminergic neurons.

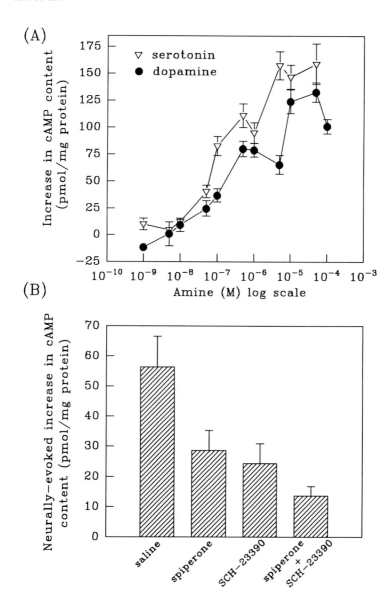

Figure 1. (A) Increase in cyclic AMP content/mg protein of locust salivary glands incubated for 10 min in various concentrations of serotonin or dopamine. (B) The effects of stimulating the salivary nerve for 5 min at 15 Hz upon the cyclic AMP content of locust salivary glands. The presence of either $10\mu M$ spiperone or $10\mu M$ SCH-23390 antagonised the increase in cyclic AMP content induced by electrical stimulation. All incubations in the presence of 0.5 mM IBMX.

ACKNOWLEDGMENTS

This work was supported by the Natural Sciences and Engineering Research Council of Canada.

REFERENCES

Ali, D.W., Orchard, I. and Lange, A.B., The aminergic control of locust (*Locusta migratoria*) salivary glands: Evidence for dopaminergic and serotonergic innervation, *J. Insect Physiol.* in press, 1993.

Altman, J.S. and Kien, J., Sub-oesophageal neurons involved in head movements and feeding in locusts, *Proc. R. Soc. Lond.* B. 205, 209-227, 1979.

Baines, R.A., Tyrer, N.M. and Mason, J.C., The innervation of locust salivary glands. I. Innervation and analysis of transmitters, *J. Comp. Physiol.* A 165, 395-405, 1989.

Gifford, A.N., Nicholson, R.A. and Pitman, R.M., The dopamine and 5-hydroxytryptamine content of locust and cockroach salivary neurones, *J. exp. Biol.* 161, 405-414, 1991.

SEROTONIN: A PUTATIVE REGULATOR OF GLYCOGEN METABOLISM IN THE CENTRAL NERVOUS SYSTEM OF THE COCKROACH, *Periplaneta americana*

J.E. Steele

Department of Zoology, The University of Western Ontario, London, Ontario, Canada N6A 5B7

INTRODUCTION

The insect nervous system must maintain a constant environment in the region of the axons to ensure normal axonal function. Achievement of this, because of the difference in the ionic composition of the extraaxonal fluid and the haemolymph surrounding the nervous tissue, implies that ion movement between these compartments is regulated. It has been suggested that ion balance is maintained by an efficient blood-brain barrier. [Treherne and Schofield, 1979] These authors postulate that ion regulation depends primarily on the presence of Na^+, K^+ pumps on the membranes of the perineurial (blood-brain barrier) and glial cells.

Na^+, K^+ pumps requires a source of energy. The perineurial cells which invest the nervous system contain major deposits of glycogen capable of serving as a source of energy. [Wigglesworth, 1960] The potential importance of glycogen is signified by the fact that its concentration in the insect nerve cord (~ 10 mg.g^{-1}) is ten fold greater than that in mammalian nervous tissue. [Steele, 1963] Metabolism of the glycogen could be affected indirectly through hormonal control of the blood-brain barrier, or by direct action on the reserves. Several hormones are known to stimulate glycogen metabolism in the nerve cord. These include crude corpus cardiacum extract which contains two neuropeptides that activate glycogen phosphorylase. [Steele, 1963] The monohydroxyphenolic amine octopamine also causes glycogenolysis in the nerve cord. [Robertson and Steele, 1973] The possibility that serotonin (5-HT) may also regulate glycogen metabolism is supported by the fact that it stimulates synthesis of the second messenger cyclic AMP in nervous tissue. [Nathanson and Greengard, 1973]

The purpose of this study was to seek evidence for an effect of 5-HT on glycogen metabolism. The study succeeded in showing that 5-HT stimulates glycogen synthesis in the nerve cord, both in vitro and in vivo, and that its action appears to be related to the activation of trehalase.

0-8493-4591-X/94/$0.00 + $.50
© 1994 by CRC Press, Inc.

MATERIALS AND METHODS

Only adult male cockroaches (*Periplaneta americana*), fed commercial dog food and provided with a source of water, were used in the study. The colony was maintained at 27°C with alternating 12h light and dark periods, the dark period beginning at 8:00 p.m.

The cockroach saline contained NaCl, 154 mM; KCl, 8 mM; $CaCl_2$, 2 mM; $MgCl_2$, 3 mM; HEPES buffer, pH 7.2, 5 mM; and trehalose, 40 mM. Nerve cords were incubated in 2 ml of saline at 30°C.

Nerve cord glycogen was determined as previously described. [Robertson and Steele, 1973] Trehalase activity was determined using nerve cord homogenates prepared in 10 mM phosphate buffer, pH 7.0. Supernatant fractions of the homogenate were incubated with 10 mM trehalose for 30 min at 37°C. The reaction was stopped by boiling the samples and the glucose determined using glucose oxidase reagent prepared with Sigma Type V glucose oxidase.

RESULTS

The glycogen concentration in abdominal nerve cords incubated in saline in vitro for 1h was 5.5 ± 0.5 $\mu g.mg^{-1}$. The amount of glycogen in the tissue increased from the resting value to 8.2 $\mu g.mg^{-1}$ as the concentration of 5-HT in the medium was raised from 10^{-8} to 10^{-6}M. This increase in glycogen was significant. Interestingly, the increase in glycogen was only half as great when the concentration of 5-HT was raised to 10^{-5}M and at higher concentrations no effect of the amine was observed.

The stimulatory effect of 5-HT on glycogen in vitro could be mimicked in vivo but with an interesting difference. No effect occurred following injection of 1 pmol of 5-HT but 10 pmol elicited a maximum effect, producing an increase of just over 100 percent. At higher concentrations of 5-HT (0.1-10 nmol) the stimulatory effect on glycogen synthesis was less marked but provoked a significant increase of approximately 20 percent.

Since the only glycogenic substrate supplied to the tissue in vitro was trehalose it seemed likely that glycogen synthesis from the disaccharide would depend on its rate of hydrolysis. To determine whether 5-HT regulates the activity of trehalase, the activity of the enzyme was determined in nerve cords incubated for 1h in vitro, with and without 10^{-6}M 5-HT. Trehalase activity in the presence of 5-HT increased from 0.7 μg to 1.7 μg glucose.mg tissue.h^{-1}. These data suggest that the increase in glycogen is the result of an increase in glucose due to the greater activity of trehalase in the presence of 5-HT.

The data prompted an interest in determining the glycogen level in the nerve cord during a 24h daily cycle. Between midnight and 4:00 p.m. the following afternoon glycogen showed little variation (5.0-5.5 μg.mg^{-1}). However, by 8:00 p.m. glycogen had fallen as much as 40 percent and within four hours had returned to the higher value. On the assumption that the return of glycogen to the higher value might have been mediated by 5-HT the insects were injected with 10 μl of 1 mM methysergide to block the 5-HT receptors. The glycogen content in the nerve cords of the methysergide treated cockroaches did not differ from that in the controls.

DISCUSSION

The data show that injection of 5-HT into the intact cockroach to yield a haemolymph concentration of not more than 10^{-7}M markedly stimulated glycogen synthesis in the nerve cord. The endogenous level of 5-HT in the nerve cord is sufficient, if released, to produce a haemolymph concentration in excess of that required for the glycogenic effect. [Sloley and Owen, 1982] This suggests that the glycogenic effect of 5-HT may be of physiological significance. Injection of larger amounts of 5-HT to produce a concentration of 10^{-3} or 10^{-4}M also cause an increase in glycogen but to a lesser extent. These results are of interest because the same concentrations of 5-HT have a potent inhibitory effect on phosphorylase in the nerve cord. [Hart and Steele, 1969] This effect would contribute to glycogen synthesis since the balance between synthesis and hydrolysis would shift in favour of the former.

Glycogen arising under the influence of 5-HT probably originates from trehalose since this is the only source of glycosyl residues in the in vitro experiment. Trehalose must be hydrolysed to be utilized and this is accomplished through the action of trehalase. This study shows that trehalase is activated by 5-HT. Since this would increase the concentration of glucose it would also provide the substrate for glycogen synthesis.

Trehalase activation by 5-HT is of special interest in view of the demonstration that 5-HT stimulates cyclic AMP production in cockroach nerve cord. [Nathanson and Greengard, 1973] The presence of protein kinases in insect nervous tissue suggests that a phosphorylation mechanism could explain the activation of trehalase. [Albin and Newburg, 1975] Trehalase activation in yeast, which is accomplished through a cyclic AMP induced phosphorylation of an inactive form of trehalase, supports that conclusion. [Thevelein, van Hollander and Shulman, 1984]

Nerve cord glycogen is depleted just before the lights go out (8:00 p.m.) in the insectary. An hour later the cockroaches become very active, while they forage for food. During the initial four hours of the dark period glycogen levels return to the higher value. It is possible that 5-HT stimulates the resynthesis of glycogen following the period of depletion.

A number of authors have suggested that insect activity can be correlated with diurnal changes in the concentration of 5-HT in the nervous system although the evidence for this is weak. [Evans, 1980] However, it is of some interest that a direct effect of 5-HT on activity can be demonstrated by injecting 5-HT into noctuid moths. This causes the duration and amplitude of night flight to be increased. [Hinks, 1967] That investigation suggests further work along the lines pursued in the present study would be of interest.

ACKNOWLEDGEMENTS

The skilled technical assistance of Heather Walker in this study is gratefully acknowledged. Financial support for the work was provided by Natural Sciences and Engineering Research Council of Canada.

REFERENCES

Albin, E.E. and Newburg, R.W., Cyclic nucleotide stimulable protein kinases in the central nervous system of <u>Manduca sexta</u>, Biochem. Biophys. Acta, 337, 389, 1975.

Evans, P.D., Biogenic amines in the insect nervous system, Advances in Insect Physiology, Vol 15, Treherne, J.E., Berridge, M.J. and Wigglesworth, V.B., Eds., Academica Press, N.Y., 1980.

Hart, D.E. and Steele, J.E., Inhibition of nerve cord phosphorylase activity by 5-hydroxyptamine, Experientia, 25, 243, 1969.

Hinks, C.F., Relationship between serotonin and the circadian rhythm in some nocturnal moths, Nature, Lond., 214, 386, 1967.

Nathanson, J.A. and Greengard, P., Octopamine-sensitive adenylase cyclase: evidence for a biological role of octopamine in the nervous tissue, Science, 180, 308, 1973.

Robertson, H.A. and Steele, J.E., Effect of monophenolic amines on glycogen metabolism in the nerve cord of the American cockroach, <u>Periplaneta americana</u>, Insect Biochem., 3, 53, 1973.

Stoley, B.D. and Owen, M.D., The effects of reserpine on amine concentrations in the nervous system of the cockroach (<u>Periplaneta americana</u>), Insect Biochem., 12, 469, 1982.

Steele, J.E., The site of action of insect hyperglycaemic hormone, Gen. Comp. Endocrinol., 3, 46, 1963.

Thevelein, J.M., den Hollander, J.A. and Shulman, R.G., Trehalose and the control of dormancy and induction of germination in fungal spores, Trends in Biochem. Sci., 9, 495, 1984.

Treherne, J.E. and Schofield, P.K., Ionic homeostasis of the brain microenvironment in insects, Trends in Neurosciences, 2, 227, 1979.

Wigglesworth, V.B., The nutrition of the central nervous system in the cockroach, <u>Periplaneta americana</u> L.: the role of perineurium and glial cells in the mobilization of reserves, J. Exp. Biol., 37, 500, 1960.

CALCIUM CHANNELS IN INSECT MUSCLE

Warburton, S.P.M.[1], Duce, I.R[1]. and Lees, G[2].

1 Life Science Department, University of Nottingham, University Park, Nottingham, U.K. 2 Department of Anaesthetics, St. Mary's Hospital Medical School, Norfolk place, London, U.K.

1. INTRODUCTION

In insect skeletal muscle, voltage clamp studies have shown that calcium is the major carrier of inward current at the non-synaptic membrane and is responsible for the observed graded potentials (Washio 1972, Fukuda et. al. 1977, Ashcroft 1981). Little is known about the calcium channels which underlie this inward current.

Single channel and whole cell data from invertebrate preparations, particularly insect neurones indicate a multitude of channel sub-types which do not fit into the vertebrate classification (Leung and Byerly 1991, Pearson et. al. 1993). Here we present evidence suggesting calcium flux through channels most similar to the L-type channel of vertebrate tissue.

2. MATERIALS AND METHODS

A. LARVAL HOUSEFLY NEUROMUSCULAR PREPARATION

Pre-pupal larvae of the housefly *Musca domestica* were opened mid-dorsally and the viscera removed to expose the musculature of the abdominal walls. The preparation was perfused with Normal saline (mM - NaCl 140; KCl 5; $MgCl_2$ 1; $NaHCO_3$ 2; $CaCl_2$ 0.75 & HEPES 5; pH 7.2) and equilibrated for 30 minutes. Recordings were made from the longitudinal muscles numbered 6A and 7A.

B. CULTURED CELLS

Primary cultures of embryonic locust muscle were prepared using a modified version of the "hanging column technique". For full details of methodology refer to Duce and Usherwood (1986).

C. ELECTROPHYSIOLOGICAL TECHNIQUES

Electrophysiology of the bodywall muscles was investigated using current clamp and two electrode voltage clamp (TEVC). In TEVC mode, fibres were held at -70mV and the membrane potential was stepped in the range -120mV to +30mV. Large outward potassium currents predominated masking the transient inward calcium current. The inward current was revealed using 10mM Ba^{2+} as the charge carrier and TEA (tetraethylammonium) in the

0-8493-4591-X/94/$0.00 + $.50
© 1994 by CRC Press, Inc.

bathing saline. "Giga Ohm Seal" single channel recordings from cultured embryonic muscle were made using the patch clamp technique in cell attached and inside-out configurations.

3. RESULTS

A. ELECTROPHYSIOLOGY

Under voltage clamp in Normal saline the major current seen is a large, fast outward potassium current which activates quickly on depolarisation and then rapidly inactivates. It is reduced on the addition of TEA (100mM) and 4-AP (1mM) and is consistent with the I_A current seen in other preparations. The slower activating, sustained I_K current appears to be masked by the very large I_A currents in Normal saline.

The inward current in these fibres inactivates swiftly and could only be studied using potassium channel blockers and Ba^{2+} as the charge carrier (Figure 1). The current is reversibly blocked by Co^{2+} and irreversibly blocked by **verapamil** and **nifedipine**.

Table 1 - Some of the passive properties of the pre-pupal larval muscles.

Resting	Potential	Time constant	Input resistance
6A	7A	6A	6A
mean s.d.	mean s.d.	mean s.d.	mean s.d.
-69mV $\pm$ 8mV	-72mV $\pm$ 7mV	35ms $\pm$ 5ms	515.3KΩ $\pm$ 73.3KΩ
n=115	n=52	n=16	n=16

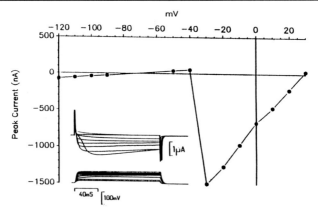

Figure 1 - I/V plot of the inward current in **10mM Ba^{2+}** with examples of currents produced by hyperpolarising pulses shown inset.

B. DEVELOPMENT OF CULTURED TISSUE

After dissociation and plating out the cells were roughly spherical in shape (5 - 20 μm diameter). Many aggregated into clumps of varying size. Differentiation occurred during the first few days. Isolated single presumptive myoblasts elongated to bipolar myocytes with radial migration of myocytes from the tissue clumps. During the second week the myocytes with extended polar processes formed head to tail chains. These aligned and apposed with other chains to form prefusion strings. In the third week the apposed strings thickened and fused to form multinuclated myotubes. Some of these branched (Figures 2 & 3).

After one month in culture the myofibres had thickened and were fibrous in nature. Several exhibited spontaneous contraction. At this time the culture as a whole was a complex network of the predominant myofibres, extensive dendritic processes of the neurocytes, combined with tracheolar, epithelial and glial cell growth.

Single channel recordings from myofibres revealed the presence of transmembrane currents with the characteristics of K^+ channels (Figure 4).

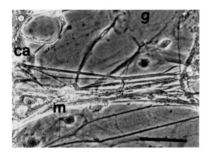
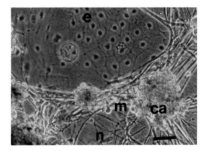

Figures 2 & 3 - Phase contrast photographs of developing myofibres (m) in culture (21 -28 days *in vitro*) suspended between cell aggregates (ca) with surrounding neurite (n), epithelial and glial (g) growth. Scale bar 50μm.

Figure 4 - Single **potassium** channel currents of a cultured myofibre (33 days *in vitro*) recorded in the cell attached mode at varying pipette potentials (180mM K^+ pipette saline; 10mM K^+ bathing saline; traces filtered at 1KHz).

Mean resting potential of the cultured myofibres was **39mV** $\pm$ 4.8mV (n=10)

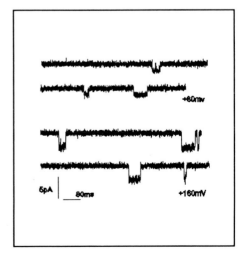

4. DISCUSSION

The kinetics of the fast outward current suggest its function as a modulator of electrical excitability in insect muscle. The timecourse of I_A activation is similar to that of the calcium current and the overlap of these currents results in a reduced inward current responsible for the graded potentials seen.

The inward current, studied using Ba^{2+} as the charge carrier, was high voltage activated, sustained and antagonised by dihydropyridines and phenylalkamines. These data suggest the inward current is due to Ba^{2+} flux through voltage sensitive calcium channels which most resemble the L-type channel of vertebrate tissue.

These studies combined with the work by Leung and Byerly (1991) and Pearson (1993) highlight the considerable biophysical and pharmacological diversity in insect calcium channels.

Although voltage clamp analyses of Dipteran larval muscle have proved useful in initial investigations, research has continued in the direction of primary tissue culture. This preparation has proved more accessible for single channel studies and for investigations using the fluorescent calcium probes fura-2 and fluo-3.

REFERENCES

Ashcroft, F.M., Calcium action potentials in the skeletal muscle fibres of the stick insect *Carausis morosus*, J.exp.Biol. 93, 257-267, 1981.

Duce, J.A. & Usherwood, P.N.R., Primary cultures of muscle from embryonic locusts (*Locusta migratoria*, *Shistocerca gregaria*): development, electrophysiological and patch-clamp studies, J.exp. Biol. 123, 307-323, 1986.

Fukuda, J., Furuyama, S. & Kawa, K., Calcium dependant action potentials in skeletal muscle of a beetle larva, *Xylotrupes dichotomus*, J.Insect.Physiol. 23, 367-374, 1977.

Leung, H-T. & Byerly, L., Characterisation of single calcium channels in *Drosophila* embryonic nerve and muscle cells, J. Neurosci. 11 (10), 3047-3059, 1991.

Pearson, H.A., Lees, G. & Wray, D., Calcium currents in neurones from locust (*Shistocerca gregaria*) thoracic ganglia, J. Exp. Biol. 177, 201-221, 1993.

Washio H., The ionic requirements for the initiation of action potentials in insect muscle fibres, J.Gen.Physiol. 59, 121-134, 1972.

Acknowledgements: S.P.M. Warburton is supported by a SERC-CASE studentship in collaboration with Roussel Uclaf, France.

CALMODULIN AND CASEIN KINASE II IN THE BRAIN AND PROTHORACIC GLANDS OF *MANDUCA SEXTA*

Qisheng Song and Lawrence I. Gilbert

Department of Biology, CB #3280 Coker Hall, University of North Carolina, Chapel Hill, NC 27599-3280

INTRODUCTION

Casein kinase II (CKII) is a multipotential serine protein kinase that mediates the phosphorylation of a variety of proteins involved in signal transduction, metabolism, transcription and translation (see Pinna, 1990). Many polycationic compounds including polyamines and basic polypeptides influence CKII activity, including polylysine which stimulates the *in vitro* phosphorylation of calmodulin (CaM), a Ca^{2+} binding protein that plays a key regulatory role in many cellular events (Klee *et al.*, 1980) e.g. prothoracicotropic hormone (PTTH) stimulation of ecdysteroidogenesis in the prothoracic gland (PG) of *Manduca sexta* (Meller *et al.*, 1988, 1990). Since phosphorylation mediated by CKII alters the biological activity of CaM (Sacks *et al.*, 1992), CKII may indirectly modulate CaM's function as a signal transducer. Because of its critical role in cell signalling, we have studied the CKII-CaM relationship in the *Manduca* brain and PGs.

RESULTS

Developmental profile of PG cytosolic CKII activity

To study any developmental alterations in CKII activity in the PG of *M. sexta*, 30,000 x g cytosolic PG fractions from days 0 to 9 fifth instar larvae (V_0-V_9) or days 0 to 2 pupae (P_0-P_2) were used as the enzyme source and partially dephosphorylated casein was the exogenous substrate. Casein phosphorylation mediated by cytosolic CKII was strikingly increased by the inclusion of 5 mM spermine in the reaction mixture i.e. approximately 3-fold above the basal level at all developmental stages tested (Table 1). Heparin, a known potent inhibitor of CKII activity decreased the apparent basal level of casein phosphorylation by an average of 3-fold, indicating that the apparent basal level actually reflected slight stimulation during the dissection and incubation processes. If one considers the true basal activity to be that occurring in the presence of heparin, then the stimulation by spermine was much greater than 10-fold. Spermine stimulated CKII activity in PG cytosolic fractions showed a developmental profile of biphasic activity during the fifth larval instar and early pupal stages (Table 1). Stimulated CKII activity was the highest between V_0 and V_2, declining thereafter to a lower plateau throughout the remaining 7 days of the fifth larval instar. Both basal and stimulated enzyme activity increased again at

0-8493-4591-X/94/$0.00 + $.50
© 1994 by CRC Press, Inc.

pupation and remained relatively high during P_1 but then decreased at P_2 and P_3. Spermine-stimulated CKII activity changed with developmental stage and at V_0 was approximately twice that of V_3-V_7 glands. Spermine-stimulated CKII activity in P_0 PGs was about 1.5 fold that of P_1-P_2.

Table 1
CKII Developmental Profile and Stimulation by 5 mM Spermine.
CPM/μg PG Cytosolic Protein   Fold Stimulation

Stage	B(Basal)	S(Spermine)	H(Heparin)	S/B	S/H	B/H
V_0	332 ± 107	747 ± 79	25 ± 3	2.3	29.9	13.0
V_1	190 ± 7	603 ± 6	40 ± 2	3.2	15.1	4.8
V_2	113 ± 20	526 ± 42	41 ± 12	4.7	12.8	2.8
V_3	121 ± 1	393 ± 11	42 ± 9	3.2	9.4	2.9
V_4	88 ± 12	317 ± 5	29 ± 6	3.6	10.9	3.0
V_5	106 ± 5	398 ± 12	31 ± 2	3.8	12.8	3.4
V_6	49 ± 8	400 ± 0	38 ± 5	8.2	10.5	1.3
V_7	62 ± 10	317 ± 52	30 ± 4	5.1	10.6	2.1
V_8	147 ± 0	478 ± 65	44 ± 11	2.8	10.9	3.3
V_9	237 ± 42	659 ± 77	43 ± 3	2.8	15.3	5.5
P_0	143 ± 17	686 ± 0	27 ± 2	4.8	25.4	5.3
P_1	102 ± 3	420 ± 12	33 ± 5	4.1	12.7	3.1
P_2	129 ± 23	482 ± 43	38 ± 10	3.7	12.7	3.3

Densitometric analysis of endogenous phosphoproteins in the PG cytosol indicated that total endogenous protein phosphorylation in the presence of spermine is high in V_1 glands, 15% lower in P_0 glands, and 20% lower in V_6 PGs, all consistent with Table 1. Western blot analysis with a specific polyclonal antibody to the human α-subunit of CKII revealed that this antibody immunostained a 39 kDa peptide, identical to the molecular weight of the α-subunit of *Drosophila* CKII (Saxena *et al.*, 1987). Although little is known about the regulation of CKII, recent evidence indicates that polyamines could act as intracellular mediators. They are present in millimolar concentrations in most eukaryotic cells and the activity of a rate-limiting enzyme in polyamine synthesis (ODC) is elevated in the fat body and brain of *Manduca* at the beginning and end of the 5th instar (Birnbaum *et al.*, 1987), a pattern similar to that of the PG CKII. Juvenile hormone titers are known to be elevated early in the instar and thus could stimulate both the CKII and ODC activities, (see Birnbaum and

Gilbert, 1990). Since one of the endogenous PG proteins phosphorylated by CKII had the same molecular weight as CaM, we wished to obtain purified *Manduca* CaM for further analysis.

Purification and characterization of *Manduca* CaM

Purification of CaM from P_0 and P_1 brains included 60% ammonium sulfate precipitation, pH 4.2 centrifugation, and phenyl-agarose column affinity chromatography (see Haiech and Capony 1988). An almost homogeneous preparation of a 19 kDa protein was obtained, and to confirm that this protein was CaM, the sample was subjected to isoelectrofocusing, calcium shifting, calcium overlay, and Western blot analyses. The purified protein yielded a single band with an IP of about 4, and Ca^{2+} caused the 19 kDa protein to migrate in SDS-PAGE with an apparent molecular weight lower than in the absence of Ca^{2+}. Calcium binding ability by the purified 19 kDa protein was detected by electroblotting to a PVDF membrane in the $^{45}Ca^{2+}$ overlay assay and Western blot analysis of the purified 19 kDa protein revealed that both monoclonal and polyclonal antibodies generated again bovine brain CaM immunoreacted with this 19 kDa protein. Thus, the putative *Manduca* CaM has similar, if not identical, properties to bovine brain CaM in terms of molecular weight, Ca^{2+} binding ability, and immunoreactivity to both monoclonal and polyclonal antibodies, suggesting that the 19 kDa protein is CaM. Our most conclusive evidence was the observation that the 19 kDa protein could activate cAMP phosphodiesterase, a widely used paradigm for identifying CaM in a functional manner (see Schachtele and Marme, 1988). The calcium overlay and Western blot analyses showed that *Manduca* brain and PG CaM shared identical properties so the purified brain CaM was used in studies on the ability of PG CKII to phosphorylate CaM.

Phosphorylation of CaM by PG CKII

The data reveal that the phosphorylation of CaM was enhanced significantly in the presence of polylysine, but that spermine was without effect. This is the first demonstration that CaM phosphorylation is enhanced by polylysine in a crude tissue preparation, presumably via the stimulation of CKII, since this is the only kinase known to be stimulated by polylysine. The increased phosphorylation of CaM in the presence of polylysine was accompanied by the decreased phosphorylation of at least 5 phosphoproteins (35 kDa, and four greater than 100 kDa) suggesting the activation of a phosphoprotein phosphatase or the inhibition of other protein kinases (see Sacks *et al.*, 1992). In addition, the phosphorylation of at least 6 other proteins (29, 39, 45, 47, 56 and 91 kDa) was enhanced, suggesting that the phosphorylation of CaM results in the activation of a CaM-dependent protein kinases(s). Although polylysine is a synthetic molecule and therefore not the physiological effector, it is known to mimic physiological relevant compounds, i.e. histone (Yamamoto *et al.*, 1978), protamine (Meggio *et al.*, 1983), basic proteins such as the k-ras protein (Gatica *et al.*, 1987) etc. Having established a role for Ca^{2+}/CaM in the action of PTTH on the PGs, it is important to establish CaM's possible role as a Ca^{2+}

binding protein in the PTTH-PG cascade e.g. activation of adenylate cyclase and CaM kinase. CaM can be phosphorylated by CKII which itself can be regulated by peptide hormones (Sacks *et al.*, 1992). The present study has provided a base for our future research that will probe the possibility that PTTH acts on the PGs by the post-translational modification of CaM via phosphorylation, perhaps through CK II stimulation.

ACKNOWLEDGEMENTS

This work was supported by NSF Grant IBN 93-00164 and NIH Grant DK 30118. We thank the J. M. McDonald Laboratory for CaM antibody, W. Combest for helpful comments and Pat Cabarga for clerical assistance.

REFERENCES

Birnbaum, M. J. and Gilbert, L. I., Juvenile hormone stimulation of ornithine decarboxylase activity during vitellogenesis in *Drosophila melanogaster*, *J. Comp. Physiol.*, B160, 145-151, 1990.

Birnbaum, M. J., Combest, W. L., Bloom, T. J. and Gilbert, L. I., Polyamine regulation of protein phosphorylation in the brain of the tobacco hornworm, *Manduca sexta*, *J. Neurochem.*, 48(3), 935-942, 1987.

Gatica, M., Allende, C. C., Antonelli, M. and Allende, J. E., Polylysine-containing peptides, including the carboxyl-terminal segment of the human c-Ki-ras 2 protein, affect the activity of some key membrane enzymes, *Proc. Natl. Acad. Sci. USA*, 84, 324-328, 1987.

Haiech, J. and Capony, J. P., Purification of calmodulin and its peptides, in *Calcium-Binding Proteins*, Vol I, Thompson, M. P., Ed., CRC Press, Inc., Boca Raton, FL, 1988, chap. 2.

Klee, C. B., Crouch, T. H., and Richman, P. G., Calmodulin, *Ann. Rev. Biochem.*, 49, 489-515, 1980.

Meggio, F., Brunati, A. M. and Pinna, L. A., Autophosphorylation of type 2 casein kinase TS at both its α-and β-subunits, *FEBS Lett.*, 160, 203-208, 1983.

Meller, V. H., Sakurai, S. and Gilbert L. I., Developmental regulation of calmodulin-dependent adenylate cyclase activity in an insect endocrine gland, *Cell Regulation*, 1, 771-778, 1990.

Meller, V. H., Combest, W. L., Smith, W. A. and Gilbert L. I., A calmodulin-sensitive adenylate cyclase in the prothoracic glands of the tobacco hornworm, *Manduca sexta*, *Mol. Cell. Endocrinol.*, 59, 67-76, 1988.

Pinna, L. A., Casein kinase 2: an 'eminence grise' in cellular regulation?, *Biochim. Biophys. Acta.*, 1054, 267-284, 1990.

Sacks, D. B., Davis, H. W., Williams, J. P., Sheehan, El L., Garcia, J. G. N. and McDonald, J. M., Phosphorylation by casein kinase II alters the biological activity of calmodulin, *Biochem. J.*, 283, 21-24, 1992.

Saxena, A., Padmanabha, R. and Glover, C. V. L., Isolation and sequencing of cDNA clones encoding alpha and beta subunits of *Drosophila melanogaster* casein kinase II, *Mol. Cell. Biol.*, 7, 3409-3417, 1987.

Schachtele, C. and Marme, D., Methods of assay of calcium-binding proteins, in *Calcium-Binding Proteins*, Vol. 1, Thompson, M. P., Ed., CRC Press, Inc., Boca Raton, FL, 1988, chap. 6.

Tabor, C. W., and Tabor, H., Polyamines, *Ann. Rev. Biochem.*, 53, 749-790, 1984.

Yamamoto, M., Criss, W. E., Takai, Y., Yamamura, H. and Nishizuka, Y., A hepatic soluble cyclic nucleotide-independent protein kinase, *J. Biol. Chem.*, 254, 5049-5052, 1979.

ACTIVITY OF AN IDENTIFIED, LIGHT-REGULATED NEURON, CONTAINING VASOPRESSIN-LIKE PEPTIDES REDUCES THE LEVEL OF cAMP IN THE LOCUST CNS.

Richard A. Baines, Kevin S.J. Thompson, Richard C. Rayne, A. Deep Alef, Ping Zhou*, David G. Watson*, John M. Midgley* and Jonathan P. Bacon.

Sussex Centre for Neuroscience, University of Sussex, Brighton, BN1 9QG U.K. *Department of Pharmaceutical Sciences, University of Strathclyde, Glasgow, G1 1XW U.K.

The suboesophageal ganglion of at least 17 grasshopper species contains the cell bodies of two large neurons (ca. 60μm in diameter) which are immunoreactive to antibodies raised against the vertebrate peptide arg-vasopressin (Tyrer et al, 1993). These two neurons, termed the vasopressin-like immunoreactive (VPLI) neurons, lie on the ventral surface of the ganglion and are particularly unusual in that they have an extensive ipsilateral arborisation in all ganglia of the CNS (Fig. 1). The VPLI neuron has been physiologically characterised in one grasshopper species, *Locusta migratoria*, where it receives excitatory input from a presynaptic neuron (Thompson and Bacon, 1991). Under bright light the presynaptic neuron is inhibited, depriving the VPLI neuron of its excitatory input. This circuit dictates that the VPLI neuron is active only during periods of darkness. The presynaptic element does not receive its light sensitivity from the compound eye or the ocelli, but from an extra-ocular photoreceptor (EOP).

Two peptides, structurally related to arg-vasopressin, have been isolated and sequenced from *Locusta* suboesophageal and thoracic ganglia (Proux et al, 1987). These are the nonapeptide CLITNCPRGamide (F1) and its antiparallel dimer (F2). The equally possible parallel dimer was not detected. To determine whether VPLI is a source of these peptides, we used HPLC separation coupled with radioimmune assay to analyse the peptide content of 200 electrophysiologically identified VPLI neurons from *Locusta migratoria*. Each VPLI neuron contains 1.2pg of F1 and 1.5pg of F2 but, in addition, contains 1.6pg of the parallel dimer (PDm). The fact that both dimers occur in equimolar amounts suggests that dimerisation is a chance (and not enzymatically driven) process that does not favour a particular dimeric configuration. This reflects exactly the situation *in vitro* where inducing dimerisation of F1 produces equal amounts of F2 and PDm. The amount of peptides found in a VPLI cell body is approximately 3% of that present in the side of the nervous system containing its arborisation.

We have also analysed VPLI neurons for aminergic transmitters. The extract of a pooled sample of 30 VPLI cells was derivatised with ditrifluoro-methylbenzoyl chloride and bis-(trimethylsilyl)-acetamide (DTFMB-TMS) and analysed by gas chromatography-negative ion chemical ionisation mass

0-8493-4591-X/94/$0.00 + $.50
© 1994 by CRC Press, Inc.

Control	cAMP	cGMP	IP3	n
(both VPLIs inactive: left side minus right)	0.4 ± 13.1	-1.0 ± 0.9	-	10
Test				
(one VPLI active: active side minus inactive)	$-38.9 \pm 16*$	-5.6 ± 3.0	2.2 ± 8.3	25

values are mean $\pm$ S. E. (pmol per mg protein)

*$P < 0.05$ (paired two-tailed t-test)

Table 1: Summary of the effect of VPLI activity on second messenger levels in the CNS of *Locusta migratoria.*

spectrometry for the DTFMB-TMS derivatives of a range of biogenic amines. Results indicate that all amines screened for (including the positional isomeric octopamines and synephrines, noradrenaline, adrenaline, dopamine, and tyramine) were not present in amounts significantly greater than background. In contrast, octopaminergic metathoracic dorsal unpaired median (DUM) neurons assayed at the same time as positive controls were found to contain between 50-200pg of octopamine per cell, as well as smaller amounts of dopamine (2-4pg) and tyramine (6-7pg). This apparent absence of aminergic transmitters in VPLI is also indicated by the lack of immunostaining with antibodies raised against tyramine, octopamine, dopamine, histamine or 5-hydroxytryptamine.

A significant step forward in understanding the function of the VPLI neuron will be the identification of post-synaptic targets. Despite previous reports that F2 (1nM) elevates fluid secretion and cAMP production by *Locusta* Malpighian tubules (Proux et al, 1987), this effect could not be replicated by another group in a direct comparison of F2 and the potent 46 residue *Locusta* diuretic peptide (Coast et al, 1993). Consequently, we have been looking for other roles for the VPLI peptides.

The extensive arborisation of the VPLI neuron within the CNS strongly suggests that its peptides may play a role within the confines of the nervous system. As a prelude to identifying this function we have measured what effect VPLI stimulation has on second messenger levels (cAMP, cGMP and inositol trisphosphate) in the CNS. The experimental approach used is outlined in Fig. 2. One VPLI neuron was impaled with a microelectrode in an isolated CNS in which the input to both VPLI neurons was suppressed by light ensuring that both neurons were inactive. The impaled neuron was stimulated by passing continuous depolarising current to promote the typical VPLI firing rate of approximately 5 action potentials per second, for a 5 minute period. Following this, the saline bathing the pro, meso and metathoracic ganglia was replaced with liquid nitrogen. These three ganglia were then cut down their midlines and each side assayed separately for its second messenger content using radioimmune assays. Because VPLI only projects ipsilaterally in the thoracic ganglia, cutting the CNS along the midline provides an ideal comparison of the experimental (active VPLI)

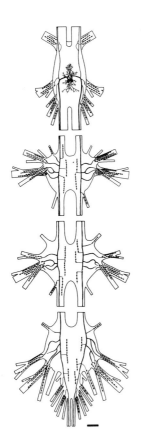

Figure 1. A diagramatic representation of the VPLI neuron in the suboesophageal and three thoracic ganglia of *Locusta migratoria*. The cell body lies in the suboesophageal ganglion. VPLI has an ipsilateral arborisation in all ganglia, some branches of which enter peripheral nerves. The endings in the nerves are not typical of neurohaemal release sites because they are not peripheral, but instead lie deep within the nerve. In addition, they are only present in one subfamily of grasshopper species (the Oedipodinae), which includes *Locusta migratoria*. Other species such as *Schistocerca gregaria* lack these endings and seemingly compensate by increasing the amount of central arborisation; see Tyrer et al, 1993 for a full morphological description. Scale bar = 200μm.

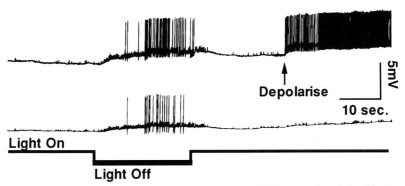

Figure 2. A simultaneous intracellular recording from the two VPLI neurons in an isolated *Locusta* CNS in which the brain is tracheated. Both VPLI neurons are active only when the preparation is in darkness. Returning to light quickly removes the excitatory input, making VPLI inactive. Under these conditions one VPLI is selectively activated by passing continuous depolarising current through the intracellular electrode. After 5 minutes of activity, the thoracic ganglia are frozen, cut through their midline and the second messenger content of both halves measured.

versus basal (inactive VPLI) conditions. Data is shown in Table 1.

Activity of VPLI is associated with a significant decrease in the amount of cAMP. Thus, the side of the CNS containing the active VPLI cell is consistently lower in cAMP (on average a 12% reduction) than the side containing its inactive counterpart. The effect of VPLI activity is limited to cAMP and does not influence cGMP or IP3 levels. In control, unstimulated, preparations both sides of the nervous system do not differ markedly in their cyclic nucleotide content.

Given that VPLI contains three, identified (vasopressin-like) peptides, each peptide was assayed for its effect on cAMP levels in both dissociated neural cultures and isolated neural membranes. The effects, which were the same for the two methodologies, show that only F2 is capable of depressing the level of cAMP, whereas F1 and PDm are ineffective (Fig. 3).

The effect of VPLI activity is particularly exciting because very few transmitters are known to depress cyclic nucleotide levels in the invertebrates and therefore this system offers us the opportunity to study this phenomenon in greater detail.

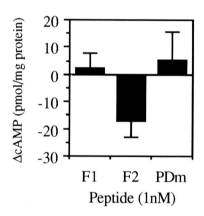

Figure 3. The effect of the three VPLI peptides on cAMP levels in isolated neural membranes stimulated with a water soluble forskolin analogue (0.1µM). Only F2 (significant at $P \leq 0.05$) is capable of reducing cAMP levels while F1 and PDm have no significant effect from control.
Values are mean ± S.E.
n = 22 (F2), 3 (F1, PDm)

REFERENCES

1. Tyrer, NM., Davis, NT., Arbas, EA., Thompson, KSJ., and Bacon, JP., Morphology of the vasopressin-like immunoreactive (VPLI) neurons in many species of grasshopper, J. Comp. Neurol. 329, 385-401, (1993).

2. Thompson, KSJ., and Bacon, JP., The vasopressin-like immunoreactive (VPLI) neurons of the locust, *Locusta migratoria*: II. Physiology, J. Comp. Physiol. A 168, 619-630, (1991).

3. Proux, JP., Miller, CA., Li, JP., Carney, RL., Girardie, A., Delaage, M., and Schooley, DA., Identification of an arginine vasopressin-like diuretic hormone from *Locusta migratoria*, Biochem. Biophys. Res. Commun. 149, 180-186, (1987).

4. Coast, GM., Rayne, RC., Hayes, TK., Mallet, AI., Thompson, KSJ., and Bacon, JP., A comparison of the effects of two putative diuretic hormones from *Locusta migratoria* on isolated locust Malpighian tubules, J. Exp. Biol. 175, 1-14, (1993).

CHANGES OF HUMORAL ANTIBACTERIAL ACTIVITY INDUCED BY HISTAMINE IN THE HAEMOLYMPH OF CABBAGE MOTH

Mirko Slovák, Alica Hučková, and Mária Kazimírová

Institute of Experimental Phytopathology and Entomology,
Slovak Academy of Sciences, 900 28 Ivanka pri Dunaji, Slovakia

INTRODUCTION

Histamine (HA) functions as a neurotransmitter in the insect nervous system, especially in the visual system. The presence of HA has been demonstrated also in the venom and/or venom system of hymenopterans. So far little is known about the functional role of HA outside the insect nervous system. The effects of HA on heart activity, blood pressure, contraction of muscles, color changes have been studied in several invertebrate species (Reite 1972). In vertebrates, HA is a well-established extracellular messenger in numerous physiological and pathophysiological conditions and also inhibits the immunoglobulin production by B lymphocytes (Reite 1972; Saxena et al. 1989; Oh et al. 1989). Since vertebrates and arthropods share analogous immunologic mechanisms (Gupta 1991) the aim of this work was to observe the possible role of histamine in immunoregulation in insects.

MATERIALS AND METHODS

Larvae and pupae were taken from a colony of *Mamestra brassicae* L. (Lep., Noctuidae) maintained in laboratory under standard conditions. All bacteria used were kept in medium LB (see: Slovák et al. 1992). The sixth instar larvae on day 2 or 3 were immobilized by CO_2 and then injected dorsolaterally between the 5th and 6th abdominal segment by 5 µl of a suspension of viable, log-phase *Escherichia coli* D21 (approx. 2×10^8 cells/ml) suspended in saline A (190mM NaCl, 19mM KCl, 7.5mM $CaCl_2$). Larvae injected with *E. coli* on day 3 were pre-injected 24 hr before (on day 2) with 5 µl of saline A (1) or with 5 µl of 3M histamine dihydrochloride (Sigma) dissolved in saline A (2). The same solutions (1, 2) were used for 1 hr pre-injection or 1 and 24 hr post-injection of larvae injected by *E. coli* on day 2. Their haemolymph was collected three days after injection of *E. coli* into ice-cooled eppendorf tubes containing a few

0-8493-4591-X/94/$0.00 + $.50
© 1994 by CRC Press, Inc.

Table 1
Survival of *Mamestra brassicae* larvae and activity against *E. coli* D22 in their cell-free haemolymph after injection of different solutions

Larvae were injected with *E.coli* D21	n	Survival %	Antibacterial activity n	x ± SEM (i:. RU)
24 hr after				
pre-injection with SA	10	80	8	53 554 ± 11 491**
pre-injection with HA	20	55	11	25 333 ± 7 289*
1 hr after				
pre-injection with SA	20	55	11	9 296 ± 1 158
pre-injection with HA	20	30	6	10 704 ± 1 560
Control (D21 only)	23	83	19	12 851 ± 1 409
1 hr before				
post-injection with SA	10	70	7	31 686 ± 12 254*
post-injection with HA	20	30	6	10 576 ± 1 377
24 hr before				
post-injection with SA	20	30	6	11 183 ± 1 397
post-injection with HA	30	37	11	5 186 ± 1 169**

RU - relative units; **$P \leq 0.01$, *$P \leq 0.05$ according to t-test; SA - Saline A, HA - Histamine.

crystals of phenylthiourea by cutting the 2nd abdominal proleg.

Diapausing pupae (2-3 weeks old) were injected ventrally between wing bases and the 1st abdominal segment with 10 μl/g pupal weight of a - saline B (130mM NaCl-5mM KCl); b - suspension of viable, log-phase *Enterobacter cloacae*, strain ß12, in saline B at approx. 3.2×10^5 cells/ml; c-g - 1M, 0.5M, 0.1M, 0.01M and 0.001M solution of histamine dihydrochloride in suspension b. The haemolymph of larvae and pupae was collected individually, centrifuged and stored as described by Slovák et al. (1992).

Antibacterial activity in haemolymph was determined by an inhibition zone assay (Slovák et al. 1992) but there were used Petri dishes with diam. 19 cm and *E.coli* D22. Diameters of clear zones were converted into relative units (RU) of antibacterial activity using the formula by Wiesner (1991). Data were evaluated using one-way ANOVA and t-test.

RESULTS
Larvae injected with *E. coli* D21 showed the highest survival rates (83%). The survival rates of larvae pre-injected or post-injected with histamine (HA) were lower (30-55%) than those of larvae pre-injected or post-injected with saline A (30-80%) (Table 1).

Three days after injecting of *E. coli* D21 the haemolymph of larvae showed a level of activity against *E. coli* D22 about 12 851 RU (Table 1). This activity was 4.2 and 2 times higher in larvae pre-injected with saline A and HA, respectively, 24 hr before injection of the bacteria. Injection of

Table 2
Survival of *Mamestra brassicae* pupae and activity against *E. coli* D22 in their cell-free haemolymph after injection of different solutions

Pupae were injected with	n	Survival %	Antibacterial activity n	Antibacterial activity x ± SEM (in RU)
Saline B	35	49	6	50.8 ± 2.0**
E. cloacae	90	22	12	632.6 ± 145.1
E. cl. + histamine				
1M	90	30	8	204.9 ± 82.0*
0.5M	60	22	6	226.1 ± 9?.6
0.1M	90	18	8	305.7 ± 151.0
0.01M	90	22	9	542.6 ± 216.1
0.001M	90	30	6	407.8 ± 147.7

RU - relative units; **$P \leq 0.01$, *$P \leq 0.05$

HA 1 hr before or after injection of bacteria decreased the levels of antibacterial activity only slightly. While the antibacterial activity of haemolymph of larvae pre-injected with saline A 1 hr before the challenge with bacteria exhibited only a tiny decrease in comparison with control, the antibacterial activity of haemolymph of larvae post-injected with saline A 1 hr after the challenge with *E. coli* D21 was 2.5 times higher. Antibacterial activity was 2.5 times lower in the haemolymph of larvae injected with HA 24 hr after injection of *E.coli* D21 in comparison with control.

The percentage of surviving pupae 3 days after injection of different solutions was considerably low (18-49%) with the highest survival rate after injection of saline B (49%) (Table 2).

The highest antibacterial activity in the haemolymph of pupae was observed after injection of *E. cloacae* (632.6 RU) (Table 2). The injection of *E. cloacae* together with 0.001-1M HA into pupae led to a lower activity than the injection of *E. cloacae* alone. This decrease of activity was dependent on HA concentration but only the injection of 1M HA induced a significant decline (3 times). The lowest antibacterial activity (50.8 RU) was exhibited by the haemolymph of pupae injected with saline B.

DISCUSSION

Our results provided evidence for the possibility of influencing the intensity of the insect humoral immune response by injecting of HA. The ability of HA to change immune responses depended on its concentration and time of application. The haemolymph of larvae injected with HA (or saline) 24 hr prior to injection with bacteria inhibited bacterial growth more strongly than the controls. HA might have been metabolized before injection of bacteria and the injury (1st injecting) provoked the synthesis of antibacterial proteins. This synthesis was enhanced after the second injection with bacteria. Rheins and Karp (1985) noticed that immune *Periplaneta americana* receiving an additional injection of antigen showed a clasic secondary response. The inhibition of antibacterial activity with HA in haemolymph of larvae injected 1 hr before or after injection with bacteria was not significant probably because of a low dose used. Experiments with simultaneous injection of HA and bacteria into pupae showed that the inhibition effect of HA was dose dependent, i.e. only a high dose of HA decreased the antibacterial activity significantly. Injection of HA 24 hr after inoculation of bacteria displayed a dramatic decline in reactivity. The fact that injection of saline only decreased the level of antibacterial activity suggests that HA injection is not totaly responsible for the activity observed.

In vertebrates, a great deal of data suggest an interrelation between the immune and the neuroendocrine systems. In insects the nervous system may influence the cellular defense mechanisms, too. Gole et al. (1982) observed the presence of octopamine-sensitive receptors on haemocytes. According to Baines et al. (1992) dopamine, octopamine and 5-HT increase phagocytosis in vitro. Ottaviani et al. (1992) stated that the same mobile cell is capable of immune and proto-stress responses using neuropeptides as mediators. So far no information is available concerning the interrelation between neuroendocrine system and humoral defense reactions.

REFERENCES

Baines, D., DeSantis, T. and Downer, R.G.H., Octopamine and 5-hydroxytryptamine enhance the phagocytic and nodule formation activities of cockroach (*Periplaneta americana*) haemocytes, *J. Insect Physiol.* 38, 905-914, 1992.

Gole, J.W.D., Downer, R.G.H. and Sohi, S.S., Octopamine-sensitive adenylate cyclase in haemocytes of the forest tent caterpillar, *Malacosoma disstria* Hubner (Lepidoptera: Lasiocampidae), *Can. J. Zool.* 60, 825-829, 1982.

Gupta, A.P., Vertebrates and arthropods share analogous immunologic mechanisms: known and predicted mediators, *J. Idaho Acad. Sci.* 27, 5-28, 1991.

Oh, C., Okamoto, H. and Nakano, K., Regulation of lymphocyte blastogenesis by histamine produced in the system per se, *Agric. Biol. Chem.* 53, 377-382, 1989.

Ottaviani, E., Caselgrandi, E., Fontanili, P. and Franceschi, C., Evolution, immune responses and stress: studies on molluscan cells, *Acta Biol. Hungarica* 43, 283-298, 1992.

Reite, O.B., Comparative physiology of histamine, *Physiol. Rev.* 52, 778-819, 1972.

Rheins, L.A. and Karp, R.D., Effect of gender on the inducible humoral immune response to honeybee venom in the american cockroach (*Periplaneta americana*), *Develop. Comp. Immunol.* 9, 41-49, 1985.

Saxena, S.P., Brandes, L.J., Becker, A.B., Simons, K.J., LaBella, F.S. and Gerrard, J.M., Histamine is an intracellular messenger mediating platelet aggregation, *Science* 243, 1596-1599, 1989.

Slovák, M., Repka, V. and Slováková, L., Evidence and partial comparison of inducible antibacterial factors in the haemolymph of *Mamestra brassicae* (Lepidoptera, Noctuidae), *Acta Entomol. Bohemoslov.* 89, 107-112, 1992.

Wiesner, A., Induction of immunity by latex beads and by hemolymph transfer in *Galleria mellonella*, *Develop. Comp. Immunol.* 15, 241-250, 1991.

INFLUENCE OF STARVATION ON HISTAMINE LEVEL IN SELECTED TISSUES OF COCKROACH *Nauphoeta cinerea*

Alica Hučková and Milan Kozánek

Institute of Experimental Phytopathology and Entomology, Slovak Academy of Sciences, 900 28 Ivanka pri Dunaji, Slovakia

INTRODUCTION

Histamine (HA) is widely distributed in vertebrates and inverte brates, where it fulfills a variety of functions (Schwartz et al. 1980, Claiborne and Selverston 1984). In insect CNS HA is considered to act as an important neurotransmitter in the optic system of some insect species (Simmons and Hardie 1988, Hardie 1989). High amounts of this biogenic amine were also found in ganglia of the retrocerebral complex and in the frontal ganglion (FG) of the stomatogastric nervous system (Hučková et al. 1992).

The present study represents a preliminary attempt to evaluate the influence of starvation and feeding on HA level in haemolymph and some tissues of cockroach *N. cinerea* dependent on FG removal.

MATERIAL AND METHODS

All experiments were performed with males of cockroach *N. cinerea* (Blattodea, Panchloridae). After last ecdysis they were taken from our laboratory colony and kept in bisexual groups at $29\pm 1^{\circ}C$, $50\pm 5\%$ RH, L:D 12:12. A semisynthetic diet and water were provided ad lib. To observe the influence of starvation and FG removal on HA concentration in brain and midgut tissue, cockroaches were divided into 4 groups: fed males with (1) and without FG (2), males 5 days after ecdysis with (3) and without (4) FG exposed to starvation for 2, 4, 6, 14 days. FG was removed from cockroaches chilled for 3 min just after last ecdysis. The influence of food supply on HA level in the brain without optic lobes, FG and haemolymph was tested in males starved for 6 and 14 days. Insects were killed 1, 6, 24 hours after feeding in a freezing mixture of dry ice and ethanol. Ganglia were dissected in a cooled 0.18 $mol.l^{-1}$ solution of NaCl and homogenized in cooled microhomogenizers containing 50-120 μl of solution A (0.2 $mol.l^{-1}$ sodium phosphate buffer pH 7.9:1 $mol.l^{-1}$ NaOH 10:1.1). The distal part of the midgut was cut, intestinal contents removed, gut wall washed in saline and water. Homogenates were centrifuged at 5000 g for 10 minutes at $4^{\circ}C$. Haemolymph was obtained after cutting antennae and the first pair of legs by centrifugation for 5 minutes at 750 g and $4^{\circ}C$, diluted with solution A 1:3 and centrifuged for 10 min at 7500 g.

0-8493-4591-X/94/$0.00 + $.50
© 1994 by CRC Press, Inc.

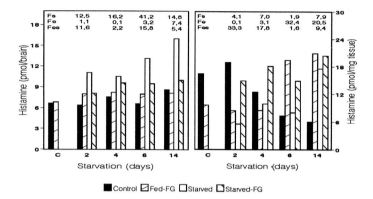

Figure 1. Influence of starvation and frontal ganglion (FG) removal on histamine concentration in the brain and midgut tissue in <u>N. cinerea</u> males. Each column represents mean of 7 individuals. F-ratio for effect of starvation (Fs), FG extirpation (Fe) and their interaction (Fes). Critical values for F are: 4.26 (p<0.05), 7.82 (p<0.01).

10 μl of the supernatant was transferred into cooled conical vials and the HA content was determined radioenzymaticly according to Brownstein et al. (1974). In all experiments O.5 ng of histamine-dihydrochloride in 10 μl 0.1 mol.l^{-1} HCl as internal standard was used. The radioactivity of samples was measured by means of a Beckman scintillation counter. The data were analyzed using two-way ANOVA, followed by multiple range test and t-test.

RESULTS

Our results showed a significant influence of starvation on the cerebral HA levels in comparison with fed individuals (Fig. 1). Both starving groups had higher amounts of HA in their brains and a tendency of HA concentration to increase with time. FG removal did not evoke significant changes in HA concentration, but our results suggest a certain interaction between starvation and FG absence (see F ratio). During the investigated period there was a significant decrease in HA concentration in the gut of control individuals. In both starving groups as well as in the fed one without FG a significant increase in HA concentration was found. A certain interaction between FG removal and starvation also occurred.

About 1.6-times higher level (p<0.05) was found in the brain of males starving for 6 days in comparison with control (Fig 2a). One hour after supplying food to the starving cockroaches the level of HA significantly increased. Feeding for 6 and 24 hours resulted in continual but not significant drop in the concentration. A similar pattern of changes in cerebral HA level was found in males deprived of food for 14 days (Fig. 2b). HA concentration in FG of re-fed animals was significantly higher (p<0.05) in comparison with starving group, but no

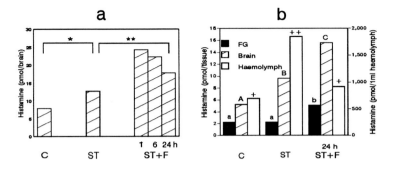

Figure 2a, b. Histamine concentration changes in the brain (a), and brain, frontal ganglion (FG) and haemolymph (b) dependent on starvation (ST) for 6 and 14 day (a, b) and feeding (ST+F) for 1, 6 and 24 hours. C - control. Each column represents the mean of 12 individuals. Columns with the same letter are not significantly different. * $p<0.05$, ** $p<0.01$.

difference occurred between fed and starved males. Compared with control, more than a twofold increase in HA level was found in haemolymph of males starving for 14 days. Twenty-four hours after feeding haemolymphatic HA concentration decreased and showed a tendency to return to the control level.

DISCUSSION

HA is present in high amounts in the gastrointestinal tract of mammals and lower vertebrates (Hakanson et al.1986). Numerous studies have also reported that HA causes weak intestine contraction, stimulates gastric secretion of HCl and is involved in the regulation of feeding. In insects the muscles of foregut are innervated from the stomatogastric nervous system (SNS), the important part of which is represented by FG. By several methods, the presence of biogenic amines in the FG neuropile of various cockroach species was revealed (Penzlin 1985) and their effects on gut are currently investigated (Luffy and Dorn 1992). In crustacea, HA was found to control rhythmic activity in the SNS in spiny lobster (Claiborne and Selverston 1984). We also found the presence of HA in FG of adults and nymphs in *N. cinerea* (Hučková et al. 1992), as well as in their midgut. That is why we hypothesized a possible involvement of HA in regulation of digestive processes. The most widely accepted function of the FG is the control of crop-emptying. After ganglionectomy undigested food is accumulated in the foregut, causing restricted passage of food into the midgut and hindgut (Davey and Therene 1963). Our observations agree well with this report. Both starving groups of cockroaches as well as fed ones without FG had shown increasing HA concentration in their guts. We suggest that this might be a response to changes in the feeding state and associated with changes in activity of some digestive enzymes. Increased HA level in fed operated group might be also associated with food pushed into the midgut in consequence of the pressure of the accumulated food from the crop.

In several insects stress stimuli have been found to increase blood and cerebral monoamine concentrations (Harris 1992). Davenport and Evans (1984) found increased haemolymph octopamine (OA) level associated with food deprivation in locust. After supplying food OA concentration remained higher up to 4 h, then declined. In starving cockroach males we observed a continual rise of HA concentration in brain dependent on the length of starvation as well as a significant increase in the haemolymph. Feeding was connected with significant increase in HA level in the brain and FG. At the same time HA concentration in haemolymph of re-fed animals decreased.

Although this is only preliminary study, we suppose a regulatory role of HA in feeding and digestive processes.

REFERENCES
Brownstein, M.J., Saavedra, J.M., Palkovits, M. and Axelrod, J., Histamine content of hypothalamic nuclei of the rat, *Brain Res.* 77, 151-156, 1974.
Claiborne, B.J. and Selverston, A. I., Histamine as a neurotransmitter in the Stomatogastric nervous system of the spiny lobster, *J. Neurosci.* 4 , 708-721, 1984.
Davenport, A. P. and Evans, P. D., Changes in haemolymph octopamine levels associated with food deprivation in the locust *Schistocerca gregaria*, *Physiol. Ent.* 9, 269-274, 1984.
Davey, K. G. and Therene, J. E., Studies on crop function in the cockroach (*Periplaneta americana* L.) II. The nervous control of crop emptying, *J. Exp. Biol.* 40, 775-780, 1963.
Hakanson, R., Böttcher, G., Ekblad, E., Panula, P., Simonsson, M. and Dohlsten, M., Histamine in the endocrine cells in the stomach., *Histochemistry* 86, 5-17, 1986.
Hardie, R.C., A histamine-activated chloride channel underlying synaptic transmission at a photoreceptor synapse, *Nature (London)* 339, 704-706., 1989.
Harris, J. W. and Woodring, J., Effects of stress, age, season , and source colony on levels of octopamine, dopamine and serotonin in the honey bee (Apis mellifera L.) brain, *J. Insect Physiol.* 38, 29-35, 1992.
Hučková, A., Kozánek, M., Vidlička, L. and Takáč, P., Histamine distribution in the nervous system of the cockroach *Nauphoeta cinerea* (Blattodea, Panchloridae) and its changes during development, in *Advances in regulation of insect reproduction*, Bennetová B., Gelbič I., Soldán T., Eds., Czech Acad. Sci., 129-134, 1992.
Luffy, D. and Dorn, A., Immunohistochemical demonstration in the stomatogastric nervous system and effects of putative neurotransmitters on the motility of the isolated midgut of the stick insect, *Carasusius morosus*. *J. Insect Physiol.* 38, 287-299, 1992.
Penzlin, H., Stomatogastric Nervous System, in *Comprehensive Insect Physiol. Biochem. and Pharmacol. vol. 5*, Kerkut, G.A., Gilbert, L.I., Eds., Pergamon Press, Oxford, 371-402, 1985.
Schwartz, J.C., Pollard, H. and Kuach, T.T., Histamine as a neurotransmitter in mammalian brain: Neurochemical evidence, *J Neurochem.* 35,26-3, 1980.
Simmons, P.J and Hardie, R.C., Evidence that histamine is a neurotransmitter of photoreceptors in the locust ocellus. *J. Exp. Biol.* 138, 205-209, 1988.

Insect Neuropeptide Research

THE QUESTION OF THE EVOLUTION OF NEUROSECRETION

Vladimír J. A. Novák

Czech Academy of Sciences
Oldřichova 39, Praha 2, 128 00, CzR

The evolutionary approach is used more and more and is important in modern Science. It consists of considering the studied question in the course of the whole existence of a given phenomenon, since its origin up to the recent time. A number of methods, both special and general ones, have been developed to determine the stages not accessible to direct observation or experiments: the comparative one, historical, unit one, deduction, synthesis, classification, etc. (Novák 1980, Novák, Leonovich 1980).

Neurosecretion may be defined as any incretory product of the nerve cells with physiological activity. It is of two types: each nerve cell including the neurosecretory cells, is concerned with the production of mediator substances of the protohormone character (working on the place of their origin), i.e., in nerve endings including synapses, active in transmitting impulses between interconnected neurons.

The second type are the typical *neurosecretory cells* producing neurohormones bound on granules of a carrier substance with specific staining properties, so that they can be visualized histologically or histochemically. The granules are transported by axonal movement into the neurohemal organs (corpora cardiaca or perisympathetic organs or neurohypophysis in vetebrates), from whence they pass into the circulatory fluid (blood, hemolymph) (Scharrer E. and Scharrer B. 1954), or, less often, to their place of action as in aphids (Johnson 1963). There the carrier substance is dissolved enzymatically and the actual neurohormone is freed for action. From this a double origin and evolution of neurosecretion follows. (Fig. 1).

The carrier substance preceded the neurohormones in phylogenesis. The nervous system and all its components are of ectodermal origin. So as all ectoderm cells are of secretory character, as seen in embryonic epidermis, their production of sheets or granules of protein-like character is not surprising. Granules with staining properties similar to those of the carrier substance were found in the glia cells of the cockroach *Periplaneta americana* by B. Scharrer (1939) and were later studied by Pipa (1961, 1962) and by Novák and Gutmann (1962). The probability that they lack any physiological function seems to favour the conclusion that the carrier substance attained its neurosecretory function only secondarily by absorbing physically or binding chemically to the actual neurohormones of polypeptide character. The structure

0-8493-4591-X/94/$0.00 + $.50
© 1994 by CRC Press, Inc.

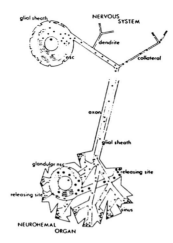

Figure 1. A neurosecretory perikaryon and its neurohemal organ. After Raabe, 1982 adapted. pk - perikaryon, nho - neurohemal organ, a - axon, gln- glandular neurosecretory cell, releasing site, d - dendrite, c - collateral, gs - glial sheath.

of their granules, their staining properties and density differentiated later together with the corresponding neurohormones and with the form, size, amount and position of the neurosecretory cells. They all, however, exhibit a positive Gomori reaction. (Fig. 2).

The nervous activity, being of electric character, is first seen at the monocellular stage (in protozoans). The first neurohormones originated, however, in the multicullular stage (in the lower coelenterates) in the form of protohormones of neurotransmitter character. Their accumulation on the pre-existing granules of what then became the carrier substance is a condition of further differentiation of both hormones and granules (Fig. 3).

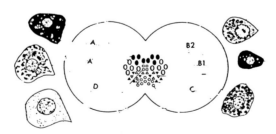

Figure 2. Six different neurosecretory cell types in the pars intercerebralis protocerebri of the bug *Roscius* sp. (A,A', B1, B2, C, D). After Raabe (1982), modified.

Differentiation is involved with the increasing number of neurosecretory functions such as conditioning and timing of growth, inducing of the moulting process, activity of endocrine glands, muscle contractions, colour change, vitellogenesis, diapause etc. (Cf. e.g. Novák 1975, Gersch 1970, Raabe 1982).

In this way the following sequence of main steps in the evolution of neurosecretion may be assumed:

(1) Nervous activity at the unimolecular stage (argentofibrils of protozoans).

(2) Origin of the carrier substance (gliosecretion, neural lamellae, trabeculae etc.).

(3) Transmitter protohormones (in nerve endings, synapses, etc.).

(4) Accumulation of neurohormones by their absorption in granules of the carrier substance.

(5) Transport of neurosecretory granules by axon movement to neurohemal organs, or, to their sites of action.

(6) Differentiation of neurohormones for various functions.

The evolution of all these various stages may be assumed to have evolved on the basis of natural selection due to their possession of better survival value in comparison with the preceding stages.

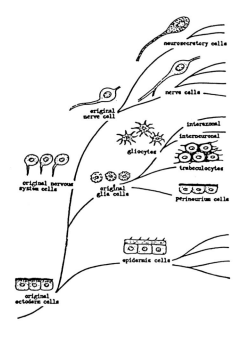

Figure 3. Supposed phylogenetical relations of Gomori positive structures in *Periplaneta americana*. After Novák, Gutmann (1962), adapted.

REFERENCES

Gersch, M., Generelle Probleme der Neuroendokrinologie wirbelloser Tiere, *Biol. Rundschau*, 8, 77-89, 1970.

Johnson, B., A histological study of neurosecretion in aphids, *J. Insect Physiol.*, 9, 727-739, 1963.

Novák, V.J.A., *Insect Hormones*, Chapman and Hall, London, IVth (IInd English) ed., 600 pp., 1975.

Novák, V.J.A., The evolutionary approach, IInd Int. Congr. Systematic and Evolutionary Biology, July 17-24, 1980. *Proc. Univ. B.C. Vancouver*, Canada, 1981.

Novák, V.J.A. and Gutmann, E., To the question and other Gomori-positive structures in the central nervous system of the cockroach *Periplaneta americana* L., *Acata Soc. ent. Cechoslov.* 59(4), 314-322, 1962.

Novak, V.J.A. and Leonovich, V., The evolutionary approach in biology and its philosophical and methodological meaning, *Teorie rozvoje vedy*, Praha, IV/2, 1980.

Pipa, R.L., Studies on the hexapod nervous system IV. A cytological and cytochemical study of neurons and their inclusion in the brain of *Periplaneta americana*, *Biol. Bull. Wood's Hole* 121, 151-535, 1961.

Pipa, R.L., A cytochemical study of neurosecretory and other neuroplasmic inclusions in *Periplaneta americana*, *Gen. Comp. Endocrinol.* 2, 44-52, 1962.

Raabe, M., *Insect Neurohormones*, Plenum Press, N. York and London, 352 pp, 1982.

Scharrer, B., The differentiation between neuroglia and connective tissue sheet in the cockroach *Periplaneta americana*, *J. Comp. Neur.*, 70, 77-88, 1939.

Scharrer, E. and Scharrer, B., Hormones produced by neurosecretory cells, *Recent Progr. Hormone Res.*, 10, 183-240, 1954.

Adipokinetic Hormone

ADIPOKINETIC HORMONES INHIBIT RNA SYNTHESIS IN THE FAT BODY OF ADULT *LOCUSTA MIGRATORIA*

Dalibor Kodrík[1] and Graham J. Goldsworthy[2]
[1]Institute of Entomology, Czech Academy of Sciences, Branišovská 31, 370 05 České Budějovice, Czech Republic
[2]Department of Biology, Birkbeck College, University of London, Malet Street, London, WC1E 7HX, U.K.

I. INTRODUCTION

There are three adipokinetic hormones (AKHs) in *Locusta migratoria*: the decapeptide AKH-I, and two octapeptides AKH-II and AKH-III (see Oudejans *et al.* 1991). In the fat body they mobilize lipid and carbohydrate, activate glycogen phosphorylase, increase intracellular cAMP (see Goldsworthy 1983; 1993; Orchard 1987), and inhibit synthesis of lipid (Gokuldas *et al.* 1988) and protein (Carlisle and Loughton 1986; Asher *et al.* 1984; Moshitzky and Applebaum 1990, Cusinato *et al.* 1991). We show here that in addition to these actions, AKHs inhibit RNA synthesis in the fat body of adult *Locusta*.

A modification of the bioassay of Kodrík and Sehnal (1991) is used with an incubation volume of $400\mu l$. Each fat body provided an experimental and a control half for direct comparison. Incubations lasted 1h, but significant inhibition of incorporation of radiolabelled uridine into total RNA could be observed within 20 min. Using this assay we identified two sources of inhibitory material for RNA synthesis in the fat body of adult *Locusta*.

II. ACTIVITY IN THE BRAIN

Inhibition of RNA synthesis by extracts of adult brain is dose-dependant, with significant responses being detected above two brain equivalents. Unlike AKHs, the activity in brain extracts is destroyed by 3 min at 100°C, which suggests that the brain factor is not AKH. Furthermore, the mechanism of action of the active factor(s) in brain differs from that for AKHs (see below).

III. ACTIVITY IN THE CORPORA CARDIACA AND AKHs

Inhibition of RNA synthesis is also shown by extracts of corpora cardiaca. The effects of the three known AKHs in corpora cardiaca from *Locusta* on RNA synthesis were studied in relationship to age and sex. No response to AKHs is detected in fat bodies from young adult males, but sensitivity them increases as locusts age to *c.* 25 days. By contrast, responses of fat bodies from females are complex: with AKH-I the situation is similar to that des-

0-8493-4591-X/94/$0.00 + $.50
© 1994 by CRC Press, Inc.

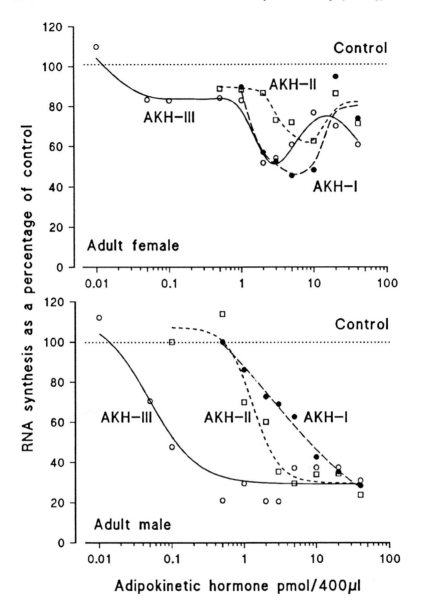

Figure 1. Dose response data for the effects of AKHs on the incorporation of ^{3}H-uridine into RNA of fat body from male and female locusts (25 days of age). The responses in fat body from females are complex in relation to dose, but all three AKHs induce their maximal effects within the range 1-20 pmol per incubation. The curves for females are fitted by eye (with some difficulty at high concentrations!), whereas those for males are computer-fitted from Hill plots. Error bars have been omitted for clarity: all points represent the mean of 5-10 observations.

cribed for male fat bodies, but AKH-II and -III are to a slight extent inhibitory on fat body from females of all ages. Nevertheless in females, as in males, fat body from 25-day old adults appears to be the most sensitive. The inhibitory effect of AKHs on RNA synthesis is dose-dependent to differing extents between males and females (Figure 1). Supramaximal doses are effective on fat bodies from males, but are less effective on fat body from females. In fat bodies from males, the inhibitory effect of the AKHs decreases in the order AKH-III > AKH-II > AKH-I in male fat body, while in those from females the order is AKH-II > AKH-I > AKH-II (Figure 1).

IV. MECHANISM OF ACTION

AKHs are thought to act in part via the production of cAMP in fat body cells, but extracellular Ca^{2+} is also involved (see Goldsworthy 1983;1990; Beenakkers *et al.* 1985; Goldsworthy *et al.* 1986; Orchard 1987). While the inhibition of RNA synthesis by AKHs requires extracellular Ca^{2+}, responses to brain extracts are unaffected by the absence of Ca^{2+} from the medium. Further, cAMP or 2'-O-monobutyryl-8-bromoadenosine 3':5'-cyclic have no effect on levels of RNA synthesis.

V. CONCLUSIONS

Control of RNA synthesis by extracts of brain has been described in insect prothoracic glands (Gersch and Brauner 1974; Aizono *et al.* 1986), and silk glands (Kodrík and Sehnal 1991). Here, we show inhibition of RNA synthesis in fat body from *Locusta* by extracts of brain, and also by the three known AKHs in this locust. The maximal levels of inhibition observed in response to tissue extracts or synthetic peptides are similar to those produced by actinomycin D. It is clear, however, that the AKHs are not identical with, and work by a different mechanism from, active material(s) in the brain. The mechanisms of action of AKHs in inhibiting lipid synthesis (Pancholi *et al.* 1991) or RNA synthesis are uncertain, but are likely to involve inositol phosphate metabolism.

The marked differences between the dose-response curves obtained for the effects of AKHs on fat body from male and female locusts are intriguing. The biphasic response in female tissue is consistent, occurring in response to all three peptides in the range 1-10 pmol/400μl of incubation medium: this would certainly 'translate' to a concentration in what is thought to be the normal physiological range for AKHs. Nevertheless, it is not clear why male tissue should be generally more sensitive and monophasic in its response.

Acknowledgements: We thank Mary Lightfoot for skilful technical assistance: supported by grants from SERC (GJG) and a Fellowship from The Royal Society (DK).

VI. REFERENCES

Aizono Y., Matsuo N., Yoshida Y., Funatsu G., Funatsu M. and Kobayashi M. *In vitro* action of prothoracicotropic hormone on RNA synthesis in the prothoracic gland of the silkworm, *Bombyx mori*. *J. Insect Physiol.* 32, 711, 1986.

Asher C., Moshitsky P., Ramachandran J. and Applebaum S.W. The effects of synthetic locusts adipokinetic hormone on dispersed locust fat body cell preparation: cAMP induction, lipid mobilization, and inhibition of protein synthesis. *Gen. Comp. Endocrinol.* 55, 167, 1984.

Beenakkers A.M.T., Van Der Horst D.J. and Van Marrewijk W.J.A. Biochemical processes directed to flight muscle metabolism, in *Comprehensive Insect Physiology, Biochemistry and Pharmacology*, G.A. Kerkut and L.I. Gilbert, Eds., Pergamon Press, Oxford, 10, 451, 1985.

Carlisle J. and Loughton B.G. The inhibition of protein synthesis in *Locusta migratoria* by adipokinetic hormone. *J. Insect Physiol.* 32, 573, 1986

Cusinato, O., Wheeler, C.H. and Goldsworthy, G.J. The identity and physiological actions of an adipokinetic hormone in *Acheta domesticus*. *J. Insect Physiol.* 37, 461, 1991.

Gersch M. and Brauer R. *In-vitro*-Stimulation der Prothoracal-drusen von Insecten als Testsystem (Prothoracaldrusentest). *J. Insect Physiol.* 20, 735, 1974.

Gokuldas M., Hunt A.P. and Candy D.J. The inhibition of lipid synthesis in vitro in the locust, *Schistocerca gregaria*, by factors from the corpora cardiaca. *Physiol. Entomol.* 13, 43, 1988.

Goldsworthy G.J. The endocrine control of flight metabolism in locusts. *Adv. Insect Physiol.* 17, 149, 1983.

Goldsworthy G.J. Hormonal control of flight metabolism in locusts, in *Biology of Grasshoppers*, R.F. Chapman and A. Joern, Eds., John Wiley & Sons, Inc., 205, 1990.

Goldsworthy G.J., Mallison K. and Wheeler C. The relative potencies of two known locust adipokinetic hormones. *J. Insect Physiol.* 32, 95, 1986.

Kodrík D. and Sehnal F. Neurohormonal stimulation of posterior silkgland in *Galleria mellonella*, in *Wild Silkmoths '90*, H. Akai and M. Kiuchi, Eds., Natl. Inst. Sericult. Insect Science, Tsukuba, 43, 1991.

Loughton B.J. and Orchard I. The nature of the hyperglycaemic factor from the glandular lobe of the corpus cardiacum of *Locusta migratoria*. *J. Insect Physiol.* 27, 383, 1981.

Moshitzky P. and Applebaum S.W. The role of adipokinetic hormone in the control of vitellogenesis in locusts. *Insect Biochem.* 20; 319, 1990.

Orchard I. Adipokinetic hormones - an update. *J. Insect Physiol.* 33, 451, 1987.

Oudejans R.C.H.M., Kooiman F.P., Heerma W., Versluis C., Slotboom A.J. and Beenakkers A.M.T. Isolation and structure elucidation of a novel adipokinetic hormone (Lom-AKH-III) from the glandular lobes of the corpus cardiacum of the migratory locust, *Locusta migratoria*. *Eur. J. Biochem.* 195, 351, 1991.

Pancholi, S., Barker, C.J., Candy, D.J., Gokuldas, M. and Kirk, C.J. Effects of adipokinetic hormones on inositol phosphate metabolism in locust fat body. *Biochem. Soc. Trans.* 19: 1045, 1991.

DIFFERENCES IN RESPONSE TO ADIPOKINETIC HORMONES BETWEEN GREGARIOUS AND SOLITARY LOCUSTS

A. Ayali, E. Golenser, and M.P. Pener

Department of Cell and Animal Biology,
The Hebrew University, Jerusalem 91904, Israel

Locust show density-dependent continuous phase polymorphism. Under crowding or under isolation they respectively develop "gregarious" or "solitary" phase characteristics and there are innumerable intermediates between the extreme phases (review by Pener, 1991). Extensive studies in many laboratories demonstrated that during the first 10-15 min of flight locusts utilize carbohydrates as flight fuel, but then adipokinetic hormones (AKHs) are released from the corpora cardiaca (CC). The AKHs induce mobilization of lipids, transformation of triacylglycerols to diacylglycerols and release of the latter from the fat body into the haemolymph. These lipids serve as the major fuel for sustained (migratory) flight. In addition, AKHs are responsible for rearrangement of carrier lipoproteins in the haemolymph resulting in an improved transport of the lipids to the flight muscles and AKHs also activate inactive glycogen phosphorylase in the fat body. AKHs, their effects and mode of action in locusts and other insects have been amply reviewed recently (Beenakkers *et al.*, 1985; Orchard, 1987; Goldsworthy and Mordue, 1989; Wheeler, 1989; Gäde, 1992). However, all research on locust AKHs and flight metabolism was carried out exclusively on the gregarious phase, that is on insects crowded in the laboratory. In the last few years we made an attempt to reveal AKH-related phase-dependent differences in the African migratory locust *Locusta migratoria migratorioides* (R. & F.). Some earlier results have already been reported and most of the relevant materials and methods described (Ayali and Pener, 1992). Only adult males were used for the study.

Following injection of graded doses of synthetic AKH I (Peninsula, lot no. 013764) to 13-19-day-old (age after fledging) or 24-30-day-old males induced consistently higher elevation of haemolymph lipids in gregarious (crowded) than in solitary (isolated) locusts (Fig. 1A). This phase-induced difference was more marked in the younger males because the level of response of the crowded locusts somewhat decreased with the age, whereas that of the isolated ones did not change within this age range. The results obtained 90 min after injections of graded doses of synthetic Lom AKH II (Sigma, lot no. 67F08451) to 12-16-day-old and 27-29-day-old males

0-8493-4591-X/94/$0.00 + $.50
© 1994 by CRC Press, Inc.

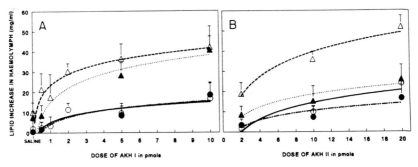

Figure 1. Increase of haemolymph lipid level in mg lipids per ml haemolymph 90-100 min after injection of graded doses of synthetic AKHs to adult males of *Locusta*. Empty triangles and evenly scattered line, crowded younger males; empty circles and continuous line, isolated younger males; full triangles and pointed line, crowded older males; full circles and unevenly scattered line, isolated older males. Average and s.d. (marked only in one direction) are shown. (A) Results with AKH I; age of younger and older males is 13-16 and 24-30 days, respectively; n=8-17 for each data point. (B) Results with Lom AKH II; age of the younger and older males is 12-16 and 27-29 days, respectively; n=7-9 for each point.

(Fig. 1B) revealed a similar trend, except that the difference between the younger and older crowded males was even more marked than that found with AKH I. Similarly, elevation of haemolymph lipid levels was higher in 12-17-day-old crowded males than in isolated ones of the same age, 90 min after injection of graded doses of CC extracts (Fig. 2), made from 10-14-day-old crowded males. Further investigations revealed that the time course of AKH-induced lipid increase in the haemolymph was rather parallel in 11-17-day-old crowded and isolated males (Fig. 3). Injection of 2 pmol AKH I induced the highest response already after 60 min and after an additional period of 30 min (=90 min after injection) haemolymph lipid levels slightly declined in both crowded and isolated locusts. Injection of 10 pmol of AKH I induced a higher response at 60 min than 2 pmol and lipid levels were maintained, or even slightly increased, during the next 30 min in both crowded and isolated males. In spite of all these similarities in the time course, AKH-induced elevation of haemolymph lipids was always consistently higher in the crowded males than in the isolated ones (Fig. 3). These findings confirmed previous results and also justified the timing in the previous experiments because if both dose and phase are considered, the best separation of the effect is obtained 90 min after injection.

The resting lipid levels (before or without any injection of AKH or saline) in the haemolymph of the isolated males were only about two-thirds of those of the crowded ones. Haemolymph volume (determined by the labelled inulin dilution method) was always lower in isolated than in crowded males of the same age. The latter finding shows that the differences in the total amount of haemolymph lipids between crowded and isolated locusts are even higher than those obtained in mg lipids per ml haemolymph.

In the experiments reported above the locust were kept under the same

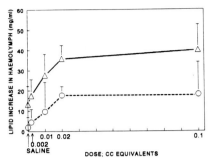

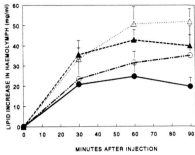

Figure 2. Increase of haemolymph lipid levels in mg lipids per ml haemolymph 90-100 min after injection of graded doses of CC extracts to 10-17-day-old adult males of crowded (triangles and continuous line) or isolated (circles and scattered line) *Locusta*. Doses are expressed in CC equivalents. Averages and s.d. (marked only in one direction) are shown; n = 6-23 for each data point. CC dissected from 10-14-day-old crowded males were used for preparation of the extract (cf. Ayali and Pener, 1992).

Figure 3. Increase of haemolymph lipid levels in mg lipids per ml haemolymph at different times after injection of AKH I to 11-17-day-old adult *Locusta* males; dose and phase interrelations. Full triangles and evenly scattered line, 2 pmol to crowded males; full circles and continuous line, 2 pmol to isolated males; empty triangles and pointed line, 10 pmol to crowded males; empty circles and unevenly scattered line, 10 pmol to isolated males. Averages and s.d. (marked only in one direction) are shown; n = 6-8 for each data point.

condition of density (either isolated or crowded) for many consecutive generations. In the next set of experiments we explored the effect of changes of density on the AKH-induced hyperlipemic response. Injection of 1 pmol of AKH I (Peninsula, lot no. 015301 which was more active than the previously used lot no. 013764) to 12-16-day-old adult *Locusta* males resulted in the following increase of haemolymph lipid levels (mg lipids per ml haemolymph, averages±s.d.) 90 min after injection:

Locusts crowded for many consecutive generations,   21.1±7.1 (n=10);

Hoppers crowded, adults isolated 0-2 h after fledging,   6.5±8.3 (n=13);

Second consecutive generation of isolated locusts,   4.9±3.4 (n=16).

These results show that the differential hyperlipemic response to AKH of crowded and isolated *Locusta* is a phase characteristic which changes rapidly with changes of density.

Ziegler *et al.* (1988) have shown that the intensity of adipokinetic response within the family of acridids is species dependent and correlates with the flight capability and capacity of the species. The present results extend this correlation to the intraspecific phase level.

AKH is known to activate inactive glycogen phosphorylase in the fat body of *Locusta*. We investigated this effect in crowded and isolated adults using the method described by Van Marrewijk *et al.* (1980). We found that although total glycogen phosphorylase activity in the fat body (μmol glycosyl/h/mg protein) is significantly higher in isolated than in crowded males at the age of 11-18 days after fledging, the proportion of the active enzyme 15-20 min after injection of 10 pmol of AKH I (Peninsula, lot no. 015301) is nevertheless the same (about 40% of the total) in crowded and isolated locusts.

This research was supported by grant no. 89-00129 from the United States-Israel Binational Science Foundation (BSF), Jerusalem, Israel. We thank Mr. Hussein Abu-Hilal for keeping the locust colony.

REFERENCES

Ayali, A. and Pener, M.P., Density-dependent phase polymorphism affects response to adipokinetic hormone in *Locusta, Comp. Biochem. Physiol.* 101A, 549-552, 1992.

Beenakkers, A.M.Th., Bloemen, R.E.B., De Vlieger, T.A., Van der Horst, D.J. and Van Marrewijk, W.J.A., Insect adipokinetic hormones, *Peptides* 6, Suppl. 3, 437-444, 1985.

Gäde, G., The hormonal integration of insect flight metabolism, *Zool. Jb. Physiol.* 96, 211-225, 1992.

Goldsworthy, G. and Mordue, W., Adipokinetic hormones: functions and structures, *Biol. Bull.* 177, 218-224, 1989.

Orchard, I., Adipokinetic hormones - an update, *J. Insect Physiol.* 33, 451-463, 1987.

Pener, M.P., Locust phase polymorphism and its endocrine relations, *Adv. Insect Physiol.* 23, 1-79, 1991.

Van Marrewijk, W.J.A., Van den Broek, A.Th.M. and Beenakkers, A.M.Th., Regulation of glycogenolysis in the locust fat body during flight, *Insect Biochem.* 10, 675-679, 1980.

Wheeler, C.H., Mobilization and transport of fuels to the flight muscles, in *Insect Flight*, Goldsworthy, G.J. and Wheeler, C.H., Eds., CRC Press, Boca Raton, Florida, 1989, chap.12.

Ziegler, R., Ryan, R.O., Arbas, E.A. and Law, J.H., Adipokinetic response of a flightless grasshopper (*Barytettix psolus*): functional components, defective response, *Arch. Insect Biochem. Physiol.* 9, 255-268, 1988.

REGULATION OF ADIPOKINETIC HORMONE SECRETION FROM LOCUST CORPORA CARDIACA: EFFECTS OF NEUROACTIVE SUBSTANCES *IN VITRO*

Henk G.B. Vullings, Paul C.C.M. Passier, Jacques H.B. Diederen, Pierre N.M. Konings and Dick J. Van Der Horst

Department of Experimental Zoology, Utrecht University, Padualaan 8, NL-3584 CH Utrecht, The Netherlands

I. INTRODUCTION

In the glandular lobes of the corpus cardiacum (CCG) of the African migratory locust, *Locusta migratoria*, intrinsic glandular cells synthesize three adipokinetic hormones (AKHs) (Oudejans et al. 1991). AKH I and AKH II coexist in the same secretory granules (Diederen et al. 1987) and are involved in the regulation of flight metabolism by activating fuel mobilization, especially lipids, for the flight muscles (Beenakkers et al. 1984). The secretory route and function of AKH III are still unknown (Oudejans et al. 1991). The release of AKH I and II is controlled by a paired group of about 15 secretomotor cell bodies situated in the lateral part of the protocerebrum, which innervate the adipokinetic cells via the nervus corporis cardiaci II (NCC II) (Rademakers 1977; Konings et al. 1989a). Electrical stimulation of the NCC II of isolated corpora cardiaca (CCs) results in the release of AKH I and II (Orchard and Loughton 1981). Stimulation of the NCC I alone does not cause release, but potentiates NCC II-stimulated AKH release. This suggests that axons of the NCC I do not initiate hormone release, but rather play a neuromodulatory role. In vivo, however, severance of both the NCC I and NCC II is required to prevent lipid mobilization during flight (Bloemen 1985). Experimental evidence exists for the involvement of octopamine (OA) in the regulation of release of AKHs from the adipokinetic cells (Orchard et al. 1993). Immunocytochemistry employing a well defined antiserum (Spörhase-Eichmann et al. 1992), however, did not reveal OA immunoreactivity in the secretomotor cells, nor in their endings in the CCG (Konings et al. 1988a). A search for OA-binding sites in the CCG equally provided no evidence for involvement of OA in AKH release (Konings et al. 1989b).

OA and several other neuroactive substances mentioned in earlier studies to be possibly involved in AKH secretion {serotonine (5HT), dopamine (DA)} were tested for their ability to induce release of AKH I from isolated CCs. Also FMRFamide was tested in view of the presence of FMRFamide-like immunopositive fibres in the CCG (Vullings et al. 1992a,b). These substances

0-8493-4591-X/94/$0.00 + $.50
© 1994 by CRC Press, Inc.

were tested alone or in combination with IBMX, an inhibitor of phosphodiesterase, or forskolin, an activator of adenylylcyclase. The latter drugs are known to potentiate cyclic AMP induced secretion processes.

II. MATERIAL AND METHODS

CCs from male locusts (*Locusta migratoria*), 12 days after imaginal ecdysis, were isolated and pooled into physiological saline consisting of (mM): NaCl 150, KCl 10, $CaCl_2$ 4, $MgCl_2$ 2, ascorbic acid 10, and HEPES buffer 10, pH 7.0 on watch glasses, each pool consisting of five CCs. After washing twice with 200 µl of fresh saline, the pools of CCs were first incubated in 200 µl of fresh saline under continuous moderate shaking in a moist chamber for 30 min at 35°C. The CCs were then incubated under the same conditions in 200 µl of fresh saline provided with or without test agent(s). Occasionally a high potassium concentration (100mM) was applied during the second and even during a third incubation period of 30 min to test the viability of the isolated CCs. Pipette tips and vessels the incubation media came in contact with, during and after the incubation experiments, were carefully coated with 1% silicone.

The incubation media were assayed with reversed-phase high-performance liquid chromatography (RP-HPLC) to determine the amounts of released AKH I. The ratio was calculated of the amount of AKH I released in the second 30 min (induced release) and that released in the first 30 min (spontaneous release). All experiments were done in 6-10 fold.

III. RESULTS AND DISCUSSION

The results are summarized in Figure 1. Isolated CCs always spontaneously release AKH into the incubation medium during both the first and the second incubation period, in the amount of about 2 pmol per CC. This may be due to injury or to denervation resulting in the loss of a possible inhibiting influence from the brain. This spontaneous release is generally higher in the second than in the first incubation period, resulting in a ratio >1. Addition of OA, 5HT, DA or FMRFamide alone to the second incubation medium did not result in AKH release above control levels. On account of these results it can be concluded that none of these substances acts as the primary signal for release of AKHs. Besides, none of these substances, except FMRFamide, could be demonstrated immunocytochemically in the NCC II fibres and their endings in the CCG. Only forskolin and IBMX were able to enhance the release of AKH on their own, suggesting that c-AMP is involved as a second messenger in the release of AKH. It is remarkable that all but one of the substances applied were able to enhance the IBMX induced release of AKH; only FMRFamide inhibits the effect of IBMX.

On account of these results, we conclude that an unknown substance acts

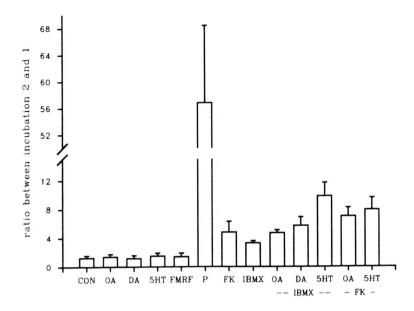

Figure 1. Effects of several neuroactive substances and high potassium (P), alone or in combination with IBMX or forskolin (FK), on adipokinetic hormone secretion *in vitro*. CON = control. For other abbreviations, see text.

as the primary neurotransmitter in the lateral secretomotor cells, at least in most of them, since a minority of these cells is FMRFamide immunopositive. Likely, the axons of the latter cells which end in the CCG are responsible for an inhibiting influence from the brain on AKH release. OA can exert its modulatory effect only via the hemolymph since it is absent from the CCG as well as from the storage part of the CC (CCS). Both 5HT and DA are present in the CCS, but not in the CCG (Konings et al. 1988b; Viellemaringe, et al. 1984). These substances may exert their modulatory influence via axo-axonal contacts between NCC I and II fibres in the CCS or via local neurons in the CCS. The electrical activity of the latter cells appeared to be influenced by NCC I and NCC II activity (Bloemen et al. 1985).

REFERENCES

Beenakkers, A.M.Th., Van Der Horst, D.J. and Van Marrewijk, W.J.A., Insect flight muscle metabolism, *Insect Biochem.*, 14, 243-260, 1984.
Bloemen, R.E.B., Regulation of the release of the adipokinetic hormone in the migratory locust, Thesis, Utrecht, 1985.
Bloemen, R.E.B., Beenakkers, A.M.Th. and de Vlieger, T.A., Influences of the NCC axons on the electrical activity in the glandular lobe of the corpus cardiacum of *Locusta migratoria*, *Comp.Biochem. Physiol.*, 84A, 319-324, 1985.

Diederen, J.H.B., Maas, H.A., Pel, H.J., Schooneveld, H., Jansen, W.F. and Vullings, H.G.B., Co-localization of the adipokinetic hormones I and II in the same glandular cells and in the same secretory granules of corpus cardiacum of *Locusta migratoria* and *Schistocerca gregaria, Cell Tissue Res.,* 249, 379-389, 1987.

Konings, P.N.M., Vullings, H.G.B., Geffard, M., Buijs, R.M., Diederen, J.H.B. and Jansen, W.F., Immunocytochemical demonstration of octopamine-immunoreactive cells in the nervous system of *Locusta migratoria* and *Schistocerca gregaria, Cell Tissue Res.,* 251, 371-379, 1988a.

Konings, P.N.M., Vullings, H.G.B., Siebinga, R., Diederen, J.H.B. and Jansen, W.F., Serotonin-immunoreactive neurones in the brain of *Locusta migratoria* innervating the corpus cardiacum, *Cell Tissue Res.,* 254, 147-153, 1988b.

Konings, P.N.M., Vullings, H.G.B., Kok, O.J.M., Diederen, J.H.B. and Jansen, W.F., The innervation of the corpus cardiacum of *Locusta migratoria*: A neuroanatomical study with the use of lucifer yellow, *Cell Tissue Res.,* 258, 301-308, 1989a.

Konings, P.N.M., Vullings, H.G.B., Van Gemert, W.M.J.B., De Leeuw, R., Diederen, J.H.B. and Jansen, W.F., Octopamine-binding sites in the brain of *Locusta migratoria, J. Insect Physiol.,* 35, 519-524, 1989b.

Orchard, I and Loughton, B.G., The neural control of release of hyperlipaemic hormone from the corpus cardiacum of *Locusta migratoria, Comp. Biochem. Physiol.,* 68A, 25-30, 1981.

Orchard, I., Ramirez, J-M. and Lange A.B., A multifunctional role for octopamine in locust flight, *Annu. Rev. Entomol.,* 38, 227-249, 1993.

Oudejans, R.C.H.M., Kooiman, F.P., Heerma, W., Versluis, C. Slotboom, A.J. and Beenakkers, A.M.Th., Isolation and structure elucidation of a novel adipokinetic hormone (Lom-AKH-III) from the glandular lobes of the corpus cardiacum of the migratory locust, *Locusta migratoria, Eur.J.Biochem.,* 195, 351-359, 1991.

Rademakers, L.H.P.M., Identification of a secretomotor centre in the brain of *Locusta migratoria*, controlling the secretory activity of the adipokinetic hormone producing cells of the corpus cardiacum, *Cell Tissue Res.,* 184, 381-395, 1977.

Spörhase-Eichmann, U., Vullings, H.G.B., Buijs, R.M., Hörner, M. and Schürmann, F-W., Octopamine-immunoreactive neurons in the central nervous system of the cricket, *Gryllus bimaculatus, Cell Tissue Res.,* 268, 287-304, 1992.

Vieillemaringe, J., Duris, P., Geffard, M., Le Moal, M., Delaage, M., Bensch, C. and Girardie, J., Immunohistochemical localization of dopamine in the brain of the insect *Locusta migratoria migratorioides* in comparison with the catecholamine distribution determined by the histofluorescence technique, *Cell Tissue Res.,* 237, 391-394, 1984.

Vullings, H.G.B., Diederen, J.H.B., Passier, P. and Veldman R.J., Regulation of adipokinetic cells in *Locusta migratoria*: opposite effects of octopamine and FMRFamide, in *Rhythmogenesis in Neurons and Networks,* Elsner, N. and Richter, D.W., Eds., Georg Thieme Verlag, Stuttgart, New York, 508, 1992a.

Vullings, H.G.B., Diederen, J.H.B., Van De Corput, M. Passier, P.C.C.M., Veldman, R.J. and Van Vuuren, M., Regulation of adipokinetic hormone secretion from locust corpora cardiaca, *Proc. Exper. & Appl, Entomol., N.E.V.* Amsterdam, 167-168, 1992b.

NOVEL ANALOGUES OF ADIPOKINETIC HORMONES: A MINIMAL CONFORMATION FOR BIOLOGICAL ACTIVITY

Graham Goldsworthy[1], Janet Thornton[2], and Alex Drake[3]

Departments of Biology[1] and Chemistry[3], Birkbeck College and
[2]Biochemistry Department, University College
University of London, London U.K.

I. INTRODUCTION

We have synthesized a series of peptides to determine the shortest sequence capable of eliciting a full adipokinetic response. Our starting molecule was the octapeptide *Acheta*-adipokinetic hormone (*Acheta*-AKH), < QVNFSTGW-NH$_2$, because this peptide is remarkably active in locusts (Fig. 1): although its C-terminus lacks the glycine[9] (Gly, G) and threonine[10] (Thr, T) residues of AKH-I, the cricket peptide has high potency compared with other octapeptides of the AKH-family (Cusinato *et al.*, 1991; Goldsworthy *et al.*, 1992). We truncated this octapeptide progressively from its N-terminus in one experiment, while in another we deleted residues sequentially starting from valine (Val, V); thus, in this latter series, always producing an N-terminally blocked peptide. All analogues were tested for hyperlipaemic activity in locusts: none of the first series of unblocked 'truncation' peptides are active.

II. BIOLOGICAL ACTIVITY OF THE DELETION PEPTIDES

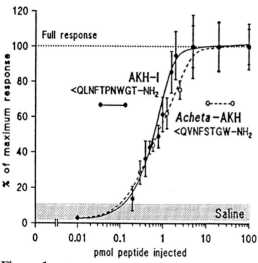

Figure 1. AKH-I and *Acheta*-AKH compared.

Only two of the deletion peptides are active: QNFSTGW-NH$_2$ and QSTGW-NH$_2$ show reduced potency but full efficacy (Fig. 2). The 'intermediate' peptide, QFSTGW-NH$_2$, is almost inactive: at the highest dose tested, it elicited a response which is 40% of the maximal response (Fig 2). N-terminally blocked peptides shorter than 5 residues are totally without activity.

0-8493-4591-X/94/$0.00 + $.50
© 1994 by CRC Press, Inc.

III. CIRCULAR DICHROISM SPECTROSCOPY OF THE DELETION PEPTIDES

Circular dichroism (CD) spectra of *Acheta*-AKH are shown in Fig. 3; essentially the peptide has a featureless spectrum in aqueous solution at room temperature (Fig. 3A), but a type C spectrum (Fig. 3B), indicative of a type-I β-turn, predominates in the presence of SDS micelles, and a type C´ spectrum (Fig. 3C), which can be indicative of a type-II β-turn (Drake *et al.* 1988), is seen at -90°C. These spectra are similar to those seen earlier for AKH-I (Goldsworthy *et al.*, 1990, 1992: Wheeler *et al.*, 1990), and may explain the close potencies of these two AKHs. The three deletion peptides shown in Fig. 2 have type C´ spectra both at low temperatures and in SDS, indicating perhaps the dominant presence of a type-II β-turn. When subjected to increasingly high temperature, however, the biologically active hepta- and penta-peptides show a shift progressively from a type C´ towards a type C spectrum. Importantly, the CD spectrum of the biologically inactive sexapeptide, <QFSTGW-NH$_2$, is relatively insensitive to increasing temperature.

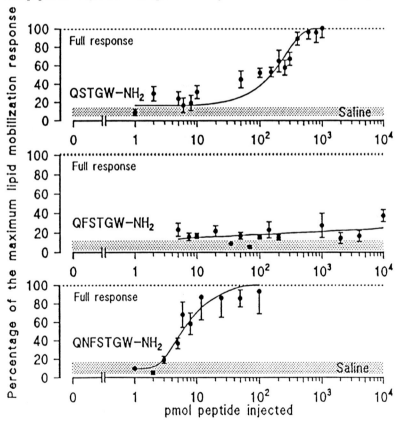

Figure 2. Dose response data for three deletion analogues.

Figure 3. The CD spectra of *Acheta*-AKH (0.3 mg/ml) in aqueous solution (A), with SDS (3 mg/ml) micelles (B), or in ethanediol/water (2:1 v/v) at -90°C (C).

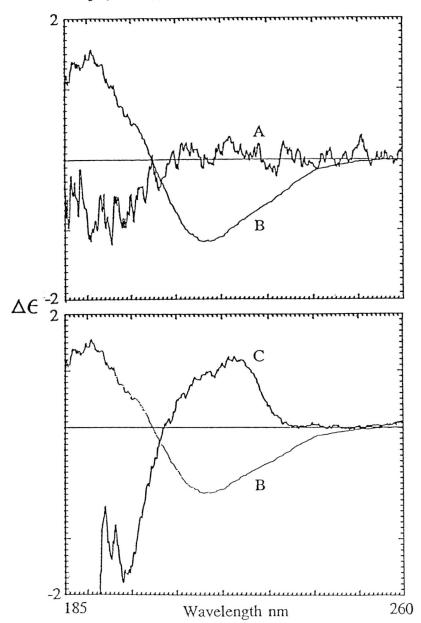

IV. CONCLUSIONS

Previously, because Phe^4 is an almost invariate feature of the AKH family, it was thought to be essential for activity. Further, we have argued that Phe^4 is an important component of a hydrophobic cluster comprising also the side chains of Leu^2 and Trp^8, which could be concerned with receptor binding (Goldsworthy *et al.* 1992). The data here suggest that neither the Phe^4, nor a hydrophobic side-chain at position two, are essential for transduction, although they may well contribute to potency.

Despite the loss of three orders of magnitude in potency, it seems remarkable that the pentapeptide $QSTGW-NH_2$ retains full efficacy in hyperlipaemic assay. This pentapeptide is the shortest deletion analogue of *Acheta*-adipokinetic hormone which can elicit a full response and, presumably, contains all the information required for signal transduction. The reduced potencies of $QNFSTGW-NH_2$ and $QSTGW-NH_2$ may be due partly to the contribution that the deleted amino acids make to potency, but may also reflect their dominant type-II turn conformation. If the assumption is made that the active conformation at the receptor site is that seen in SDS (a type-I β-turn), then the activity of the deletion peptides would be determined by their capacity to assume such a conformation. There is good agreement between the CD data and the biological activity to support such a hypothesis.

Acknowledgments. We are grateful to Mary Lightfoot and Gita Panchal for skilful technical assistance, and Colin Wheeler for helpful discussions. Supported by grants from the SERC (UK).

V. REFERENCES

Cusinato, O., Wheeler, C.H. and Goldsworthy, G.J. The identity and physiological actions of an adipokinetic hormone in *Acheta domesticus. J. Insect Physiol.* 37, 461, 1991.

Drake, A.F., Siligardi, G. and Gibbons, W.A. Reassessment of the electronic circular dichroism criteria for random coil conformations of poly-L-lysine and the implications for protein folding and denaturation studies, *Biophys. Chem.*, 31, 143, 1988

Goldsworthy, G.J., Coast, G.M., Wheeler, C.H., Cusinato, O., Kay, I. and Khambay, B. The structure and functional activity of neuropeptides, in *Insect Molecular Science*, J.M. Crampton and P. Eggleston, Eds., Academic Press, London and San Diego. 205, 1992.

Goldsworthy, G.J., Wheeler, C.H., Cusinato, O. and Wilmot, C.M. Adipokinetic hormones: structures and functions, in *Progress in Comparative Endocrinology*, A. Epple, C.G. Scanes and M.H. Stetson, Eds., Alan R. Liss. Inc, New York. 28, 1992.

Wheeler, C.H., Drake, A.F., Wilmot, C.M., Thornton, J.M. and Goldsworthy, G.J. Structures in the AKH family of neuropeptides, in *Insect Neurochemistry and Neurophysiology 1989*, A.B. Borkovec and D.B. Gelman, Eds., Humana Press, Clifton, New Jersey, 235, 1990.

NOVEL ANALOGUES OF ADIPOKINETIC HORMONES: MODIFICATIONS AT THE N-TERMINUS

Graham J. Goldsworthy and Ornella Cusinato
Department of Biology, Birkbeck College, University of London,
Malet Street, London, WC1E 7HX, U.K.

I. INTRODUCTION

African migratory locusts, *Locusta migratoria*, produce three adipokinetic hormones (AKHs): a decapeptide AKH-I, and two octapeptides AKH-II and AKH-III (see Oudejans *et al.* 1991). In a lipid mobilization assay *in vivo*, AKH-I is the most potent (Fig. 1). We have previously proposed a model for AKH-I (<QLNFTPNWGT-NH$_2$) in which the molecule takes up a folded conformation, and Circular Dichroism spectroscopy provides firm evidence that it adopts a type-I β-turn when in an environment similar to that of a biological membrane (Goldsworthy *et al.* 1990, 1992; Wheeler *et al.* 1990).

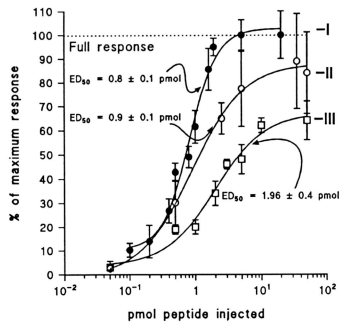

Figure 1. Dose response data for the three known AKHs in *Locusta*.

0-8493-4591-X/94/$0.00 + $.50
© 1994 by CRC Press, Inc.

A further feature of our model is that one face of the molecule presents a hydrophobic cluster comprising the side chains of tryptophan (Trp, W), phenylalanine (Phe, P) and leucine (Leu, L). In this study we have used the lipid mobilization assay *in vivo* to test whether the spacing of the Leu from the Phe and the Trp is critical for potency and/or efficacy, and investigated the importance of the blocked N-terminus. In this assay, lipid mobilization is measured in male *Locusta* 12-20 days after the imaginal moult. The responses

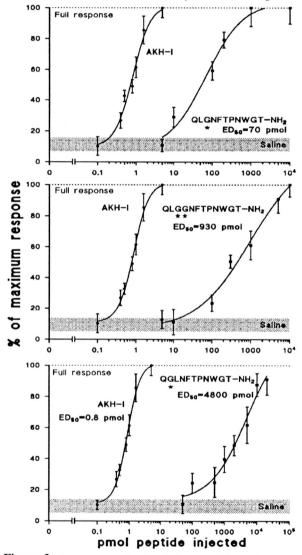

Figure 2. Dose response data for glycine-extended peptides. Note the different scale on lower graph.

of individual groups of locusts are calculated as percentage of that to a supramaximal dose of AKH-I in a similar group of locusts assayed on the same day, and expressed as a mean of at least 10 locusts $\pm$ SE. In this way the assays are standardised to the extent that inter-assay comparisons can be made reliably.

II. THE SPACING OF LEU RELATIVE TO PHE AND TRP

Analogues have been synthesized in which one ($<$QL<u>G</u>NFTPNWGT-NH$_2$) or two ($<$QL<u>GG</u>NFTPNWGT-NH$_2$) glycine (Gly, G) spacers are interposed between Leu and asparagine[3] (Asn, N). These peptides retain full efficacy, but there is a progressive loss of potency as Gly spacers are introduced (Fig. 2).

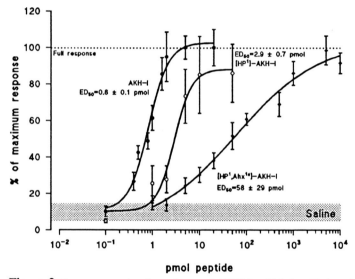

Figure 3. Dose response data for analogues of AKH-I in which the pGlu is replaced.

III. THE IMPORTANCE OF THE N-TERMINAL RESIDUE

In a further peptide, $<$QGLNFTPNWGT-NH$_2$, in which a single Gly is positioned **after** the pyroglutamyl (pGlu, $<$Q) residue and **before** the Leu, the position of Leu relative to Phe and Trp remains unchanged. Nevertheless, this peptide shows an even more pronounced loss of potency (Fig. 3) than seen previously even with two Gly spacers after the Leu.

The pGlu can be replaced by hydroxyphenyl propionate (HP) with only a 2-3 fold loss of potency. Moving HP away from the remainder of the molecule using an aminohexanoic acid (Ahx) spacer causes a further loss in potency, but importantly, efficacy remains unaffected in these analogues (Fig. 3).

V. CONCLUSIONS

The introduction of Gly spacers between the Leu and Asn[3] leads to a 70-900 fold loss in potency, and it is tempting to suggest that the relative spacing of the cluster of Leu, Phe and Trp could be critical in determining receptor binding. However, the even more marked loss of potency with a single Gly spacer between the pGlu and Leu argues against such an interpretation; rather, the relative spacing of the blocked N-terminus from the remainder of the molecule is important. It seems that pGlu itself is not critical at the N-terminus; it can be replaced by HP with only slight loss of potency. Nevertheless, distancing the N-terminal residue from the remainder of the molecule reduces potency with both the natural pGlu residue and the hydroxyphenol propionate substitute. The latter analogue is markedly more active than the former, and the slopes of the dose-response curves are very different: with [HP[1],Ahx[1a]]-AKH-I, it is not simply a displacement to the right as is observed more or less with the other analogues discussed here. The apparent importance of a hydrophobic (blocked) N-terminal residue in close proximity to the remainder of the molecule deserves further study.

The changes which have been introduced into the N-terminus of the AKH-I molecule described here all reduce potency without affecting efficacy. These findings re-enforce our view that hormonal transduction is associated largely with the β-turn region of the peptide (Goldsworthy *et al.* 1992; Cusinato and Goldsworthy, this volume).

Acknowledgements: We thank Mary Lightfoot and Gita Panchal for skilful technical assistance, and Colin Wheeler and Janet Thornton for helpful discussions: supported by grants from the SERC (UK).

VI. REFERENCES

Goldsworthy, G.J., Wheeler, C.H., O. Cusinato and Wilmot, C.M. Adipokinetic hormones: structures and functions, in *Progress in Comparative Endocrinology*, A. Epple, C.G. Scanes and M.H. Stetson Eds., Alan R. Liss Inc., New York, 28, 1990.

Goldsworthy, G.J., Coast, G.M., Wheeler, C.H., Cusinato, O., Kay, I. and Khambay, B. The structure and functional activity of neuropeptides, in *Insect Molecular Science*, J.M. Crampton and P. Eggleston Eds., Academic Press, London and San Diego. 205, 1992.

Oudejans R.C.H.M., Kooiman F.P., Heerma W., Versluis C., Slotboom A.J. and Beenakkers A.M.T. Isolation and structure elucidation of a novel adipokinetic hormone (Lom-AKH-III) from the glandular lobes of the corpus cardiacum of the migratory locust, *Locusta migratoria, Eur. J. Biochem.* 195, 351, 1991.

Wheeler, C.H., Drake, A.F., Wilmot, C.M., Thornton, J.M. and Goldsworthy, G.J. Structures in the AKH family of neuropeptides, in *Insect Neurochemistry and Neurophysiology 1989.* A.B. Borkovec and D.B. Gelman, Eds., Humana Press, Clifton, New Jersey, 235, 1990.

Allatoregulatory Peptides

STRUCTURE AND IMMUNOLOGICAL PROPERTIES OF ALLATOREGULATING PEPTIDES FROM A CRICKET, *Gryllus bimaculatus*

K.H. Hoffmann[1], T. Neuhäuser[1], M.W. Lorenz[1] and B. Stay[2]

[1]Department of General Zoology, University of Ulm, 89069 Ulm, Germany; [2]Department of Zoology, The University of Iowa, Iowa City, Iowa 52242, USA

Insect corpora allata (CA) synthesize juvenile hormones (JH) and the activity of the CA may be controlled by stimulatory and inhibitory signals which reach the glands either via the haemolymph or via nervous connections (Goodman 1990). Neuropeptides that either stimulate (allatotropin) or inhibit (allatoinhibin, allatostatin) JH synthesis have been described in several insect species. So far, the chemical nature of 11 allatostatins (Pratt et al. 1989, Woodhead et al. 1989, Kramer et al. 1991, Pratt et al. 1991, Duve et al. 1993) and one allatotropin (Kataoka et al. 1989) has been published. Ten of the allatostatins show structural similarities in the C-terminus amino acids. Antibody raised to one of the five allatostatins of the cockroach *Diploptera punctata* (Dip-AS1) was used to identify the nerve cells of the brain that produce this allatostatin and carry it to the CA (Stay et al. 1992). Cricket (*G. bimaculatus*) CA synthesize and release JH III (Koch and Hoffmann 1985). The biosynthetic activity of the glands has been investigated under a variety of physiological and experimental conditions (Espig and Hoffmann 1985, Wennauer et al. 1989, Klein et al. 1993).

We describe the isolation, and partial purification, of a number of allatoregulating factors from adult female *G. bimaculatus*. We also demonstrate Dip-AS1 immunoreactivity in the brain and retrocerebral complex of adult crickets.

EXTRACTION AND PARTIAL PURIFICATION OF ALLATO-INHIBITORY BRAIN PEPTIDES

Fractionation of the 16-40% acetonitrile post Seppak C18 eluate prepared from a methanolic extract of 800 adult female brains was carried out using a three-step C18, C8 and C4 reversed-phase HPLC procedure. Columns were eluted with an increasing gradient of acetonitrile in 0.1% TFA and aliquots of these portions were bioassayed for their effects upon JH III biosynthesis (Tobe and Pratt 1974) using CA taken from 3-days old virgin females. Using this technique, we obtained at least four fractions which inhibited JH III biosynthesis significantly (Fig. 1). The elution times of the allatoinhibitory factors differ from those for *D. punctata* allatostatins 1-4.

0-8493-4591-X/94/$0.00 + $.50
© 1994 by CRC Press, Inc.

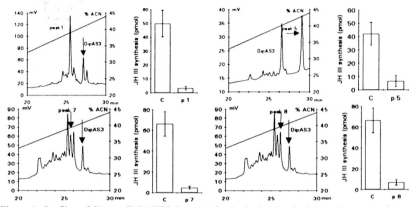

Figure 1. Profiles of Kromasil C4 HPLC step 3 in the isolation of cricket allatostatins from brains. Optical density at 220 nm (mV). The solid line indicates the composition of the acetonitrile gradient. Dip-AS3, marker compound allatostatin 3 from *D. punctata*. The biological activities of peaks 1, 5, 7 and 8 are shown in histogram form. They were tested at 20 brain equivalents per CA pair (n=10; ± SEM).

Preliminary amino acid sequences and molecular weights were so far recorded for two peptides from peaks 1 (8 residues, Mr 1128) and 7 (11 residues, Mr 1326), respectively. Both peptides do not share a resemblance to the *D. punctata* allatostatin family. As yet, however, we have insufficient analytical information to confirm the chemical identity of any of the presumptive cricket allatostatins.

IMMUNOCYTOCHEMICAL DISTRIBUTION OF ALLATOSTATINS IN THE CRICKET BRAIN

G. bimaculatus brain-retrocerebral complexes immunoreacted with a monoclonal antibody against Dip-AS1. A summary of the immunoreactivity is shown in Figure 2. There are four strongly immunoreactive medial cells in the pars intercerebralis, and several lateral cells of the protocerebrum, as well as other groups of cells in the protocerebrum and tritocerebrum also have reactive cell bodies (Fig. 2A). All the nerve tracts from the brain to the CA through the corpora cardiaca (CC) were immunoreactive (NCC 1, 2, 3; Fig. 2B). Branches of immunoreactive nerves are seen in the CC as well as in the CA. The distribution of Dip-AS1 immunoreactive material in the cricket brain is similar to that reported for *D. punctata* itself (Stay et al. 1992). The immunological responds are in harmony with our previous findings that cricket CA were sensitive to the presence of Dip-AS1 in the incubation medium ($ED_{50} = 3 \times 10^{-9}$ M) (Neuhäuser et al. submitted) and let suggest that at least one of the allatostatic cricket petides bears close homology to the *D. punctata* allatostatins.

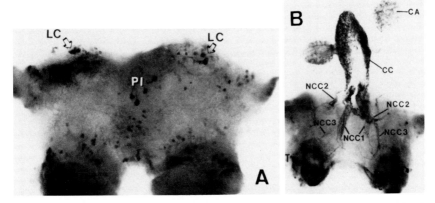

Figure 2. *G. bimaculatus* brain-CC-CA (3-day virgin female) immunoreacted with monoclonal antibody (5F10) against *D. punctata* allatostatin 1. A) Dip-AS1 immunoreactive cells in the brain, anterior view. PI, four large cells in the pars intercerebralis; LC, lateral cells of the protocerebrum. Desheathed brain; magnification x70. B) Back of brain-CC-CA complex viewed from bottom showing immunoreactivity in nervus corporis cardiaci 1 (NCC 1), NCC 2 , NCC 3, CC and CA. T, immunoreactive cells in the tritocerebrum. Magnification x40.

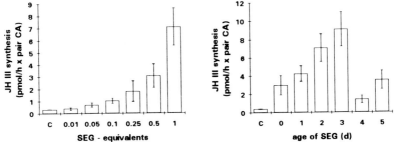

Figure 3. A) Dose response of a SEG allatotropic factor. C, control synthesis in CC-CA-complexes from newly emerged males. B) JH III biosynthesis of CC-CA-complexes from newly emerged males after addition of a methanolic SEG extract from 0 to 5-days old donor females into the incubation medium (1 SEG equivalent per assay). Error bars represent SEM; n=30.

ALLATOTROPIC ACTIVITY FROM CRICKET SUBESOPHAGEAL GANGLION

The subesophageal ganglia of *G. bimaculatus* contain a factor which stimulates spontaneously inactive CA from newly emerged males to synthesize JH III (15-fold increase in JH III synthesis in the presence of one SEG equivalent; Fig. 3A). The allatotropic factor appears to be a peptide based on its proteinase K sensitivity and heat stability. A maximal allatotropic activity was present in SEG of 3-days old females (Fig. 3B), at the time when JH III synthesis *in vivo* was also highest. Further studies on the isolation and identity of the cricket SEG allatotropin are in progress.

REFERENCES

Duve, H., Johnsen, A.H., Scott, A.G., Yu, C.G., Yagi, K.J., Tobe, S.S. and Thorpe, A., Callatostatins: Neuropeptides from the blowfly *Calliphora vomitoria* with sequence homology to cockroach allatostatins, *Proc. Natl. Acad. Sci. USA* 90, 2456-2460, 1993.

Espig, W. and Hoffmann, K.H., Juvenile hormone and reproduction in the cricket. II. Effect of rearing temperature on corpus allatum activity (*in vitro*) in adult females, *Experientia* 41, 758-759, 1985.

Goodman, W.G., Biosynthesis, titer regulation, and transport of juvenile hormone, in *Morphogenetic hormones in arthropods*, Vol. 1, Gupta, A.P., Ed., Rutgers University Press, New Brunswick, 1990, chap. 5.

Kataoka, H., Toschi, A., Li, J.P., Carney, R.L., Schooley, D.A. and Kramer, S.J., Identification of an allatostatin from adult *Manduca sexta*, *Science* 243, 1481-1483, 1989.

Klein, P.M., Lorenz, M.W., Huang, D. and Hoffmann, K.H., Age dependency and regulatory properties of juvenile hormone III biosynthesis in adult male crickets, *Gryllus bimaculatus*, *J. Insect Physiol.* 39, 315-324, 1993.

Koch, P.B. and Hoffmann, K.H., Juvenile hormone and reproduction in crickets, *Gryllus bimaculatus* de Geer: Corpus allatum activity (*in vitro*) in females during adult life cycle, *Physiol. Entomol.* 10, 173-182, 1985.

Kramer, S.J., Toschi, A., Miller, C.A., Kataoka, H., Quistad, G.B., Li, J.P., Carney, R.L. and Schooley, D.A., Identification of an allatostatin from the tobacco hornworm *Manduca sexta*, *Proc. Natl. Acad. Sci. USA* 88, 9458-9462, 1991.

Neuhäuser, T., Sorge, D., Stay, B. and Hoffmann, K.H., Responsiveness of the adult cricket (*Gryllus bimaculatus* and *Acheta domesticus*) retrocerebral complex to allatostatin-1 of a cockroach, *Diploptera punctata*, *J. Comp. Physiol.*, submitted.

Pratt, G.E., Farnsworth, D.E., Siegel, N.R., Fok, K.F. and Feyereisen, R., Identification of an allatostatin from adult *Diploptera punctata*, *Biochem. biophys. Res. Commun.* 163, 1243-1247, 1989.

Pratt, G.E., Farnsworth, D.E., Fok, K.M., Siegel, N.R., McCormack, A.L., Shabanowitz, J., Hunt, D.F. and Feyereisen, R., Identity of a second type of allatostatin from cockroach brains: An octadecapeptide amide with a tyrosine-rich address sequence, *Proc. Natl. Acad. Sci. USA* 88, 2412-2416, 1991.

Stay, B., Chan, K.K. and Woodhead, A.P., Allatostatin-immunoreactive neurons projecting to the corpora allata of adult *Diploptera punctata*, *Cell Tissue Res.* 270, 15-23, 1992.

Tobe, S.S. and Pratt, G.E., The influence of substrate concentration on the rate of insect juvenile hormone biosynthesis by corpora allata of the desert locust *in vitro*, *Biochem. J.* 144, 107-113, 1974.

Wennauer, R., Kassel, L. and Hoffmann, K.H., The effect of juvenile hormone, 20-hydroxyecdysone, precocene II and ovariectomy on the activity of the corpora allata (*in vitro*) in adult female *Gryllus bimaculatus*, *J. Insect Physiol.* 35, 299-304, 1989.

Woodhead, A.P., Stay, B., Seidel, S.L., Khan, M.A. and Tobe, S.S., Primary structure of four allatostatins: neuropeptide inhibitors of juvenile hormone synthesis, *Proc. Natl. Acad. Sci. USA* 86, 5997-6001, 1989.

We are grateful to Dr. R. Kellner (EMBL Heidelberg, Germany) for peptide sequencing and mass determinations. This study was supported by a Landesforschungsschwerpunkt Baden-Württemberg and the Fonds der Chemischen Industrie.

FEEDBACK REGULATION OF THE ALLATOTROPIC HORMONE IN THE *GALLERIA MELLONELLA* (LEPIDOPTERA) LARVAL BRAIN

Malgorzata Muszynska-Pytel[1], Piotr Mikolajczyk[1], Ewa Szolajska[2] and Maciej A. Pszczolkowski[1]

[1]Department of Invertebrate Physiology, Warsaw University, Xwirki i Wigury 95, 02-089 Warszawa, Poland
[2]Institute of Biochemistry and Biophysics, Polish Academy of Science, Rakowiecka 26, 02-536 Warszawa, Poland

INTRODUCTION, MATERIALS AND METHODS

Allatotropic hormone (ATTH) stimulates juvenile hormone (JH) synthesis in *corpora allata* (CA) of several insect species. Numerous studies have shown that high titre of JH (or JH analog) inhibits synthetic activity of CA (Kramer and Staal 1981, Khanetal 1982, Edwards et al. 1987). It has been postulated that decrease in CA activity is not due to direct inhibition of the glands by JH but is rather mediated by the brain (Tobe and Stay 1979). In this paper we present data on feedback regulation of ATTH by JH in *G. mellonella*.

The experiments were performed on one day old last instar (7th) larvae. On the day of the moult into the last larval stadium the larva is regarded as one day old and is designated day-1 larva. The larvae were chilled for 3 h at 0°C. Hydroprene and JH-I were applied topically at a dose of 10 μg and 20 μg on dorsal abdomen in 1 μl of acetone. Brain ATTH activity has been determined by the brain implantation assay according to Sehnal and Granger (1975) and expressed in allatotropic activity units (ATU) (Muszynska-Pytel et al. 1992). *In vitro* assay for ATTH has been performed according to Muszynska-Pytel et al. (1991).

RESULTS AND DISCUSSION

Day 1 last instar larvae were chilled for 3 h at 0°C, and were then transferred to 30°C. The ATTH of the brain from chilled larvae was determined during the two day post-chilling period, using a brain implantation assay. It was found that brain ATTH decreased from 0.5±0.09 ATU to 0.28±0.1 ATU during the 3 h chilling period. Gradual elevation in brain activity occurred during the first 12 h and reached the maximum 1.14±0.16 ATU, at 18 h post chilling. Brain ATTH decreased gradually thereafter, and 48 hr it was nearly the same as the activity exhibited by the brain before chilling (Tab.1). At the beginning of the post chilling period the *in vitro* rates of JH synthesis by CC-CA were about 50% lower when compared with that

0-8493-4591-X/94/$0.00 + $.50
© 1994 by CRC Press, Inc.

of the unchilled control. The glands from chilled larvae synthesized 12.0±5.5 fmole/h x gland pair, and the glands from unchilled control 25.5±6.2 fmole of JH. The JH synthetic activity increased thereafter and reached a peak 24 h after chilling. These glands synthesized approximately 61% more JH than unchilled control (41.1±5.1 fmole of JH/h x gland pair) (Tab.1). Application of 10 μg of hydroprene caused nearly a three-fold decrease in the brain allatotropic activity of day-1 last instar larva. Within 6 h after treatment this activity dropped from 0.51±0.1 ATU to 0 ATU. Low allatotropic activity of the brain was recorded for the next 18 h of the post-treatment

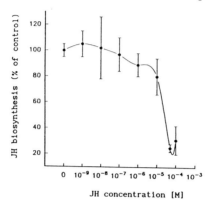

Figure 1. The rate of JH synthesis by CC-CA form day-1 last instar larvae *in vitro* incubated at different concentration of JH-I and 10^{-6} M EPPAT (JH esterase inhibitor).

Table 1.

Time dependent changes in brain allatotropic activity (ATTH) and *in vitro* rate of JH synthesis by CC-CA after chilling or topical application of 10μg of hydroprene on day-1 last instar larvae.

hours after treatment	chilling		hydroprene	
	allatotropic activity of the brain (ATU)	JH synth. in vitro (% of control)	allatotropic activity of the brain (ATU)	JH synth. in vitro (% of control)
0	0.275 ± 0.08	46.26 ± 18.3	0.511 ± 0.1	25.1 ± 6.0
6	0.411 ± 0.02	73.72 ± 6.2	0.0	6.2 ± 0.1
18	1.146 ± 0.16	127.53 ± 5.7	-	-
24	1.052 ± 0.04	160.57 ±11.3	0.083 ±0.01	6.4 ± 3.1
36	0.733 ± 0.11	139.34 ± 6.8	-	-
48	0.583 + 0.04	124.16 ± 9.6	0.448 ± 0.02	4.9 ± 3.8
control	0.511 ± 0.09	100	0.483 ± 0.04	100

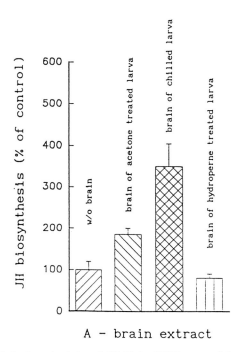

A - brain extract

Figure 2. *In vitro* stimulation of CC-CA by ATTH extracted from the brain. The ATTH activity was tested at 2 brains equivalent on the glands from head cap slippage, 6th instar larva. Glands incubated in plain Grace's medium synthesized approximately 11.0±2.75 fmole/h/gland pair.

period (Tab.1). Thereafter it slowly increased and at the end of the second day the brain of the treated larvae exhibited the same level of allatotropic activity as the brain of untreated control, i.e. 0.48-0.04 ATU.

Exogenous hydroprene also caused considerable changes in the activity of CC-CA. Six hours after topical application of 10 μg hydroprene/larva the rate of JH synthesis decreased from 25.0±5.6 fmole to 6.9±1.6 fmole/h x gland pair and was approximately 74% lower than that of the glands before treatment. Similar time-dependent changes in the rate of JH synthesis by CC-CA and in brain allatotropic activity were found after application of 20 μg of exogenous JH-I on young, last instar larva (data not shown). Drop in JH synthesis is not caused by direct inhibition of CA by JH since the rate of JH production did not decrease when this gland was incubated *in vitro* in Graces's medium containing a physiological concentration of JH-I (Fig.1).

The observed decrease in the rate of JH synthesis by CA after hydroprene application may be due to the inhibition of allatotropin synthesis in the brain. This hypothesis was verified in the next experiment. The brains of chilled or hydroprene-treated larvae were extracted with methanol, allatotropic activity of the extracts was tested at 2 brain equivalents/100 μl medium. It was found that the extracts from brains of chilled larvae synthesis by the glands incubated with extract from the brain of hydroprene-treated larvae was inhibited by 30% as compared with controls incubated in plain medium (Fig.2). Our data indicate that inhibition of G. mellonella CA, as measured by the subsequent decrease in the in vitro rate of JH synthesis by exogenous JH cannot be related to its direct inhibitory effect on CA because the physiological concentrations of JH did not inhibit of JH synthesis by CC-CA *in vitro*. The data suggest that JH titre regulates synthetic activity of CA indirectly, possibly by inhibition of allatotropin synthesis via a long-loop negative feedback mechanism.

REFERENCES

Kramer, S. J., and Staal, G. B., *In vitro* studies on the mechanism of action of anti-juvenile hormone agents in larvae of *Manduca sexta*, in *Juvenile Hormone Biochemistry*, Pratt, G. E., Brooks, G. T., Eds., Elsevier/North Holland, Amsterdam, pp 425-437, 1981. Bogus, M. and Cymborowski, B., Chilled *Galleria mellonella* larvae: mechanism of supernumerary moulting, *Physiol. Entomol.* 6, 343, 1981.

Khan, M. A., Koopmanschap, A. B., and de Kort, C. A. D., The effects of juvenile hormone, 20-hydroxyecdysone and precocene II on activity of *corpora allata* and the mode of negative-feedback regulation of these glands in the adult Colorado potato beetle, *J. Insect Physiol.* 28, 995, 1982.

Edwards, J. P., Chambers, J., Price, N. R., and Wilkins, J. P. G., Action of a juvenile hormone analogue on the activity of *Periplaneta americana corpora allata in vitro* and on juvenile hormone III levels *in vivo*, *Insect Biochem.* 17, 1115, 1987.

Tobe, S. S., and Stay, B., Modulation of juvenile hormone synthesis by an analogue in the cockroach, *Nature* 281, 481, 1979.

Sehnal, F., and Granger, N. A., Control of *corpora allata* function in *Galleria mellonella. Biol. Bull.* 148, 106, 1975.

Muszynska-Pytel, M., Pszczolkowski, M. A., Mikolajczyk, P., and Cymborowski, B., Strain specificity of *Galleria mellonella* larvae to juvenilizing treatments. *Comp. Biochem. Physiol.* 103A, 119, 1992.

Muszynska-Pytel, M., Szolajska, E., Pszczolkowski, M. A., Michalik, J., and Mikolajczyk, P., *In vitro* bioassay for the brain allatotropic hormone of *Galleria mellonella, Folia Biol.* 39, 58, 1991.

A NEUROPEPTIDE FAMILY THAT INHIBITS THE MANDIBULAR ORGAN OF CRUSTACEA AND MAY REGULATE REPRODUCTION

Hans Laufer[1], Lei Liu[1] and Francois Van Herp[2]

[1]Department of Molecular and Cell Biology, The University of Connecticut, Storrs, CT 06269-3125, USA and [2]Department of Experimental Zoology, Catholic University, Nijmegen, The Netherlands

I. INTRODUCTION

Mandibular organs (MOs) of crustacea produce methyl farnesoate (MF) [Laufer et al., 1987]. This unepoxidated form of JHIII has JH activity in insects. Crustacea have physiological processes that are similar to the events controlled by JH in insects. The JH-like compound synthesized and secreted by the MO functions in adult crustaceans in a manner similar to its function in adult insects, that is, JHs seem to stimulate reproduction in both males and females [Laufer and Borst 1988; Homola et al., 1991]. Synthesis of MF in MOs is negatively regulated by factors from sinus glands (SG) [Laufer et al., 1987 and Landau et al., 1989]. These factors are called mandibular organ inhibiting hormones (MO-IHs).

The sinus gland X-organ neurosecretory complex (SG-XO), located in the eyestalks, is considered to be the most important neuroendocrine regulating and integrating center of crustaceans. A number of crustacean neuropeptides from the SG have been completely identified with regard to their structure, function and distribution [reviewed by Keller, 1992]. A novel family of relatively large peptides from the sinus gland X-organ complex has been elucidated which include several isoforms of crustacean hyperglycemic hormone (CHH), a molt-inhibiting hormone (MIH) and vitellogenesis or gonad-inhibiting hormone (VIH or GIH). CHHs regulate glycogen metabolism in crustacea. They have molecular weights of about 8 kDa, containing about 72 amino acid residues, including 6 cysteins which form 3 disulfide bridges. Evidence is accummulating which shows multiple physiological activities among this peptide family. The MIH of *Homarus americanus* also causes hyperglycemia [Chang et al., 1990] and one group of the lobster CHH isoforms has a stimulatory effect on oocyte growth when injected into the prawn, *Palaemonetes varians*, a heterologous bioassay performed in vivo [Tensen et al., 1989].

0-8493-4591-X/94/$0.00 + $.50
© 1994 by CRC Press, Inc.

In this report, CHH isoforms from the crayfish *Procambarus clarkii* were tested in vitro on MOs from the same species and were shown to have MO-IH activity. This discovery is the first characterization of any MO-IHs and classifies the MO-IHs as members of the CHH/MIH/VIH peptide family. A fully determined structure of the MO-IH will reveal the relationship between the hormones and give a new model for the control of reproduction in crustacea.

II. MATERIALS AND METHODS

The crayfish *Procambarus clarkii* were from the Louisiana State University, kept in fresh water in "Living Stream" tanks and were fed twice a week with shrimp pellets.

Neuropeptide samples from *Procambarus clarkii* sinus glands were prepared by a two-step purification procedure on HPLC from crude sinus gland extracts according to Tensen (1991). Four isoforms of CHH and a CHH precursor related peptide (CPRP) were isolated. The neuropeptides were lyophilized and redissolved in freshwater crustacean saline [Van Harreveld, 1936] for bioassay.

Paired MOs were removed and incubated in 400ul of freshwater crustacean saline containing 62.5 uCi ^{3}H-methyl-methionine for 4 hours at room temperature (20-22 °C) according to Landau et al. (1989). Each isoform was tested on 6-8 pairs of MOs, the ipsilateral gland from a single organism serving as a control and the contralateral gland receiving the experimental treatment. The results were averaged and then converted to the percentage of MO activity relative to the control.

III. RESULTS

The HPLC fractions containing the individual CHH isoforms, called CHH I, II, VI, and VII, and the CPRP were collected from both female and male SGs. Assays were conducted at a dose of 0.2 sinus gland equivalents. The results show CHHI had very active inhibition and CHHII was similarly active. CHHVI gave less inhibition while CHHVII and CPRP did not show any inhibition. The results of the present experiment on *P. clarkii* CHHs provide evidence that peptides from the SG may have multiple functions affecting MF synthesis by inhibiting it. They appear to have different target tissues, affecting blood sugar, MF synthesis, molting and vitellogenin synthesis. Furthermore, and most significant, is the fact that MO-IHs have been identified and characterized as members of the CHH-related peptide family.

IV. DISCUSSION

Since the juvenile hormones in insects are produced by the corpora allata (CA) and the MOs produce a juvenile hormone-like compound in crustacea, we consider the juvenoids homologues and their secretory endocrine organs to be homologous [LeRoux, 1968]. The JH synthesis in CA is inhibited by neuropeptides, so-called allatostatins. A number of allatostatins have been isolated and sequenced [Woodhead et al., 1989]. MO-IH can be considered to be a functional homologue of allatostatins. So the question arises whether these are also structurally related.

Isoforms of CHHs have been isolated from several crustaceans and were found to have an effect on oocyte growth [Van Herp, 1992]. The CHHs and related neuropeptides from *Procambarus bouivier* have CHH, MIH and VIH activity [Huberman et al., 1992].

Since MF has effects on vitellogenesis, MO-IHs can be considered to have actions as a VIH as well. This action may be either directly on the vitellogenin synthetic tissue or less directly by affecting the MO which in turn may affect vitellogenin synthesis. The endocrine control of female reproduction in crustacea has been studied for several decades, but the regulatory pathway and the target tissues of VIH are still unknown. From our results the following can be suggested: 1) MO-IHs may be the same hormones as CHH I and II and one of the target tissues of CHH appears to be the MO. CHH may function through inhibition of the MF synthesis of the MO. Although, we cannot exclude a direct interaction of the CHHs on the gonads by these experiments. 2) The interaction of the SG neuropeptides and the MO seems to be the same in male and female animals since we found no differences in the inhibition assay on MO.

The relationships among the CHH-related peptide family and the primary physiological function of the different peptides are still uncertain. A fully identified structure of MO-IH will help to clarify this situation and provide understanding of the mechanism of control of crustacean reproduction [Van Herp, 1992].

Acknowledgements: The research reported here was supported in part by the Sea Grant College Program (NOAA). The crayfish used in this study were supplied through the generosity of Dr. Robert Romaire, Louisiana State University Agricultural Center, Baton Rouge, LA.

REFERENCES

Chang, E., Prestwich, G.D. and Bruce, M.J., Amino acid sequence of a peptide with both molt-inhibiting and hyperglycemic activities in the lobster *Homarus americanus*, *Biochem. & Biophy. Res. Commun.*, 171(2), 818-826, 1990.

Homola, E., Sagi, A. and Laufer, H., Relationship to claw form and exoskeleton condition to reproductive system size and methyl farnesoate in the male spider crab, *Libinia emarginata*, *Invert. Reprod. & Devel.*, 20, 245-251, 1991.

Huberman, A., Aguilar, M.B., and Brew, K., A neuropeptide hormone family from the Mexican crayfish *Procambarus bouvieri* (*Ortmann*): intraspecies and interspecies comparisons, Abst. to 1st European Crustacean Conference, Paris, Sept. 2, 1992.

Keller, R., Crustacean neuropeptides: Structure, functions and comparative aspects, *Experientia*, 48, 439-448, 1992.

Landau, M., Laufer, H., and Homola, E., Control of methyl farnesoate synthesis in the mandibular organ of the crayfish, *Procambarus clarkii*: Evidence for peptide neurohormones with dual functions, *Invert. Reprod. & Devel.*, 14, 165-168, 1989.

Laufer, H., Borst, D., Baker, F.C., Carrasco, C., Sinkus, M., Reuter, C.C., Tsai, L.W. and Schooley, D.A., Identification of a juvenile hormone-like compound in a crustacean, *Science*, 235, 202-205, 1987.

Laufer, H. and Borst, D., Juvenile hormone in crustacea, in *Endocrinology of Selected Invertebrate Types*, Vol. 2, Laufer, H. and Downer, R., Eds., Alan R. Liss Inc., New York, 1988,.

Le Roux, A., Description d'organes mandibularies nouveaux chez les crustacean Decapodes, CR hebd Acad. Sci., *Ser. D. Sci. Nat.*, 266, 1414-1417, 1968.

Tensen, C.P., Janssen, K.P.C., Van Herp, F., Isolation, characterization and physiological specificity of crustacean hyperglycemic factors from the sinus gland of the lobster, *Homarus americanus* (Milne-Edwards), *Invert. Reprod. & Devel.*, 16, 135-164, 1989.

Tensen, C.P., Hyperglycemic neuropeptides in crustaceans: A biochemical and molecular biological study, PhD Thesis, Catholic University, Nijmegen, 1991.

Van Harreveld, A., A physiological solution for fresh water crustaceans, *Proc. Soc. Exp. Biol. Med.*, 34, 428-432, 1936.

Van Herp, F., Inhibiting and stimulating neuropeptides controlling reproduction in crustacea, *Invert. Reprod. & Devel.*, 22, 21-30, 1992.

Woodhead, A.P., Stay, B., Seidel, S.L., Khan, M.A. and Tobe, S.S., Primary structure of four allatostatins: neuropeptide inhibitors of juvenile hormone synthesis, *Proc. Natl. Acad. Sci. USA*, 86, 5997-5999, 1989.

FMRFamide

ROLE OF A FMRFAMIDE-LIKE PEPTIDE IN THE ADULT FEMALE LOCUST, *LOCUSTA MIGRATORIA*

V.M. Sevala, V.L. Sevala and B.G. Loughton
Department of Biology, York University
North York, Ontario, Canada M3J 1P3

FMRFamide, first described by Price and Greenberg (1977) as a cardioexcitatory peptide in *Macrocallista nimbosa* is one of a family of N-terminally extended RFamides termed FaRPs (Greenberg et al., 1988). They have been recognised in a wide range of invertebrates and appear to act as neurotransmitters, neuromodulators and neurohormones (Evans et al., 1989). In the locust one such FaRP has been shown to inhibit oviducal contractions (Lange et al., 1991) and in *Rhodnius prolixus*, another FaRP was found to cause ovulation (Sevala et al., 1992). In the present study we have used immunocytochemistry and radioimmunoassay to investigate the role of a FaRP in oviposition in the locust. FMRFamide-like immunoreactivity was detected in the brain and suboesophageal ganglion of adult female locusts (Fig. 1A). The distribution of immunoreactive cell bodies and axons was similar to that described by Myers and Evans (1987). When staining intensity of median neurosecretory cells of adult female locust (MNC) was measured in successive days of the oviposition cycle a striking pattern of changes became evident. Immediately after oviposition FMRFamide-like staining of MNC was weak (Fig. 2A). Staining increased slightly on days one and two (Fig. 2B and C). Staining increased dramatically on day 3 (Fig. 2D) but declined once more on days 4 and 5 (Fig. 2E and F). Radioimmunoassay of FMRFamide-like material in the hemolymph showed that the titre was low on days one and two but increased sharply on the third day, reaching a peak on day five (Fig. 1D). On the sixth day, immediately after oviposition, the titer of FMRFamide-like material returned to "day 0" levels. An immunoblot of hemolymph after SDS-PAGE revealed an FMRFamide immunoreactive peptide of approximately 8 kDa in the hemolymph of adult female locusts.

Hemolymph taken from adult female locusts during the oviposition cycle showed increasing levels of myotropic activity as the time of oviposition approached (Fig. 1B). Gel permeation chromatography yielded an FMRFamide immunoreactive fraction (MW approximately 8 kDa). Figure 2G shows the contractile activity of isolated oviduct after application of increasing concentrations of the FMRFamide immunoreactive fraction. When anti-FMRFamide serum was added to the bath together

0-8493-4591-X/94/$0.00 + $.50
© 1994 by CRC Press, Inc.

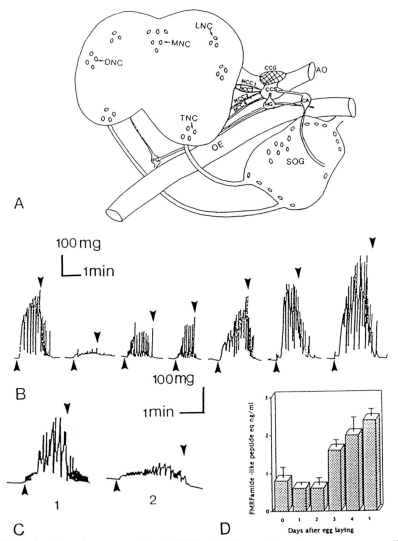

Fig 1: A. Schematic representation of FMRFamide-like peptide positive neurosecretory cell bodies in the locust brain (AO-aorta; CA-corpus allatum; CCG-glandular lobe of corpus cardiacum; CCS-storage lobe of corpus cardiacum; DNC-deutocerebral neurosecretory cells; FG-frontal ganglion; HG-hypocerebral ganglion; LNC-lateral neurosecretory cells; MNC-median neurosecretory cells; NCC-nervous corpus cardiacum; OE-oesophagus; SOG-suboesophageal ganglion; TNC-tritocerebral neurosecretory cells).

B. The effect of hemolymph during oviposition cycle on the contractions of isolated locust oviduct. In figure B and C: ▲ indicates the application of peptide and ▼ indicates times of flushing.

C. The effect of incubating the FMRFamide-like fraction (1) separated by gel permeation with anti-FMRFamide serum(2).

D. Hemolymph titre of FMRFamide-like peptide of mated adult female locusts during the oviposition cycle.

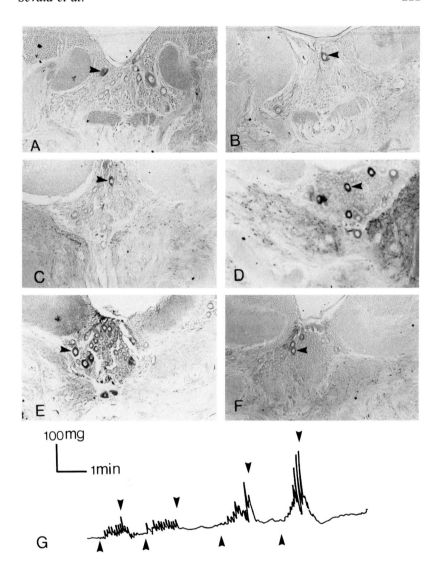

Fig 2: FMRFamide-like immunoreactivity in the median neurosecretory cells (arrowhead) of adult female locusts stained on successive days of oviposition cycle. **A.** Day of oviposition; **B.** Day 1; **C.** Day 2; **D.** Day 3; **E.** Day 4; **F.** Day 5 after oviposition; **G.** Contractions of the locust oviduct showing the effect of addition of increasing doses of FMRFamide-like material (gel permeation) to the incubation medium. From the left 10, 25, 50, 100 ul hemolymph equivalents per assay (bath volume 2.5 ml).

with the FMRFamide-like fraction, the myotropic effect was blocked (Fig. 1C). The data are consistent with the hypothesis that an 8 kDa FMRFamide-like peptide is synthesized in the MNC and released in increasing amounts into the hemolymph as the oviposition cycle progresses. Its presence in the hemolymph then stimulates contractions of the lateral oviduct and initiates oviposition.

REFERENCES

Evans, P.D., S. Robb and C.A. Cuthbert., Insect neuropeptides-identification, establishment of functional roles and novel target sites for pesticides, *Pestic. Sci.*, 25, 71, 1989

Greenberg, M.J., K. Payzee., R.J. Nachman., G.M. Holman and D.A. Price., Relationships between the FMRFamide-related peptides and other peptide families, *Peptides*, 9, 125, 1988.

Lange, A.B., I. Orchard and V.A. TeBrugge., Evidence for the involvement of a schisto-FLRFamide-like peptide in the neural control of locust oviduct, *J. Comp. Physiol.*, 168A, 383, 1991.

Myers, C.M., and P.D. Evans., An FMRFamide antiserum differentiates between populations of antigens in the brain and retrocerebral complex of the locust, *Schistocerca gregaria*, *Cell Tissue Res.*, 250, 93, 1987.

Price, D.A., and M.J. Greenberg., Structure of a molluscan cardioexcitatory neuropeptide, *Science*, 197, 670, 1977.

Sevala, V.L., V.M. Sevala, K.G. Davey and B.G. Loughton., A FMRFamide-like peptide is associated with the myotropic ovulation hormone in *Rhodnius prolixus*, *Arch. Insect Biochem. Physiol.*, 20, 193, 1992.

ISOLATION AND BIOACTIVITY OF FMRFAMIDE-RELATED PEPTIDES FROM THE LOCUST CENTRAL NERVOUS SYSTEM

Angela B. Lange, Ian Orchard and Neda M. Peeff

Department of Zoology, University of Toronto,
Toronto, Ontario, Canada, M5S 1A1.

INTRODUCTION

Peptides based upon the molluscan cardioexcitatory peptide Phe-Met-Arg-Phe-NH$_2$ (FMRFamide) are now known to be widely distributed in the animal kingdom (Price and Greenberg, 1989). In insects, several distinct groups of these FMRFamide-related peptides (FaRPs) have been sequenced, including a number of extended FMRFamides in *Drosophila* and *Calliphora* (Nichols, 1992; Duve *et al.*, 1992) and extended FLRFamides in *Leucopheae*, *Schistocerca*, *Manduca*, *Drosophila* and *Nuobellieria* (see Nichols, 1992; Fonagy *et al.*, 1992; Robb *et al.*, 1989). In locusts, as with other insects species, FMRFamide-like immunoreactivity has been described in neurons present throughout the central nervous system, some of which project to neurohaemal organs, others to peripheral targets. FaRPs have been shown to be involved in the control of locust oviducts with a peptide chromatographically similar to SchistoFLRFamide (PDVDHVFLRFamide) being localised to the innervated regions of the oviduct (Lange *et al.*, 1991). SchistoFLRFamide, and the native peptide, inhibit contractions of the locust oviducts (Lange *et al.*, 1991). Other FaRPs however, stimulate contractions (Peeff *et al.*, 1993) leading to the suggestion that locust oviducts may be under excitatory control by blood borne FaRPs and inhibitory control by FaRPs delivered via the nerve supply. In order to better understand the physiological control of locust oviducts by FaRPs, we set out to isolate and sequence *Locusta* FaRPs from the central nervous system and to examine the structure / activity properties of FaRPs using the locust oviduct bioassay.

RESULTS AND DISCUSSION

Several hundred brains or ventral nerve cords were dissected from adult *Locusta migratoria* and processed through Sep-Pak C$_{18}$ cartridges, followed by several reversed phase HPLC systems using C$_{18}$ and phenyl columns. Fractions containing FMRFamide-like immunoreactivity, as determined by RIA, were purified to a single UV peak and then sent for sequencing to the

0-8493-4591-X/94/$0.00 + $.50
© 1994 by CRC Press, Inc.

TABLE 1

Putative sequences of peptides from locust central nervous system

PDVDHVFLRFamide
ADVGHVFLRFamide
XXERNFLRFamide
ADDRNFIRFamide
XFIRFamide
AFIXRFamide

X denotes some uncertainty in the amino acid residue

Core Facility for Protein and DNA Chemistry, Queen's University, Ontario. The putative sequences obtained are shown in Table 1. Of the six peptides sequenced, one is identical to SchistoFLRFamide whereas a second differs only in positions 1 and 4. The extension of the third -FLRFamide has no structural similarity with the first two. Two other peptides are extended FIRFamides, whereas the sixth peptide is an extended IXRFamide.

FaRPs were examined for their ability to alter contractions of the oviducts of *Locusta migratoria*. SchistoFLRFamide and the closely related peptide sequenced in the present study (ADVGHVFLRFamide) both decreased the amplitude and frequency of myogenic contractions and reduced basal tonus, with thresholds at approximately 10^{-9} M (Figure 1). In addition, both peptides also reduced the amplitude of proctolin-induced contractions (Figure 1) in a dose-dependent manner (threshold approximately 5×10^{-8} M and maxima at approximately 10^{-5} M). Interestingly, these two inhibiting peptides are structurally related to leucomyosuppressin (pQDVDHVFLRFamide)(Holman *et al.*, 1986), an inhibitor of cockroach hindgut and to neomyosuppressin (TDVDHVFLRFamide)(Fonagy *et al.*, 1992) which inhibits cockroach hindgut and locust oviduct. This sub-family of visceral muscle inhibitory peptides was further examined following truncation from the N-terminus. When the truncated version, HVFLRFamide, was examined on locust oviduct it was found to retain its inhibitory properties whereas the truncated version, VFLRFamide, did not. Furthermore, removal of the amide, as in PDVDHVFLRF also resulted in a loss of inhibitory activity. The necessary requirements for inhibition appear to be the histidine group and an amidated C-terminus.

In contrast to the inhibitory effects described above other FaRPs are capable of stimulating contractions of locust oviducts (Peeff *et al.*, 1993), although with thresholds and maxima considerably higher than those required by the inhibitory peptides. These stimulatory peptides include the other four peptides isolated in the present study, as well as the non-native FaRPs, FLRFamide, FMRFamide, TNRNFLRFamide and YGGFMRFamide, with thresholds at about 10^{-6} M and maxima at about 10^{-4} M (Peeff *et al.*, 1993).

ADVGHVFLRFamide   PDVDHVFLRFamide

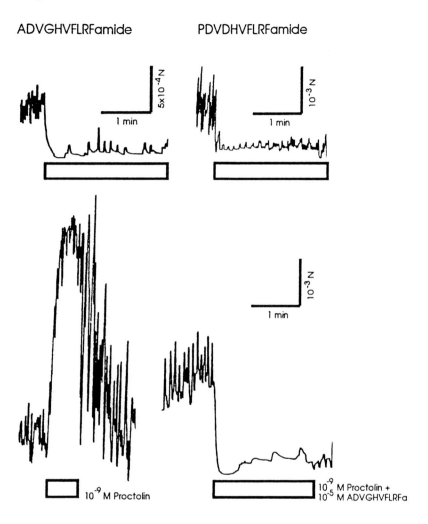

Figure 1. Effects of ADVGHVFLRFamide and PDVDHVFLRFamide upon locust oviduct contractions. Upper traces: Both peptides reduce the amplitude and frequency of spontaneous contractions and relax basal tonus. Lower traces: ADVGHVFLRFamide inhibits proctolin-induced contractions. Peptides applied during bar.

The results indicate that, as with other insects the locust central nervous system contains several FaRPs, six of which have been sequenced in the present study. In addition the results suggest the presence of at least two receptor sites for FaRPs on locust oviducts. One receptor site leads to inhibition of contraction whereas the other leads to stimulation of contraction. The presence of a histidine is critical for inhibition since HVFLRFamide is inhibitory, whereas VFLRFamide is stimulatory.

ACKNOWLEDGEMENTS

This work was supported by the Natural Sciences and Engineering Research Council of Canada, and by Insect Biotech Canada, a Federal Network of Centres of Excellence.

REFERENCES

Duve, H., Johnsen, A.H., Sewell, J.C., Scott, A.G., Orchard, I., Rehfeld, J.F. and Thorpe, A., Isolation, structure, and activity of -Phe-Met-Arg-Phe-NH$_2$ neuropeptides (designated calliFMRFamides) from the blowfly *Calliphora vomitoria*, *Proc. Natl. Acad. Sci. USA* 89, 2326-2330, 1992.

Fonagy, A., Schoofs, L., Proost, P., Van Damme, J., Bueds, H. and De Loof, A., Isolation, primary structure and synthesis of neomyosuppressin, a myoinhibiting neuropeptide from the grey fleshfly, *Neobellieria bullata*, *Comp. Biochem. Physiol.* 102C, 239-245, 1992.

Holman, G.M., Cook, B.J. and Nachman, R.J., Isolation, primary structure and synthesis of leucomyosuppressin, an insect neuropeptide that inhibits spontaneous contractions of the cockroach hindgut, *Comp. Biochem. Physiol.* 85C, 329-333, 1986.

Lange, A.B., Orchard, I. and Te Brugge, V.A., Evidence for the involvement of a SchistoFLRF-amide-like peptide in the neural control of locust oviduct, *J. Comp. Physiol. A* 168, 383-391, 1991.

Nichols, R., Isolation and structural characterization of *Drosophila* TDVDHVFLRFamide and FMRFamide-containing neural peptides, *J. Mol. Neurosci.* 3, 213-218, 1992.

Peeff, N.M., Orchard, I. and Lange, A.B., The effects of FMRFamide-related peptides on an insect (*Locusta migratoria*) visceral muscle, *J. Insect Physiol.* 39,207-215, 1993.

Price, D.A. and Greenberg, M.J., The hunting of FaRPs: The distribution of FMRFamide-related peptides, *Biol. Bull.*177, 198-205, 1989.

Robb, S., Packman, L.C. and Evans, P.D., Isolation, primary structure and bioactivity of SchistoFLRF-amide, a FMRF-amide-like neuropeptide for the locust, *Schistocerca gregaria. Biochem. biophys. Res. Commun.* 160, 850-856, 1989.

Pheromone Biosynthesis Activating Neuropeptide

PBAN/MRCH CONTENT AND VARIATIONS IN COLOR IN POLYMORPHIC *SPODOPTERA LITTORALIS* MOTHS

Miriam Altstein, Ezra Dunkelblum, Orna Ben-Aziz, Jacob Meisner and Yoav Gazit

Institute of Plant Protection, Volcani Center, ARO, 50 250, Israel

Some insect species exhibit morphological color changes during larval and pupal stages in response to environmental factors, such as background color, temperature, photoperiod and population density [for review see Pener, 1991]. Among Lepidopteran species, larvae of some Noctuidae moths such as *Luecania separata, Anticarsia gemmatalis, Spodoptera litura* and *Spodoptera littoralis* display significant differences in their morphometrics, appearing bright in color when individually reared and dark black when reared under crowded conditions. The black coloration in these moths is due to deposition of melanin in the cuticle [Ikemoto, 1971].

Insect cuticular melanization is regulated by endocrine and neuro-endocrine factors [for review see Pener, 1991; Raabe, 1989; Riddiford and Hiruma, 1988]. Various neurohormones originating in the brain-corpora cardiaca-corpora allata complex or suboesophageal ganglion have been reported to be associated with cuticular melanization in moths. One of them, termed melanization and reddish colouration hormone (MRCH), was characterized [Matsumoto et al., 1981, 1988], and its primary structure was elucidated in *Bombyx mori* [Matsumoto et al., 1990]. MRCH was found to be identical to the pheromone biosynthesis activating neuropeptide (Bom-PBAN) isolated from the same moth [Kitamura et al., 1989] and structurally related to the primary structure of Hez-PBAN isolated from *Helicoverpa zea* [Raina et al., 1989].

The ability of PBAN/MRCH to induce melanization suggests that it may play a role in color polymorphism. A direct correlation, however, between the content of endogenous PBAN/MRCH and color polymorphism has not been demonstrated. In the present study we have tested whether such a correlation exists by performing a comparative analysis of the content of endogenous PBAN/MRCH levels in crowded and individually reared *S. littoralis*, which exhibit morphometric color variations - black and light brown, respectively [Rivnay and Meisner, 1966]. Since PBAN exhibits immunochemical and biological cross

0-8493-4591-X/94/$0.00 + $.50
© 1994 by CRC Press, Inc.

of the neuropeptide in head extracts of dark (crowded) and light (individually reared) larvae of *S. littoralis*. Analysis of PBAN-like immunoreactivity (IR) in head extracts of 6th instar crowded larvae revealed the presence of 313.3 ± 84.1 fmol/head (±SE; n=3) (Fig. 1A). The IR in the 6th larvae instar of the singly reared insects showed a 2-fold higher content (634.3 ± 134.8 fmol/head; ±SE; n=4), (Fig. 1A). The results indicated that PBAN is present in both populations, and raised the assumption that the differences in melanization result from a differential release of the neuropeptide rather than control of its biosynthesis or degradation in the neurosecretory cells.

To test this hypothesis, the content of PBAN in the hemolymph of 6th instar larvae of both populations was analyzed. Since determination of PBAN-like IR in larval or adult moths hemolymph has not been reported so far, we have worked out an extraction procedure aimed to obtain maximal recovery. The use of a variety of extraction protocols revealed that the combination of acid extraction followed by TCA precipitation and partial purification of the nonprecipitable fraction on a Sep Pak column resulted in a recovery of 78%.

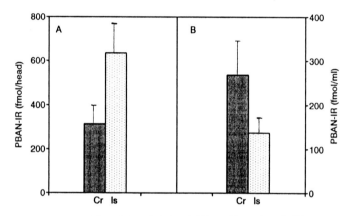

Figure 1. Comparison of PBAN-like IR in head (A) and hemolymph (B) extracts of 6th instar crowded (shaded bars) and isolated (open bars) Spodoptera littoralis larvae. PBAN-like IR was determined using YG-16 antiserum in the two step competitive ELISA as described by Gazit et al., [1992]. PBAN content in head extracts (prepared as previously described by Gazit et al., [1990; 1992]) was determined using six serial dilutions, ranging from 0.23-7.5 head equivalents. PBAN content in hemolymph extracts (prepared as described above) was performed using four serial dilutions, ranging from 100-800 μl. Each value represents the mean ± SE of 3-5 experiments.

Injection of a cocktail of protease inhibitors (bacitracin, aprotinin, N,α-p-tosyl-L-lysine chloromethyl ketone - TLCK, and phenylmethyl-sulfonyl fluoride - PMSF) to the larvae was routinely performed, 2 hours prior to hemolymph extraction, to avoid possible degradation of the neuropeptide during the extraction procedure. Following this procedure the PBAN-IR content in the hemolymph of the crowded larvae was found to be 267.8 ± 59.8 fmol/ml (±SE; n=5) PBAN. The content of the neuropeptide in isolated larvae was 1.9-fold lower (137.8 ± 28.4 fmol/ml; ±SE, n=5) (Fig. 1B).

Confirmation of the presence of PBAN in hemolymph and head extracts of *S. littoralis* larvae was performed by the pheromonotropic bioassay [Gazit et al., 1990]. The results indicated that both head and hemolymph extracts contain a bioactive peptide capable of stimulating sex pheromone biosynthesis in *H. peltigera* (Fig. 2).

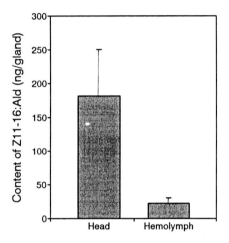

Figure 2. Pheromonotropic activity of head and hemolymph extracts of 6th instar crowded Spodoptera littoralis larvae. Activity was determined using the pheromonotropic bioassay described by Gazit et al., [1990] using 5-7 female. H. peltigera female was injected with 0.5 head equivalents or hemolymph extracts collected from eight larvae. Results are expressed as activity per larvae after correction for extractable vs. total hemolymph volume and recovery (78 and 66% for hemolymph and heads, respectively).

The higher levels of PBAN in the hemolymph of the dark larvae supports our assumption that the difference in the content of the neuropeptide in head extracts is probably due to a higher release of PBAN to the hemolymph, and raises the possibility that one of the steps involved in the complex process of color polymorphism is regulation of the release of PBAN/MRCH from the head neurosecretory cells to the circulatory system.

ACKNOWLEDGEMENT

This research was supported by the Endowment Fund for Basic Research in Life Sciences: Charles H. Revson Foundation, administered by the Israel Academy of Sciences and Humanities.

REFERENCES

Altstein, M. Gazit, Y. Dunkelblum, E., Neuroendocrine control of sex pheromone biosynthesis in Heliothis peltigera, Arch. Insect Biochem. Physiol., 22, 153, 1993.
Gazit, Y. Dunkelblum, E. Benichis, M. and Altstein, M., Effect of synthetic PBAN and derived peptides on sex pheromone biosynthesis in Heliothis peltigera (Lepidoptera :Noctuidae), Insect Biochem., 20, 853, 1990.
Gazit, Y. Dunkelblum, E. Ben-Aziz, O. and Altstein, M., Immunochemical and biological analysis of pheromone biosynthesis activating neuropeptide in Heliothis peltigera, Arch. Insect Biochem. Physiol., 19, 247, 1990.
Ikemoto, H., On the black pigment of the larval integument of the armyworm, Leucania separata. Botyu-Kagaku, 36, 128, 1971.
Kitamura, A. Nagasawa, H. Kataoka, H. Inoue, T. Matsumoto, S. Ando, T and Suzuki, A., Amino acid sequence of pheromone-biosynthesis-activating neuropeptide (PBAN) of the silkmoth Bombyx mori, Biochem. Biophys. Res. Commun., 163, 520, 1989.
Matsumoto, S. Isogai, A. Suzuki, A. Ogura, N. and Sonobe, H., Purification and properties of the melanization and reddish colouration hormone (MRCH) in the armyworm, Leucania separata (Lepidoptera), Insect Biochem., 11, 725, 1981.
Matsumoto, S. Isogai, A. and Suzuki, A., Purification and characterization of melanization and reddish coloration hormone (MRCH) in lepidopteran insects, Adv. Pigment Cell Res., 437, 1988.
Matsumoto, S. Kiramura, A. Nagasawa, H. Kataoka, H. Orikasa, C. Mitsui, T.K and Suzuki, A., Functional diversity of a neurohormone produced by the suboesophageal-ganglion: Molecular identity of melanization and reddish colouration hormone and pheromone biosynthesis activating neuropeptide, J. Insect Physiol., 36, 427, 1990.
Pener, M. P., Locust phase polymorphism and its endocrine relations, Adv. Insect Physiol., 23, 1, 1991.
Raabe M. Pigment synthesis and breakdown - color changes, in Recent Developments in Insect Neurohormones, Raabe M. Ed., Plenum, New York. pp 289-312, 1989.
Raina, A. K. Jaffe, H. Kempe , T. G. Keim, P. Blacher, R. W. Fales, H. M. Riley, C. T. Klun, J. A. Ridgway, R. L. and Hayes, D. K., Identification of a neuropeptide hormone that regulates sex pheromone production in female moths, Science, 244, 796, 1989.
Riddiford, L.M and Hiruma, K., Regulation of melanization in insect cuticle, Adv. Pigment Cell Res., 423, 1988.
Rivnay, E. and Meisner, J., The effect of rearing conditions on the immature stages and adults of Spodoptera littoralis (Boisd.), Bull. Entomol. Res., 56, 623, 1966.

SECOND MESSENGER INTERACTIONS IN RESPONSE TO PBAN STIMULATION OF PHEROMONE GLAND CULTURES

Ada Rafaeli[1] and Victoria Soroker[2]

[1]Department of Stored Products, ARO, The Volcani Center, Institute for Technology & Storage of Agricultural Products, P. O. Box 6, Bet Dagan, Israel
[2]Department of Zoology, Tel Aviv University, Ramat Aviv, Israel

PBAN, pheromone biosynthesis activating neuropeptide, has been identified as a pheromonotropic neuropeptide produced in the suboesophageal ganglion of a number of lepidopteran species (Raina, 1993). Utilizing *in vivo* studies on *Helicoverpa armigera* and *H. zea*, we have established that the pheromone gland, situated between the 8th and 9th abdominal segments of the ovipositor tip, is capable of *de novo* pheromone biosynthesis when challenged with Hez-PBAN (Soroker and Rafaeli, 1989; Rafaeli *et al.*, 1990a,b; 1991a,b; 1992; Rafaeli, 1993; Rafaeli and Soroker, 1989). Our objective was to study the transduction of the PBAN response in the target tissue to gain information on receptor interaction by PBAN in its role as a pheromonotropic neuropeptide.

Cyclic-AMP has been shown to act as a second messenger of the PBAN response based on evidence that forskolin (adenylate cyclase activator), 8-bromo-cAMP (cAMP analog) and isobutyl methyl xanthine (phosphodiesterase inhibitor) mimicked the pheromonotropic response (Rafaeli *et al.*, 1990b; Rafaeli, 1993; Jurenka *et al.*, 1991). In addition, a 28 fold increase in intracellular cAMP levels occurred as a result of pheromonotropic stimulation (Rafaeli *et al.*, 1991b; Rafaeli and Soroker, 1989). The importance of extracellular Ca^{2+} for PBAN stimulation was also demonstrated (Jurenka *et al.*, 1991). We observed a discrepancy between the doses of Hez-PBAN required to induce pheromonotropic activity (the physiological response) and those required to induce an increase in intracellular cAMP levels (the cellular response) (Rafaeli and Soroker, 1989). We therefore hypothesized dual receptor mechanisms for PBAN: response to low PBAN levels (a high affinity receptor), and a response to higher PBAN levels (a low affinity receptor).

We subsequently implicated the involvement of another transducer by demonstrating an increase in pheromonotropic activity as a result of the phorbol ester PMA (phorbol-12-myristate 13 acetate) and the diacylglycerol analog OAG (1,2 dioleolyl *sn* glycerol), both of which activate protein kinase-C (PKC). Lithium chloride, responsible for inhibiting the recycling of inositol, caused a significant drop in the normal pheromonotropic response to PBAN. In addition, the intracellular Ca^{2+} ionophore, ionomycin, caused an increase in pheromonotropic activity (Rafaeli, 1993).

To learn more about the importance of different second messenger systems and their mediating protein kinases during pheromone biosynthesis, we

0-8493-4591-X/94/$0.00 + $.50
© 1994 by CRC Press, Inc.

used several pharmacological agents to interfere with the transducing pathways. The Ca^{2+} dependency of these effects was also tested using Ca^{2+}-free physiological saline which was augmented with EGTA (1 mM). Pheromone gland cultures were prepared from virgin 2-3-day-old females during the photophase as previously reported (Rafaeli *et al.*, 1990; Rafaeli and Soroker, 1989). They were incubated in the presence of [^{14}C]-sodium acetate (NEN, Boston, USA) in the absence or presence of the test substances and analyzed on radio-TLC as reported previously (Soroker and Rafaeli, 1989). The percentage stimulation of [^{14}C] incorporation into the pheromone fraction, (Z)-11 hexadecenal, (Z11-16:Ald) was calculated from values obtained by control cultures.

At low PBAN concentrations the pheromonotropic response was dependent on Ca^{2+}. However, at higher concentrations (20 pmoles) a loss of Ca^{2+}-dependency became apparent and a significant stimulation above control values was detected (Table 1). On the other hand the PMA response was found to be Ca^{2+} dependent (Table 1). Moreover, in the presence of ionomycin and in the absence of Ca^{2+}, PMA caused a significant stimulation of pheromone biosynthesis. Thus, ionomycin's effect on free intracellular Ca^{2+} was sufficient to mediate the stimulation of PKC by PMA. In addition, the cAMP analog, Sp-cAMP (cyclic adenosine 3',5' monophosphothionate, Sp diasteromer, Biolog Life Science, CA, USA), was tested for pheromonotropic activity in the presence and absence of Ca^{2+} and it was apparent that the response to Sp-cAMP was not dependent on the extracellular Ca^{2+} levels (Table 1). On the other hand, the diasteromer Rp-cAMP, an antagonist of Sp-cAMP, did not

Table 1
Effect of Calcium on the Action of PBAN and Pharmacological Agents

Treatment[#]	Percentage Stimulation of [^{14}C] Incorporation into Z11-16:Ald	
	Calcium-Rich	**Calcium-Free**
PBAN 0.25pmoles	21.6± 13.6 (19)	46.4± 36.9 (6)
PBAN 2.00pmoles	608.0± 82.5 (14)[*]	70.4± 30.8 (7)
PBAN 5.00pmoles	597.0±137.0 (19)[*]	14.7± 13.1 (5)
PBAN 20.0pmoles	758.0±136.2 (14)[*]	238.6± 74.9 (8)[*]
PMA 0.01mM	433.7± 97.7 (23)[*]	53.5± 20.4 (14)
Ionomycin 0.001mM	306.3± 136.1 (7)[*]	79.5± 28.1 (14)
PMA+Ionomycin	not tested	183.6± 52.9 (14)[*]
Sp-cAMP 1mM	209.0± 44.4 (5)[*]	399.6±140.9 (8)[*]

[#]Test substances added in 10µl incubation medium containing 0.5µCi [^{14}C]-sodium acetate per pheromone gland

Data represented as means ± SEM, numbers in brackets indicate the number of replicates

[*] Denotes a significant stimulation above control values.

inhibit the response obtained both at PBAN concentrations of 1 pmole and 20 pmoles (data not shown).

Since Ca^{2+} was implicated as a mediator of the pheromonotropic responses we tested, the involvement of a Ca^{2+}-calmodulin dependent kinase, Calmidazolium, an inhibitor of Ca^{2+} calmodulin, slightly increased the pheromonotropic activity of gland cultures. In the presence of 1 pmole PBAN an enhanced stimulation of the pheromonotropic response was observed (Table 2). On the other hand, in the presence of 20 pmoles PBAN, no significant difference in the normal response to PBAN was observed (Table 2).

Table 2
Effect of Calmidazolium on the Pheromonotropic Action of PBAN

[PBAN] (pmoles)	Percentage Stimulation of $[^{14}C]$ Incorporation into Z11-16:Ald	
	PBAN	PBAN+Calmidazolium
0	0	188.5± 46.6(21)
1	496.0± 61.9(15)	979.6±145.4(16)*
20	1026.7±218.6(7)	1079.6±115.7(7)

Data represented as means ± SEM, numbers in brackets indicate the number of replicates
* Denotes a significant stimulation above control values.

Since no inhibitory effect was found in the presence of calmidazolium, Ca^{2+}-calmodulin is probably not involved in pheromonotropic stimulation by PBAN. However, it is evident that at low PBAN concentrations an enhancement of the normal pheromonotropic response was obtained. This may be due to a release of bound Ca^{2+} as a result of calmidazolium inhibition of Ca^{2+}-calmodulin. Thus, any internal changes to intracellular Ca^{2+} levels will indirectly affect the response to PBAN. Since this effect is only apparent at low concentrations of PBAN, this may mean that the high affinity receptor is most sensitive to Ca^{2+} levels.

A study of the minimum essential time required to expose pheromone glands to PBAN revealed that within 0.25 minutes a significant increase in pheromone production was already evident and maximum stimulation occurred after 0.5 minute exposure to PBAN. Nevertheless, no increase in intracellular cAMP was detected after 0.5 minutes exposure time thereby indicating that another pathway is responsible for initial stimulation (data not shown). Our findings therefore suggest a more modulating role for protein kinase A where the dominant function relies on the PKC pathway which results in an increase in intracellular Ca^{2+} whilst cAMP acts as a modulator. Indeed, in *B. mori* it was also found that an initial stimulation was required for triggering pheromone production before IBMX or forskolin could enhance pheromone production (Fonagy *et al.*, 1992). This bifurcating signal pathway provides the versatility necessary to introduce subtle variations in a control mechanism which consists of a complex array of biochemical events.

Acknowledgments: This research as supported by a Binational Agriculture Research Development Grant (BARD Research Project No. IS 1645-89).

REFERENCES

Raina, A. K. Neuroendocrine control of sex pheromone biosynthesis in Lepidoptera. *Ann. Rev. Entomol.* 38, 329-349, 1993.

Soroker, V. and Rafaeli, A. *In vitro* hormonal stimulation of acetate incorporation by *Heliothis armigera* pheromone glands. *Insect Biochem.* 67, 1-5, 1989.

Rafaeli, A., Soroker, V., Raina, A. ., and Klun, J. A. Stimulation of *de novo* biosynthesis of (Z)-11 hexadecenal by *in vitro* pheromone glands of *Heliothis spp.*, in *Insect Neurochemistry and Neurophysiology*, Borkovec, A. B. and Masler, E. P., Eds. Humana Press, Clifton, New York, 1990a, p. 309-312.

Rafaeli, A. Soroker, V., Kamensky, B., and Raina, A. K. Action of pheromone biosynthesis activating neuropeptide on *in vitro* pheromone glands of *Heliothis armigera* females. *J. Insect Physiol.* 36, 641-646, 1990b.

Rafaeli, A., Hirsch, J., Soroker, V., Kamensky, B., and Raina, A. K. Spatial and temporal distribution of PBAN in *Helicoverpa (Heliothis) armigera* using RIA and *in vitro* bioassay. *Arch. Insect Biochem. Physiol.* 18, 119-129, 1991a.

Rafaeli, A., Kamensky, B., and Soroker, V. The use of cryopreservation for convenient bioassay of insect organs *in vitro*. *Physiol. Entomol.* 16, 457-461, 1991b.

Rafaeli, A., Soroker, V., Hirsch, J., Kamensky, B., and Raina, A. K. Influence of photoperiod and age on the competence of pheromone glands and on the distribution of immunoreactive PBAN in *Helicoverpa spp. Arch. Insect Biochem. Physiol.* 22, 169-180, 1992.

Rafaeli, A. Pheromonotropic stimulation of moth pheromone gland cultures *in vitro*. *Arch. Insect Biochem. Physiol.* 1993 (in press).

Rafaeli, A. and Soroker, V. Cyclic-AMP mediation of the hormonal stimulation of ^{14}C-acetate incorporation by *Heliothis armigera* pheromone glands *in vitro*. *Mol. Cell. Endocrinol.* 65, 43-48, 1989.

Jurenka, R. A., Jacquin, E., and Roelofs, W. L. Stimulation of pheromone biosynthesis in the moth *Helicoverpa zea*: Action of a brain hormone on pheromone glands involves Ca^{2+} and cAMP as second messengers. *Proc. Natl. Acad. Sci. USA* 88, 8621-8625, 1991.

Fonagy, A., Matsumoto, S., Uchiumi, K., and Mitsui, T. Role of calcium ion and cyclic nucleotides in pheromone production in *Bombyx mori. J. Pest. Sci.* 17, 115-121, 1992.

Proctolin

BIOLOGICAL ACTIVITIES OF PROCTOLIN
AND ITS ANALOGUES

Gregorz Rosiński[a], Andrzej Plech[b], Angela B. Lange[c], Ian Orchard[c], Hubert Bartosz-Bechowski[d], Wiesław Sobótka[e] and Danuta Konopińska[d].

[a] Dept. Animal Physiol., A. Mickiewics Univ., Poznań, Poland; [b] Dept. Pharmacol., Medical University, Zabrze, Poland; [c] Dept. Zoology, Univ. Toronto, Toronto, Canada; [d] Institute of Chemistry, Univ. Wrocław, Wrocław, Poland; [e] Institute Industrial Organic Chemistry, Warsaw, Poland

INTRODUCTION

To explain the role of the amino acid residues in position 1 to 4 of the proctolin (RYLPT) chain, the myotropic activity of series (A-E) of 42 of its analogues were obtained (Konopinska *et al.* 1992): X-Tyr-Leu-Pro-Thr (A), where X = Lys(1), His(2), homo-Arg(3), Can(4), Cit(5), τ-Abu (6), Phe(p-NH$_2$) (7), Phe(p-guanidino) (8), Gac (9), Gav (10), Orn (11), Gap (12); Arg-X-Leu-Pro-Thr (B) where X = Phe(p-NH$_2$) (13), Phe(pNMe$_2$) (14), Phe(p-NO$_2$) (15), Phe(p-OMe) (16), Cha(4'-OMe) (17), Phe(p-OEt) (18), Phe(p-Guanidino) (19), L-Dopa (20), Tyr(3'-NH$_2$) (21), Tyr(3'-NO$_2$) (22), Afb(p-OH) (23), Afb(p-NH$_2$) (24), Afb(p-NO$_2$) (25) and Tyr(OSO$_3$H)(26); Arg-Tyr-X-Pro-Thr (C) where X = Val (27), Gly (28), Thr (29), Pro (30), Acp (31) and Ach (32); Arg-Tyr-Leu-X-Thr (D) where X = Hyp (33), Hyp(4-OMe) (34), Thz (35), homo-Pro (36), Sar (37), Ach (38); and X-Leu-Pro-Thr (E), where X = Tyr(3'-NH$_2$) (39), Phe(p-NH$_2$) (40), Phe(p-NO$_2$) (41) and Phe(3'-CO$_2$-4'-OMe) (42).

Bioassays were carried out on the semi-isolated hearts of *Peiplaneta americana* and *Tenebrio molitor* using a cardioacceleratory test. The ability of selected proctolin analogues (1,3,14,20,39) to mimic the basal contraction induced by proctolin on locust (*Locusta migratoria*) oviduct was also examined (Orchard *et al.* 1989; Lange *et al.* 1993). Moreover, the influence of some of these analogues (1,2,5,13,15-17,20) on rat behaviour *in vivo* was investigated.

RESULTS AND DISCUSSION

Among the 42 analogues studied for myotropic activity, 12 peptides were found to increase the heartbeat frequency of *P. americana* and *T. molitor* with effects similar to or greater than those of proctolin (Table 1). Most of

0-8493-4591-X/94/$.00 + $.50
© 1994 by CRC Press, Inc.

229

TABLE 1

Myotropic activity of proctolin analogues on insect heart or oviduct bioassays relative to proctolin (%).

Peptide	Contraction activity (*in vitro*)		
	Heart		Oviduct
	P. americana[a]	*T. molitor*[b]	*L. migratoria*[b]
Proctolin	100	100	100
1	110	50	100
2	50	100	91
3	100	80	-
13	300	183	69
14	100	320	71
15	370	170	-
16	390	155	-
18	100	0	-
19	20	0	-
20	500	50	115
21	110	100	-
22	120	0	-
26	100	100	-
27	25	55	-
28	10	30	-
33	50	40	-
36	25	20	-
39	inhib (10%)	inhib (10%)	0

[a] at concentration 10^{-9} M
[b] at concentration 10^{-8} M
- not tested
inhib, inhibitory

the active compounds were analogues modified in position 2 of the peptide chain. In the case of the tetrapeptide 39, a decrease in heartbeat frequency was observed.

Proctolin derivatives modified in position 1 (series A) retained their full bioactivity when the N-terminal amino acid residue included a guanidine group, as in [homo Arg]-proctolin (3), or [Can]-proctolin (4). Alterations in position 3 or 4 (series C and D) resulted in compounds (27, 28, 33, 36) with low cardioacceleratory properties. Agonistic effects on locust oviduct of five pentapeptides modified in positions 1 or 2 (series A and B) (1,3,13,14,20) were also observed (Lange *et al.* 1993).

From the analysis of the myotropic effects of proctolin analogues on these various preparations (Table 1) the following conclusions were drawn: 1) their cardiotropic effects depend upon the presence and type of substituted "para" position 2 of the proctolin chain; 2) additional substitutions in position 3' of the Tyr phenyl ring facilitates the interaction with the receptor site in *P. americana*; 3) antagonistic activity of the tetrapeptide, 39, with respect to proctolin may be inferred from its ability to decrease heartbeat frequency; 4) the presence of an N-terminal Arg moiety as well as Leu[3] and Pro[4] residues in the pentapeptide chain are of critical importance for myotropic activity; 5) the myotropic effects of selected analogues on the locust oviduct bioassay suggests that there are sub-types of proctolin receptors in insects, and further emphasises that these analogues may be useful tools for examining proctolin receptors. Other biological investigations of proctolin and some of its analogues (1,2,5,13,15-17,20) concerned their influence on rat behaviour (*in vivo*). An evaluation of cardiovascular and antinociceptive effects after icv injections revealed long-term analgesic properties of proctolin and three analogues (13,16,17) which were prevented by naloxone. Two of these pentapeptides, 13 and 17, increased atrial blood pressure. Once again these effects were blocked by naloxone. From these results we may infer that the behavioural effects in rats, induced by proctolin and selected analogues, may be associated with opiate receptors.

CONCLUSION

The results presented here indicate that the insect neuropeptide proctolin and some of its analogues, besides having biological activity in arthropods, showed significant biological effects in mammals. Proctolin analogues modified in position 2 revealed neuromodulatory action both in vertebrates and invertebrates.

ACKNOWLEDGMENT

Financial support from KBN, grant 4 0688 01 (for G.R., A.P., W.S., H.B.-B. and D.K.) and NSERC (for I.O. and A.B.L.) is gratefully acknowledged.

REFERENCES

Konopinska, D., Rosinski, G. and Sobotka, W., Insect peptide hormones, an overview on the present literature, *Int. J. Peptide Protein Res.* 39, 1-11, 1992.

Lange, A.B., Orchard, I. and Konopinska,D. The effects of selected proctolin analogues on contractions of locust (*Locusta migratoria*) oviducts. *J. Insect Physiol.* 39, 347-351, 1993.

Orchard, I., Belanger, J.H. and Lange, A.B., Proctolin: A review with emphasis on insects. *J. Neurobiol.* 20, 470-496, 1989.

A COMPARISON OF THE POTENCY OF PROCTOLIN ANALOGUES AND THEIR RELATIVE AFFINITY FOR THE PROCTOLIN BINDING SITES ON LOCUST OVIDUCT

V.M. Sevala, L. King, J. Puiroux, A. Pedelaborde and B.G. Loughton
Department of Biology, York University, North York
Ontario, Canada M3J 1P3.

Proctolin is an unblocked pentapeptide which has been shown to function as a neuromodulator of muscle activity in a variety of insects and other arthropods (see Orchard et al., 1989). Though there have been reports of proctolin in other phyla all of these claims have yet to be substantiated. In our own crude experiments we have failed to alter the contractile activity of a variety of muscle types in earthworms or in snails. The restricted distribution of proctolin has made its receptor an attractive target for insecticides. It was with this in mind that we set out to characterize the proctolin receptor of *Locusta*.

We have already succeeded in demonstrating proctolin binding to crude membrane preparations of hindgut and oviduct. In hindgut a single binding site was detected. In oviduct, which had a greater density of receptors, two binding sites were apparent (Puiroux et al. 1992a & 1992b).

We have attempted to characterize the proctolin receptors in two ways. First, by determining the relative affinity of the receptor for proctolin and its analogues and secondly, by comparing the relative binding affinity of an analogue with its biological activity. In this way we hope to be able to recognize those residues important for binding and those important for biological activity.

Several laboratories have investigated the effects of modifications of the amino acid residues at each position. These studies have shown that modification at every position alters biological activity. However, few modifications of the tyrosine residue at position two have been shown to enhance proctolin activity. Starratt and Brown (1979) showed that the addition of a methoxy group at the para position enhanced activity on the cockroach hindgut and Konopinska and collaborators have shown that other modifications of the tyrosine at this position enhance the analogue's biological activity on *Periplaneta* and *Tenebrio* heart (Konopinska et al., 1986). However, when several of these analogues were applied to the locust oviduct no enhancement of activity was detected (Lange et al. 1993).

An easily detected radio-ligand is an essential element in recognising a receptor and so we set out to determine the biological activity and

0-8493-4591-X/94/$0.00 + $.50
© 1994 by CRC Press, Inc.

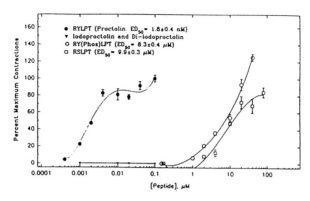

<u>Figure 1</u>   The effect of proctolin analogues on hindgut
contractions. The degree of contraction is expressed
relative to that produced by 0.1μM proctolin.

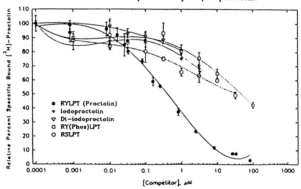

<u>Figure 2</u>   Competitive displacement of [³H]- proctolin bound to
hindgut membranes by proctolin analogues.

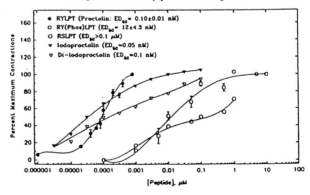

<u>Figure 3</u>   The effect of proctolin analogues on oviduct contractions.
The degree of contraction is expressed relative to that
produced by 0.8nM proctolin.

relative binding affinity of proctolin analogues iodinated or phosphorylated at position two. For comparison we investigated the methoxy-tyrosine analogue of Starratt & Brown and a proctolin analogue in which serine was submitted for tyrosine.

Figure one shows the relative potency of the analogues to cause contractions of the locust hindgut. It is evident that the analogues are several orders of magnitude less potent the proctolin. This lack of potency is matched by the ability of these analogues to compete with proctolin for hindgut binding sites (Figure 2).

When the same analogues were tested against the locust oviduct it became evident that the iodoproctolins possessed significant stimulating activity (Fig. 3). By contrast, the serine substituted analogue and the phosphotyrosine analogue were much less potent. The biological activity of the iodotyrosine analogues was not matched by their ability to compete with proctolin for oviduct binding sites. In fact, their competitive ability was similar to that of the phsphotyrosine and serine analogues (Fig. 4).

In an attempt to resolve this discrepancy we examined the degradation of proctolin and the iodoproctolins by hindgut and oviduct membrane preparations since Konopinska et al. (1986) had suggested that the potency of the analogue they tested might be due to increased resistance to hydrolytic enzymes. Figure 5 shows proctolin and iodoproctolin degradation by a series of hindgut and oviduct membrane preparations. No significant difference between the degradation of iodoproctolin and proctolin could be detected. In fact if anything iodoproctolin was degraded more quickly than proctolin itself. Thus, it is unlikely that the potency of the iodoproctolin analogue is due to its resistance to degradation at the target membrane.

These studies further demonstrate the differences between hindgut and oviduct proctolin receptors. In our earlier work we demonstrated a single binding site on the hindgut and two apparent binding sites on the oviduct. However additional data on the potentiation of binding by enkephalins suggested there might only be a single binding site (Puiroux et al. 1992b). In addition only a single molecular type was detected when receptors were solubilized from hindgut and oviduct membranes (Puiroux et al. unpublished).

The differences in response to stimulation by hindgut and oviduct argue for differences in receptor types. It is possible that a second receptor type will become evident in membrane extracts when analyses are performed using sufficient starting material.

Thus we believe that not only are there species differences in proctolin receptors as has been shown by Konopinska and coworkers (1986) and Lange et al. (1993) but that there are at least two proctolin receptor subtypes in the locust.

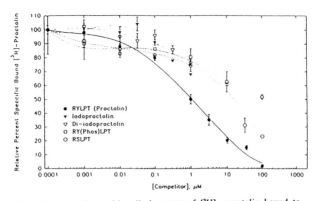

Figure 4   Competitive displacement of [³H]- proctolin bound to oviduct membranes by proctolin analogues.

Figure 5   Degradation of proctolin and iodoproctolin by locust oviduct and hindgut membranes in vitro. Proctolin (1mM) was incubated with 100μg membrane. Aliquots of the medium were assayed by RPHPLC for proctolin. The figure shows the results of several experiments. Open symbols represent proctolin, closed symbols represent iodoproctolin.

REFERENCES

Orchard I., Belanger J.H., and Lange A.B., Proctolin: a review with emphasis on insects. J. Neurobiol. 20, 470, 1989.

Konopinska D., Sobotka W., Lesicki A., Rosinski G., and Sujak P., Synthesis of proctolin analogs and their cardioexcitatory effect on cockroach, *Periplaneta americana L.*, and yellow mealworm, *Tenebrio molitor L.* Int. J. Peptide Protein Res. 27, 597, 1986.

Puiroux J., Pedelaborde A. and Loughton B.G., Characterization of a proctolin binding site on locust hindgut membranes. Insect Biochem. Molec. Biol. 22, 547, 1992(a).

Puiroux J., Pedelaborde A., and Loughton B.G., Characterization of proctolin binding sites on locust oviduct membranes. Insect Biochem. Molec. Biol. 22, 859, 1992.

Starrat A.N. and Brown B.E., Analogs of the pentapeptide proctolin - synthesis and structure-activity studies. Biochem. Biophys. Res. Commun. 90, 1125, 1979.

Lange A.B., Orchard I., and Konopinska D., The effects of selected proctolin analogues on contractions of locust (*Locusta migratoria*) oviducts. J. Insect Physiol. 39, 347, 1993.

PROCTOLIN-ANTAGONISTIC EFFECTS OF TOLYPIN (EFRA-PEPTIN F) ON THE AUTONOMIC NERVOUS SYSTEM OF INSECTS

Karel Sláma, and Jaroslav Weiser

Institute of Entomology, Academy of Sciences, Branišovská 31, 37005 České Budějovice, Czech Republic

Autonomic physiological functions in insects are regulated by special parasympathetic-like, cholinergic nervous system (coelopulse), with the centers in the thoracic ganglia. The functions can be exactly measured by monitoring special extracardiac pulsations in hemocoelic pressure. The system controls homeostatic equilibrium by affecting circulation of hemolymph, exchange of gases through selected spiracles, or affecting osmolarity and water balance (for more details and references see Sláma 1988, Sláma and Coquillaud 1992).

In the foregoing paper (Sláma et al. 1993) we have reported that administration of proctolin in 100nM to 5μM concentrations increases the amplitude of hemocoelic pressure pulsations in *Tenebrio* pupae up to six-fold within a few minutes. With dosages corresponding to 1μM concentrations, the effects of proctolin culminated between the 2nd and 3rd hour post-treatment, with a successive decline to the original level during the next 15-20hrs. It has been concluded that proctolin may be a valid neurohormone that is involved in homeostatic physiological regulations.

Tolypin is a naturally occurring peptide with insecticidal properties. It has been found as a mosquitocidal principle in *Tolypocladium* (Weiser and Maťha 1988a). Chemically it belongs to the group of antibiotics with a high content of α-aminoisobutyric acid, the so called peptaibiotics. Its structure may be identical with the previously isolated efrapeptins (cf. Maťha et al. 1992). In this study, tolypin has been subjected to general toxicological studies in pupae of the mealworm, *Tenebrio molitor*, with special attention to its possible effects on the autonomic nervous system, the coelopulse.

The methods of recording the extracardiac pulsations in hemocoelic pressure were described elsewhere (Sláma and Miller 1987). Toxicological studies with tolypin revealed LD-50 of 0.34 μg/spec., as it is shown in Figure 1. Dosages around 1 μg/spec. caused paralysis of neuromuscular

0-8493-4591-X/94/$0.00 + $.50
© 1994 by CRC Press, Inc.

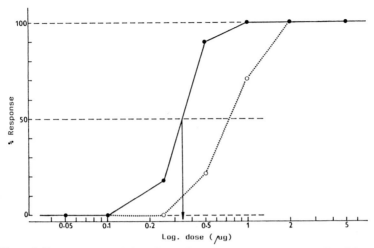

Figure 1. Dose-response relationships for Tolypin in 4-day-old pupae of *Tenebrio molitor*. Solid line indicates mortality 24 hr after injection, broken line shows immobilization and paralysis of irritational responses at 3hr (dotted line). The arrow indicates LD-50 (0.34 μg, 25°C, n - 10).

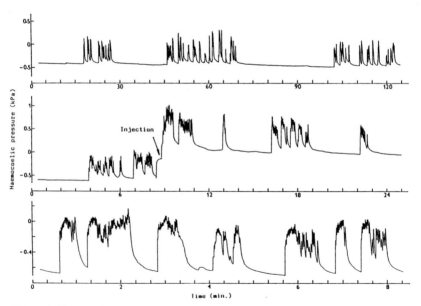

Figure 2. Hemocoelic pulsations in a normal, 4-day-old pupa of *Tenebrio*. Regular series of the pulsations in a control pupa are on the top. Middle portion exemplifies the effect of 2 μl control injection of insect Ringer. Lower portion provides a more detailed record of one series of the pulsations (from the top) in enlarged scale.

activity within 3 hr after the injections. Larger doses were lethal to all individuals. Extracardiac pulsations in hemocoelic pressure were recorded in 4-day-old pupae. Examples of these pulsations in the control pupae can be seen in Figure 2, which also shows the effect of Ringer injection.

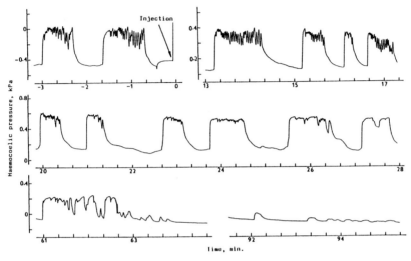

Figure 3. Typical example of the reduced pulsation amplitude after injection of 1 µg of Tolypin into 4-day-old pupa of *Tenebrio*. Total paralysis of the coelopulse system has occurred approximately at 100 min. after the injection.

Injections of 1 µg of tolypin caused opposite effects to proctolin, e.g. rapid diminution of the pulsation amplitude, which can be clearly seen in Figure 3. Large concentrations (more than 2 µg/spec.) of tolypin prevented all hemocoelic pulsations within a few min., causing tetanic-like contractions with temporary peaks of high hemocoelic pressure. In all these studies we have found a clearcut physiological antagonism between tolypin and proctolin, but a real nature of this effect remains to be elucidated.

After complete paralysis of somatic musculature by tolypin, the pupae still exhibited very fine pulsations in hemocoelic pressure (see Figure 4). They were identified as forward oriented beatings of the dorsal vessel. Finally, we have also investigated the effects of another peptide, cyclosporin A, which is less toxic than tolypin (Weiser and Maťha 1988b). In contrast to tolypin, this compound induced very frequent, long-lasting pulsations (see Figure 5), which usually indicate suffocation.

These results suggest that functions of the coelopulse system may serve as a good criterion for the *in vivo* effectiveness of a wide range of the biologically active peptides or neuropeptides.

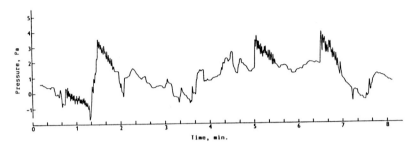

Figure 4. A fraction of record taken with enlarged sensitivity and expanded scale in a completely paralysed, 4-day-old pupa of *Tenebrio*, 12hr after injection of 0.5 μg of Tolypin. The series of pulsations with extremely small amplitude (smaller than 1 Pa) reveal the periods of the forward oriented beating of the dorsal vessel.

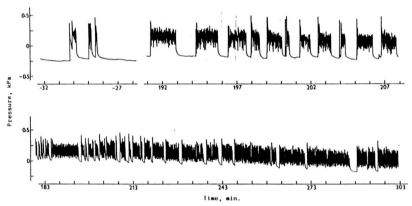

Figure 5. Hemocoelic pulsations recorded 30 min. before and 3 - 4 hr after the injection of 10 μg of Cyclosporin A in the 4-day-old pupa of *Tenebrio* (upper part). Lower portion shows almost uninterrupted pulsations recorded in another pupa from the same experimental series.

REFERENCES

Maťha, V., Jegorov, A.,Kiess, M., and Brückner, H., Morphological alterations accompanying the effects of peptaibiotics, α-aminoisobutyric acid-rich secondary metabolites of filamentous fungi, on *Culex pipiens* larvae, *Tissue and Cell,* 24, 559-564,1992.

Sláma, K.,A new look at insect respiration, *Biol.Bull.*,175, 289-300,1988.

Sláma, K. and Coquillaud, M.-S., Homeostatic control of respiratory metabolism in beetles, *J. Insect Physiol.*, 38, 783-791, 1992.

Sláma, K. and Miller, T.A., Insecticide poisoning:disruption of a possible autonomic function in pupae of *Tenebrio molitor. Pestic. Biochem. Physiol.* 29, 25-34, 1987.

Sláma, K., Konopińska, D. and Sobótka, W., Effects of proctolin on autonomic physiological functions in insects. *Eur. J. Entomol.*,90, 23-35, 1993.

Weiser, J. and Maťha, V., Tolypin, a new insecticidal metabolite of fungi of the genus *Tolypocladium. J. Invertebr. Pathol.*, 51, 949-946, 1988a.

Weiser, J. and Maťha, V., The insecticidal activity of cyclosporines on mosquito larvae, *J. Invertebr. Pathol.*, 51, 92-93, 1988b.

Peptides Affecting Development

BRAIN FACTORS CONTROL THE ACTIVITY OF PROTHORACIC GLAND NERVES IN THE COCKROACH *PERIPLANETA AMERICANA*

Klaus Richter

Sächsische Akademie der Wissenschaften zu Leipzig,
Forschungsgruppe Jena; 07703 Jena, Postfach100322, Germany

Anatomical evidences of prothoracic gland innervation in Blattodea, Coleoptera, Diptera and Lepidoptera [3,6] as well as some physiological indications suggest that the control of the moulting glands, at least in these insects, is a considerable complex process including endocrine, neuroendocrine and neuronal regulation.

The prothoracic gland of *Periplaneta americana* is innervated by neurons of the prothoracic ganglion [2]. In the prothoracic gland nerve two types of efferent action potentials are to be differentiated in the pattern of discharges [4,8]. The electrophysiological features of these two types correspond to motor (amplitude 225 µV, duration 1.2 ms) and neurosecretory axons (amplitude 30 µV, duration 2.3 ms).

The first peak of ecdysteroids produced by the prothoracic gland between ld (larval day) 18 and ld 23 of the last larval instar is accompanied by an important increase of the small potential type in the gland nerve. The large increase in ecdysteroid titre at the end of the moulting period, on the other hand, is in no correlation to the small type of action potentials. The large type of spike potentials does not show significant correlations to the secretory activity of the prothoracic gland [8]. The spike activities of the neurons of the prothoracic gland nerve in *Periplaneta* are inhibited by a GABAergic mechanism, connected to the subesophageal ganglion [4,6-8]. GABA causes inhibition of ecdysone secretion in prothoracic glands as well as inhibition of discharges of small action potentials in the gland nerve. During the first peak of ecdysteroid production this inhibition seems to be omitted, resulting in an increase of the small potentials.

Except for the fact that the function of the first ecdysone peak in cockroaches is unknown as yet, the gland is obviously dependent on nervous connections to produce the first ecdysteroid peak, as could be shown with transection of the prothoracic gland nerve *in vivo* [5].

In *Periplaneta*, the small axons of the prothoracic gland nerve contain neurosecretory peptide granules [1], whose content could not be identified as yet. It is the aim of this investigation to look for further factors, especially peptide factors, of the central nervous system which are implicated in nervous regulation of the prothoracic gland.

0-8493-4591-X/94/$0.00 + $.50
© 1994 by CRC Press, Inc.

243

The mean length of the moulting interval under these conditions was 30.5 ± 0.6 days (n=366). The efferent activity of the prothoracic gland nerve was obtained extracellularily [4]. The homogenates or fractions were applied to the perfusion solution (*in situ*) or to bath volume (*in vitro*). Crude brain and corpora cardiaca extracts were prepared as described previously [10].

Results

1. Effect of brain homogenates *in situ*

There are two characteristic small and transient peaks in spike activity of the prothoracic gland nerve during the last larval instar [8]. The first peak appears between ld 12 and ld 15 and the second one between ld 20 and ld 22. Both are characterized by a large increase of discharges of the small type of action potentials.

As a first experimental series, effects of brain homogenates on the prothoracic gland nerve of larvae of the ld 9 or ld 10 *in situ* were measured. The effect of homogenates consisted of increases of efferent spike activity in the prothoracic gland nerve. The strongest effect was observed with brain homogenates from larvae between ld 13 and ld 15 and of the ld 20. Prior to ld 13 or later to ld 20 no substantial effect of brain homogenates was observed. 15 min after application of a homogenate corresponding to 3 brain equivalents, the increase of spike activity was about 100%.

Application of homogenates of corpora cardiaca also resulted in an increase of prothoracic gland nerve activity. The main effect was induced by corpora cardiaca from ld 15 and ld 16. The increase induced by a homogenate corresponding to 3 pair corpora cardiaca equivalents was

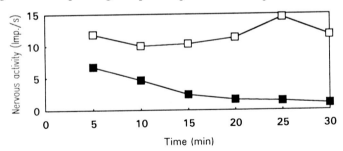

Figure 1.   Effect of brain homogenate (3 brain equivalents) from 14th/15th days of the last larval instar on the nervous activity of the prothoracic gland nerve *in vitro*. Preparation consisting of the prothoracic ganglion with the gland connected only. Black squares: untreated control, open squares: administration of brain homogenate, 5 min before recording .

comparable to that of active brain homogenates. Homogenates of corpora cardiaca prepared from larvae of other days of the moulting interval, did not show any neurotropic effect.

2. Effect of brain homogenates *in vitro*

An isolated preparation (last instar larvae of ld 9 or 10), consisting of the prothoracic ganglion together with the prothoracic gland, both connected by the prothoracic gland nerve was used under *in vitro* conditions. Administration of a homogenate of 3 brain equivalents from larvae of ld 14/15 of the moulting interval resulted in a duplication of spike activity within two minutes. This increase remained rather constant during the following thirty minutes (fig. 1).

In preparations in which the functional connection between the prothoracic ganglion and the subesophageal ganglion was intact, the spike activity in untreated controls remained on a low level, as a result of the inhibitory effect of the subesophageal ganglion [5]. Administration of 3 brain equivalents to this kind of preparation resulted in an increase of spike activity only during the first 15 min of the experimental interval. Later on the effect was suppressed by about 50%. This inhibiting effect of the subesophageal ganglion could be shown by comparison of the spike activity after administration of homogenate to preparations with and without connection to the subesophageal ganglion, respectively. The increase of spike activity in preparations without subesophageal ganglion differed from those with this ganglion by a factor of 103 (fig. 2).

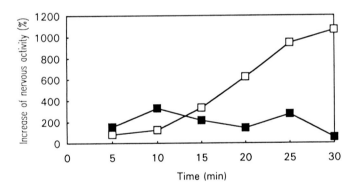

Figure 2. Increase of nervous activity in the prothoracic gland nerve of a prothoracic ganglion separated (open squares) and in functional connection to the subesophageal ganglion (black squares) in comparison to untreated controls (%) (*in vitro*). Administration of a homogenate corresponding to 3 brain equivalents, 5 min before recording.

The increase of spike activity after administration of brain homogenates was mainly due to an increase of the small type of action potentials. Course and amount of the entire nervous activity are identical with course and amount of the small type of spikes in the prothoracic gland nerve during the experimental interval.

An extract containing the equivalent of 0.75 brain was sufficient to stimulate spike activity by 200%, 20 min following administration. The effect reached a saturation with 3 brain equivalents (about 1000%, 20 min after administration). Higher concentrations did not lead to a further increase of nervous activity.

3. Preliminary experiments to characterize the brain factor

The neurotropic activity of brain homogenates was lost after extraction with ethanol as well as by treatment with the enzyme pronase. Pronase inactivated by boiling (3 min at 100°C) did not destroy the neurotropic activity of the brain homogenate. Extraction of brain homogenates with methanol resulted also in a loss of neurotropic activity. On the other hand, heating for 2 min to 100°C was without influence on the activity of the homogenate.

In a preliminary experiment, a homogenate of 100 brains including corpora cardiaca was separated by gel filtration on Sephadex G 100 (Columns 1.3 x 75cm, eluted with 0.2 M acidic acid; 9 ml/h [10]). Of all fractions neurotropic activity was detected in two peaks, one near the upper exclusion limit and the other before the lower exclusion limit. When tested on the prothoracic gland nerve activity both fractions showed about the same effectiveness.

Our experiments lead to the conclusion that the prothoracic gland of *Periplaneta* is inhibited by neuronal pathways from the subesophageal ganglion [7] and disinhibited by brain factors effective on neurosecretory neurons of the prothoracic gland nerve.

References

1. **Birkenbeil, H.**, Ultrastrukturelle und immuncytochemische Untersuchungen der Ecdyson bildung in Häutungsdrnüsen von Crustaceen und Insekten. Thesis, Friedrich-Schiller Universität Jena, 1991.
2. **Birkenbeil, H., and Agricola, H.** , Die Innervierung der Prothorakaldrünse von *Periplaneta americana* L. *Zool. Anz.* 204, 331-336, 1980.
3. **Herman, W.S.**, The ecdysial glands of arthropods. *Int. Rev. Cytol.* 22, 269-347, 1967.
4. **Richter, K.**, Experimentelle Untersuchungen der nervösen Steuerung der Prothorakaldrünse bei *Periplaneta americana*. *Zool. Jb. Physiol.* 87, 65-78, 1983.
5. **Richter, K.**, Physiological investigations on the role of the innervation in the regulation of the prothoracic gland in *Periplaneta americana*. *Arch. Insect Biochem. Physiol.* 2, 319-329, 1985.
6. **Richter, K.**, Zur Frage der Innervierung der Häutungsdrnsen bei Arthropoden. *Zool. Jb. Physiol.* 90, 43-63, 1986.
7. **Richter, K.**, The involvement of GABA in the regulation of the nerve activity on the prothoracic gland in *Periplaneta americana*. *Zool. Jb. Physiol.* 93, 375-387, 1989.
8. **Richter, K.** , Prothoracicotropic activity in the brain of the cockroach *Periplaneta americana*. *J. Insect Physiol.* 38, 349-355, 1992.
9. **Richter, K.**, Further physiological evidence for nervous regulation of the prothoracic gland in the cockroach *Periplaneta americana*. *Zool. Jb. Physiol.* 97, 31-46, 1993 .
10. **Richter,K.**, Brain factors control the activity of prothoracic gland nerves in the cockroach *Periplaneta americana*. *J. Insect Physiol.* 1993 (in press).

ACTIVATION OF THE PROTHORACICOTROPIC HORMONE-LIKE ACTIVITY IN PRE-HATCH EGGS OF THE GYPSY MOTH, *LYMANTRIA DISPAR*

Thomas J. Kelly, Robert A. Bell, Belgaum S. Thyagaraja*,
and Edward P. Masler

Insect Neurobiology and Hormone Laboratory, USDA, ARS, Beltsville, MD
20705, and *Department of Entomology,
University of Maryland, College Park, MD 20742, USA.

Except for *Bombyx mori*, little is known about the endocrine regulation of embryonic development in lepidopterous insects. The role of neurohormones, such as the prothoracicotropic hormone (PTTH), is even less well understood. Only recently has PTTH activity been demonstrated in lepidopterous eggs and embryos (Chen *et al.* 1986, 1987; Dorn *et al.* 1987; Fugo *et al.* 1987; Masler *et al.* 1991). Following egg hatch, the primary role of PTTH is to stimulate the developmental processes associated with insect molting. Failure of PTTH synthesis or secretion is often associated with diapause (Denlinger 1985). In the univoltine gypsy moth, *Lymantria dispar*, development is interrupted within the egg by an intensive and obligatory diapause at a mature, pharate larval stage (Leonard, 1968; Bell, 1989).

We have initiated an analysis of PTTH in the embryonated egg of *L. dispar* and its possible role in the regulation of diapause, since the neuroendocrine system is fully-formed in these mature, pharate larvae and may function in a manner similar to that following egg hatch. Results to date establish the existence of PTTH-like activity in embryonated eggs of *L. dispar* (Bell *et al.* 1990; Masler *et al.* 1991, and references therein). This activity has been partially characterized as stable to heat treatment, organic solvent extraction, and labile to protease treatment (Masler *et al.* 1991). In eggs with developing embryos, PTTH activity appears to peak when embryonic development is complete, around day 13 to 15, in conjunction with a peak in whole egg ecdysteroids, primarily 20-hydroxyecdysone (Bell *et al.* 1990). Subsequently, the pharate larva enters diapause around day 20 (Bell 1989; Bell *et al.* 1990), where it remains for many months. PTTH activity and whole egg ecdysteroid titers fall in conjunction with entry into diapause (Bell 1989). In the post-diapause egg, PTTH-like activity has been demonstrated in whole egg extracts (Masler *et al.* 1991, and references therein) and activity is associated with the brain (Table 1).

0-8493-4591-X/94/$0.00 + $.50
© 1994 by CRC Press, Inc.

Table 1

Activation of Brain PTTH-like Activity Following Diapause-
break in Pharate-First-Instar Larvae of *Lymantria dispar*

Net Synthesis
(pg ecdysone equiv./gland)

Brain[1] extract	25°C Treatment	Brain Equivalent Dose (per 25 μl)			
		0.25	1	2	4
unboiled[2]	-	0,0,0[3]	0,0,0	9,39,8	8,150,5
unboiled	+	0,0,0	31,23,21	180,227,137	517,524,419
boiled	-	0,0,0	0,0,0	19,33,74	36,304,263
boiled	+	309,3,6	898,79,101	290,136,157	295,126,153

[1] Newly laid eggs from mated females were maintained at 25°C, 16:8 light/dark cycle for 30-35 days and then at 5-7°C for ca. 160 days to allow diapause development, as described by Bell (1989). Following this treatment, brains were either dissected immediately in *Bombyx* saline and frozen at -80°C until use (indicated above as 25°C treatment "-"), or brains were dissected following maintenance at 25°C (3-5 days) when the eggs in each egg mass began to hatch (indicated above as 25°C treatment "+").

[2] Brains were homogenized in Grace's medium in 1.5 ml tubes (50 brains/312.5μl), centrifuged, and supernatants assayed on day-5, last-instar (5th) female prothoracic glands, as previously described (Kelly *et al.* 1992). Brain extracts were either placed in boiling water for two minutes or left unboiled and dilutions prepared in Grace's medium to final concentrations of 0.25,1,2, and 4 brain equivalents/25μl, as indicated above.

[3] Represents the net synthesis for three separate glands.

In contrast to the PAF-positive, presumably neurosecretory material, which packs the neurosecretory cells of brains from chilled eggs of *L. dispar*, but decreases to nearly undetectable levels after incubation at 25°C (Loeb and Hayes 1980), the PTTH-like activity increases when chilled eggs are maintained at 25°C (Table 1). The PTTH-like activity in the pharate-first-instar larval brain is heat stable (Table 1), and there is apparently little of the ecdysteroid ketoreductase activity associated with unboiled whole egg extracts (Kelly *et al.* 1990a; Kelly *et al.* 1990b). Such ecdysteroid ketoreductase activity is unstable to heat treatment and results in an *ca.* 75% increase in immunodetectable ecdysteroids following incubation of prothoracic glands with unheated whole egg extract as compared to heat-treated extract (Kelly *et al.* 1990a; Kelly *et al.* 1990b). Whether or not this brain activity present in pharate-first-instar *L. dispar* larvae represents primarily the large-form of

PTTH, such as that present in *B. mori* eggs containing mature embryos (Chen *et al.* 1987), or the smaller form "bombyxin" present throughout embryonic development (Fugo *et al.* 1987) is unknown. However, our preliminary analysis presented in Table 1 suggests greater variability in the response of day-5 glands to these pre-hatch brain extracts as compared to their response to brain extracts from day-5, last-instar larval females which contain the larger and smaller forms of PTTH (Masler *et al* 1986; Kelly *et al.* 1992). Preliminary analysis of post-diapause, pre-hatch *L. dispar* whole egg extracts suggests the predominance of a very small form of PTTH (Kelly *et al.* 1990a). Further analysis is necessary to understand the variable response of the glands to pre-hatch brain extract, and to characterize the forms, function and locations of the PTTHs in the egg of the gypsy moth.

Thanks to Carol A. Masler and Carol H. Robinson for technical assistance, Tuananh T. Vu for collection and assay of samples, and B. Irene Bedard for typing the manuscript.

REFERENCES

Bell, R. A., Respiratory activity during embryonic development in a diapausing and a selected non-diapausing strain of the gypsy moth, *Lymantria dispar* L., *Comp. Biochem. Physiol.* 93A, 767-771, 1989.

Bell, R. A., Kelly, T. J., Masler, E. P., Thyagaraja, B. S., DeMilo, A. B. and Borkovec, A. B., Endocrinology of embryogenesis and late embryonic diapause in the gypsy moth, *Lymantria dispar*, in *Insect Neurochemistry and Neurophysiology·1989*, Borkovec, A. B. and Masler, E. P., Eds., Humana Press, Clifton, New Jersey, 1990, pp. 341-344.

Chen, J. H., Fugo, H., Nakajima, M., Nagasawa, H. and Suzuki, A., Neurohormones in developing embryos of the silkworm, *Bombyx mori*: the presence and characteristics of prothoracicotropic hormone B, *J. Insect Physiol.* 33, 407-411, 1987.

Chen, J.-H., Fugo, H., Nakajima, M., Nagasawa, H. and Suzuki, A., The presence of neurohormonal activities in embryos of the silkworm, *Bombyx mori*, *J. Seric. Sci. Jpn.* 55, 54-59, 1986.

Denlinger, D. L., Hormonal control of diapause, in *Comprehensive Insect Physiology, Biochemistry and Pharmacology*, Vol 8, Kerkut, G. A. and Gilbert, L. I., Eds., Pergamon Press, Oxford, 1985, chap. 11.

Dorn, A., Gilbert, L. I, and Bollenbacher, W. E., Prothoracicotropic hormone activity in the embryonic brain of the tobacco hornworm, *Manduca sexta*, *J. Comp. Physiol. B.* 157, 279-283, 1987.

Fugo, H., Chen, J. H., Nakajima, M., Nagasawa, H. and Suzuki, A., Neurohormones in developing embryos of the silkworm, *Bombyx mori*: the presence and characteristics of prothoracicotropic hormone-S, *J. Insect Physiol.* 33, 243-248, 1987.

Kelly, T. J., Masler, E. P., Thyagaraja, B. S., Bell, R. A., Gelman, D. B., Imberski, R. B. and Borkovec, A. B., Prothoracicotropic hormone and ecdysteroid ketoreductase from pre-hatch eggs of the gypsy moth, *Lymantria dispar*, in *Insect Neurochemistry and Neurophysiology·1989*, Borkovec, A.B. and Masler, E. P., Eds., Humana Press, Clifton, New Jersey, 1990a, pp. 357-360.

Kelly, T. J., Masler, E. P., Thyagaraja, B. S., Bell, R. A. and Imberski, R. B., Development of an *in vitro* assay for prothoracicotropic hormone of the gypsy moth, *Lymantria dispar* (L.) following studies on identification, titers and synthesis of ecdysteroids in last-instar females, *J. Comp. Physiol. B*, 162, 581-587, 1992.

Kelly, T. J., Thyagaraja, B. S., Bell, R. A., Masler, E. P., Gelman, D. B. and Borkovec, A. B., Conversion of 3-dehydroecdysone by a ketoreductase in post-diapause, pre-hatch eggs of the gypsy moth, *Lymantria dispar*, *Arch. Insect Biochem. Physiol.*, 14, 37-46, 1990b.

Leonard, D. E., Diapause in the gypsy moth, *J. Econ. Entomol.* 61, 596-598, 1968.

Loeb, M. J. and Hayes, D. K., Neurosecretion during diapause and diapause development in brains of mature embryos of the gypsy moth, *Lymantria dispar*, *Ann. Ent. Soc. Amer.*, 73, 432-436, 1980.

Masler, E. P., Bell, R. A., Thyagaraja, B. S., Kelly, T. J. and Borkovec, A. B., Prothoracicotropic hormone-like activity in the embryonated eggs of gypsy moth, *Lymantria dispar* (L.), *J. Comp. Physiol. B*, 161, 37-41, 1991.

Masler, E. P., Kelly, T. J., Thyagaraja, B. S., Woods, C.W., Bell, R. A. and Borkovec, A. B., Discovery and characterization of prothoracicotropic hormones of the gypsy moth, *Lymantria dispar*, in *Insect Neurochemistry and Neurophysiology·1986*, Borkovec, A. B. and Gelman, D. B., Eds., Humana Press, Clifton, New Jersey, 1986, pp. 331-334.

MULTIPLE PEPTIDE EXPRESSION BY THE L-NSC III IN THE TOBACCO HORNWORM, *MANDUCA SEXTA*

Rosemary S. Gray, David P. Muehleisen, and Walter E. Bollenbacher

Department of Biology, University of North Carolina at Chapel Hill, Chapel Hill, NC 27599-3280, USA

The co-expression of multiple peptide phenotypes by neurons is being discovered with increasing frequency. Research on these multiple phenotypic neurons is presenting insight into the molecular dynamics of the synthesis and physiological roles of the co-expressed peptides. In the tobacco hornworm, *Manduca sexta*, at least three different CNS neurons appear to express multiple peptides. The lateral neurons of the abdominal ganglia express both cardioacceleratory peptides and bursicon (Tublitz and Sylwester 1990), and the cerebral ventral medial neurons (VMN) that express the behavioral eclosion hormone (EH) (Hewes and Truman 1991) also express the developmental neurohormone, prothoracicotropic hormone (PTTH) (Westbrook *et al.* 1993). A third group of *Manduca* CNS neurons that appears to express multiple secretory peptide phenotypes is the lateral neurosecretory cell group III (L-NSC III) located in the protocerebrum. These cells express one molecular variant of PTTH, big PTTH (O'Brien *et al.* 1988), which drives insect postembryonic development by stimulating the prothoracic glands (PG) to synthesize ecdysteroids (Bollenbacher and Granger 1985). In addition to expressing big PTTH, the *Manduca* L-NSC III express a peptide (Gray 1992) immunologically similar to the PTTH from the commercial silkmoth, *Bombyx mori* (Kawakami *et al.* 1990). Recent evidence has revealed that the *Bombyx* peptide does not have amino acid sequence similarity with the *Manduca* PTTH (Muehleisen *et al.* 1993). A third neuropeptide, a 28 kDa peptide, having chromatographic properties similar to *Manduca* big PTTH but lacking PTTH activity, is also expressed by the L-NSC III. The relationships of these peptides to one another and the development of *Manduca* are discussed.

The generation of a MAb to *Manduca* big PTTH (O'Brien *et al.* 1988) identified the L-NSC III as the principal site of big PTTH synthesis in this insect (Fig. 1A). An anti-*Bombyx* PTTH MAb immunostains similarly located neurons in the *Bombyx* brain (Mizoguchi 1990). To investigate the similarity between these two PTTHs, the *Bombyx* PTTH MAb was used to immunostain *Manduca* brains. Immunological staining with the *Manduca* big PTTH and the *Bombyx* PTTH MAbs was quite similar (Fig. 1A, B), although in contrast to the *Manduca* big PTTH MAb, the *Bombyx* PTTH MAb did not immunostain the VMN. This data initially suggested that the *Manduca* PTTH and the *Bombyx* PTTH were possibly immunologically similar. However, purification of *Manduca* big PTTH resulted in the determination of a partial amino acid sequence that had no similarity to the *Bombyx* PTTH sequence (Muehleisen *et*

0-8493-4591-X/94/$0.00 + $.50
© 1994 by CRC Press, Inc.

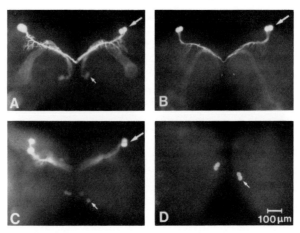

Figure 1. Immunofluorescent staining of *Manduca* day 1 pupal brain whole mounts with *Manduca* big PTTH MAb (A), *Bombyx* PTTH MAb (B), 28 kDa antiserum (C), and EH MAb (D). Large arrows indicate L-NSC III, small arrows indicate VMN.

al. 1993). To investigate whether *Bombyx* PTTH possessed prothoracicotropic activity in *Manduca*, it was used in a *Manduca in vitro* PG assay for PTTH (Bollenbacher *et al.* 1979, 1984). Three different stages of purified *Bombyx* PTTH (generously provided by Dr. J. Nishiitsutsuji-Uwo, Shionogi & Co. Ltd., Kyoto, Japan) failed to activate the PG (Muehleisen *et al.* 1993), supporting the finding that the *Manduca* PTTH and the *Bombyx* PTTH peptides are not similar.

Initial attempts to purify *Manduca* big PTTH by conventional means suggested it was a 28 kDa peptide (Bollenbacher *et al.* 1981, 1984). Purification through three HPLC columns (C_{18}, C_4, and biphenyl) ultimately resulted in the separation of the 28 kDa peptide from PTTH as determined by ELISA and bioassay (Gray *et al.* 1993). An immunoaffinity chromatographic approach to purifying *Manduca* big PTTH has revealed it is a 25.5 kDa dimer consisting of ~16 kDa subunits (Muehleisen *et al.* 1993), whereas the 28 kDa peptide is a monomer (Gray *et al.* 1993). Murine antiserum to the 28 kDa peptide was obtained and immunocytochemistry of *Manduca* brains revealed that the protein is expressed by the L-NSC III and the VMN (Gray *et al.* 1993; Fig. 1C).

The expression by the L-NSC III of at least three different peptide phenotypes, a 28 kDa peptide, *Manduca* big PTTH, and *Bombyx* PTTH-like peptide, raises questions about the genomic organization of these peptides. The chemical data available suggest that they are not processed from a single polyprotein. The *Bombyx* PTTH cDNA has revealed that this PTTH and two other peptides are processed from a polyprotein precursor (Kawakami *et al.* 1990) which does not have any deduced amino acid similarity to *Manduca* big PTTH or the 28 kDa peptide. While *Manduca* big PTTH, *Bombyx* PTTH, and the 28 kDa peptide appear to come from separate genes, this does not preclude the possibility that they function in a coordinated manner. For example, the 28 peptide could act synergistically with big PTTH to elicit molting, possibly at the level of the PG. In this context, it is interesting that the 28 kDa peptide is expressed by the L-NSC III only during stages of the life cycle when the PG are present (see Herman 1967). In addition, the recent suggestion that *Manduca* big PTTH has functions other than as a prothoracicotropin, including paracrine and neuromodulatory roles (Westbrook *et al.* 1993), presents more possibilities for 28 kDa peptide involvement. This is supported by the distribution of the 28 kDa peptide in the dendritic fields and in the retrocerebral CA, which is similar to the distribution of big PTTH (O'Brien *et al.*,1988). Evidence that the 28 kDa peptide is a secretory product, and thus possibly having a regulatory function, has come from preliminary electron microscopy data that indicate the peptide is packaged in neurosecretory granules (K. Hartfelder, R. Gray, W. Bollenbacher, unpublished).

Insight into the physiological significance of the co-expressed L-NSC III peptides may be obtained from comparable systems in other organisms. For example, in humans one co-expressed peptide may alter the half life of another (Brain and Williams 1988) or in mice it may change the expression or activity of the other's receptor (New and Mudge 1986; Mulle *et al.* 1988). Therefore, the 28 kDa peptide could affect the action of *Manduca* big PTTH in a similar manner. Alternatively, the peptide could be released independently to exert control over targets not regulated by *Manduca* big PTTH or the *Bombyx* PTTH-like peptide.

As with the 28 kDa peptide, the *Bombyx* PTTH-like peptide may have related and/or independent functions. Since it does not activate the *Manduca* PG, its possible function in driving molting and metamorphosis may be indirect, possibly acting in concert with big PTTH. Alternatively, it may have a unique function(s) not associated with molting.

Recent research in this laboratory is raising the possibility that PTTH has functions in addition to its role in the control of molting and metamorphosis (Westbrook *et al.* 1993). The expression of *Manduca* big PTTH by the L-NSC III throughout the *Manduca* life cycle suggests it functions from early embryogenesis through adult life. In addition, big PTTH's presence in multiple CNS neurons presents an even broader view of the peptide's functions (Westbrook *et al.* 1993). Correspondingly, co-expression of big PTTH and the 28 kDa peptide with EH by the cerebral VMN (Fig. 1; see Hewes and Truman 1990) poses the possibility that both peptides have multiple functions. Therefore, the combination of EH, big PTTH, and 28 kDa peptide in the VMN may be necessary for eclosion behavior. Whereas, the combination of *Manduca* big PTTH, *Bombyx*-like PTTH, and the 28 kDa peptide in the L-NSC III may be necessary for molting. In this light, *Manduca* L-NSC III and VMN may be suitable models for investigating the relationships of co-expressed peptide phenotypes during the life history of a neuron.

Acknowledgments: The authors would like to thank Dr. Anne L. Westbrook and Ms. Susan Whitfield for assistance with the illustration. This work was supported by a grant to WEB from the National Institutes of Health (NIH) (DK-31642) and a NIH Postdoctoral Training Fellowship to DPM (NS-07166).

REFERENCES

Bollenbacher, W. E., Agui, N., Granger, N. A. and Gilbert, L. I., *In vitro* activation of insect prothoracic glands by the prothoracicotropic hormone, *Proc. Natl. Acad. Sci. USA* 76, 5148-5152, 1979.

Bollenbacher, W. E., Katahira, E. J., O'Brien, M. A., Gilbert, L. I., Thomas, M. K., Agui, N. and Baumhover, A. H., Insect prothoracicotropic hormone: Evidence of two molecular forms, *Science* 224, 1243-1245, 1984.

Bollenbacher, W. E. and Gilbert L. I., Neuroendocrine control of postembryonic development in insects: The prothoracicotropic hormone, in *Neurosecretion: Molecules, Cells, Systems*, Farner, D. S., and Lederis, K., Eds., Plenum Press, New York, pp. 361-379, 1981.

Bollenbacher, W. E. and Granger, N. A., Endocrinology of the prothoracicotropic hormone, in *Comprehensive Insect Physiology Biochemistry and Pharmacology*, Vol. 7, Kerkut, G. A., and Gilbert, L. I., Eds., Pergamon Press, Oxford, 1985, pp. 109-151.

Brain, S. D. and Williams, T. J., Substance P regulates the vasodilatory activity of calcitonin gene-related peptide, *Nature* 335, 73-75, 1988.

Gray, R. S., Muehleisen, D. P., Katahira, E. J. and Bollenbacher, W. E., A 28 kDa cerebral neuropeptide from *Manduca sexta*: Relationship to the insect prothoracicotropic hormone, *Cell. Mol. Neurobiol.* 13, 39-58, 1993.

Gray, R. S., Multiple peptide expression by the prothoracicotropic hormone producing neurons, the L-NSC III, in the tobacco hornworm, *Manduca sexta*, Ph. D. Dissertation, University of North Carolina at Chapel Hill, 1992.

Herman, W. S., The edysial gland of arthropods, in *International Review of Cytology*, Vol. 22, Bourne, G. H. and Danielli, J. F., Eds., Academic Press, New York, 1967, pp. 269-347.

Hewes, R. S. and Truman, J. W., The roles of central and peripheral eclosion hormone release in the control of ecdysis behavior in *Manduca sexta*, *J. Comp. Physiol.* 168, 697-707, 1991.

Kawakami, A., Kataoka, H., Oka, T., Mizoguchi, A., Kimura-Kawakami, M., Adachi, T., Iwami, M., Nagasawa, H., Suzuki, A. and Ishizaki, H., Molecular cloning of the *Bombyx mori* prothoracicotropic hormone, *Science* 246, 1333-1335, 1990.

Mizoguchi, A., Immunological approach to synthesis, release and titre fluctuation of bombyxin and prothoracicotropic hormone of *Bombyx mori*, in *Molting and Metamorphosis*, Ohnishi, E., and Ishizaki, H., Eds., Springer-Verlag, Berlin, 1990, pp. 17-32.

Muehleisen, D. P., Gray, R. S., Katahira, E. J. and Bollenbacher, W. E., Immunoaffinity purification of a cerebral prothoracicotropic peptide from *Manduca sexta*, *Peptides* 14, 531-541, 1993.

Mulle, C., Benoit, P., Pinset, C., Roa, M. and Changeux, J.-P., Calcitonin gene-related peptide enhances the rate of desensitization of the nicotinic acetylcholine receptor in cultured mouse muscle cells. *Proc. Natl. Acad. Sci.* 85, 5728-5732, 1988.

New, H. V. and Mudge, A. W., Calcitonin gene-related peptide regulates muscle acetylcholine receptor synthesis, *Nature* 323, 809-811, 1986.

O'Brien, M. A., Katahira, E. J., Flanagan, T. R., Arnold, L. W., Haughton, G. and Bollenbacher, W. E., A monoclonal antibody to the insect prothoracicotropic hormone, *J. Neurosci.* 8, 3247-3257, 1988.

Schotzinger, R. J. and Landis, S., Cholinergic phenotype developed by noradrenergic sympathetic neurons after innervation of a novel cholinergic target *in vivo*, *Nature* 335, 637-639, 1988.

Tublitz, N. J. and Slywester, A. W., Postembryonic alteration of transmitter phenotype in individually identifed peptidergic neurons, *J. Neurosci.* 10, 161-168, 1990.

Westbrook, A. L., Regan, S. A. and Bollenbacher, W. E., Developmental expression of the prothoracicotropic hormone in the CNS of the tobacco hornworm, *Manduca sexta*, *J. Comp. Neurol.* 327, 1-16, 1993.

EXPRESSION OF THE PROTHORACICOTROPIC HORMONE IN THE TOBACCO HORNWORM, *MANDUCA SEXTA*

Anne L. Westbrook, Sheila A. Regan,
and Walter E. Bollenbacher

Department of Biology, The University of North Carolina at Chapel Hill,
Chapel Hill, NC, 27599-3280, USA

Neuropeptides in vertebrates and invertebrates are now known to elicit a wide range of functional responses that are dependent upon the developmental stage of the organism and site of release (Scharrer 1990). In insects, despite the extensive knowledge about the neuroendocrine actions of several peptides, relatively little is known about their potential for multiple functions, i.e., neuroendocrine, neuromodulatory, or paracrine.

Recent investigations of one insect developmental neuropeptide, the prothoracicotropic hormone (PTTH), have raised the possibility that this peptide is multifunctional (Westbrook *et al.* 1991, 1993). For several decades, the sole known function for PTTH was as a regulator of postembryonic molting and metamorphosis (Bollenbacher and Granger 1985). In the tobacco hornworm, *Manduca sexta,* a group of two bilaterally paired cerebral neurosecretory cells, the group III lateral neurosecretory cells (L-NSC III), was originally identified as the source of PTTH (Fig. 1) (Agui *et al.* 1979). The axons of these peptidergic neurons terminate in the retrocerebral organ, the corpus allatum (CA), where PTTH is released into the hemolymph to stimulate ecdysteroid synthesis by the prothoracic glands (PG). Immunocytochemical studies using a monoclonal antibody (MAb) to *Manduca* big PTTH enabled the morphology of the L-NSC III to be analyzed in detail (O'Brien *et al.* 1988; Westbrook *et al.* 1991) as well as the development of these neurons to be traced during *Manduca*'s embryogenesis (Westbrook and Bollenbacher 1990).

During the course of those immunocytochemistry studies, several additional groups of neurons were identified in the brain, frontal ganglion, and subesophageal ganglion which appear to express the big PTTH phenotype (Fig. 1) (Westbrook *et al.* 1993). Interestingly, one of these PTTH-expressing groups of cells, the ventromedial neurons (VMN), are also the principal source of eclosion hormone in *Manduca* (Truman and Copenhaver 1989). Thus, the VMN appear to coexpress two neuropeptides that are critical for the insect's development. Each of the cell groups that express the PTTH phenotype does so in a distinct temporal pattern during *Manduca*'s development. Similar to the L-NSC III, the VMN, frontal ganglion neurons, subesophageal ganglion lateral neurons, and subesophageal ganglion medial neurons begin expressing PTTH during embryonic development. The far lateral neurons, on the other hand, are not

0-8493-4591-X/94/$0.00 + $.50
© 1994 by CRC Press, Inc.

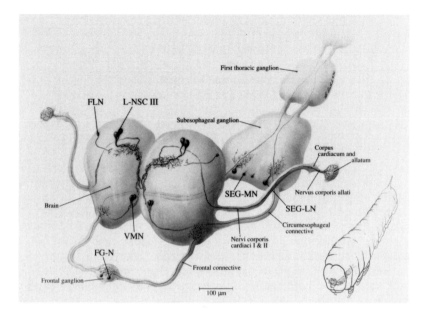

Figure 1. Schematic illustration of the CNS neurons that express the big PTTH phenotype during larval development of *Manduca sexta*. (Reprinted from Westbrook, A.L., Regan, S.A. and Bollenbacher, W.E., *The Journal of Comparative Neurology* 327, 1-16, 1993.)

detected until the first larval instar. Only the L-NSC III and VMN remain immunoreactive during the pupal, pharate adult, and adult stages of the life cycle. The distinctive architectures of these neurons and their precise temporal pattern of peptide expression support a multifunctional role for PTTH during the life cycle. Depending on the developmental stage of the insect, PTTH could exert neuroendocrine effects on a target tissue(s) other than the PG. In addition, it could fulfill local functions, either neuromodulatory or paracrine, through its release into specific microenvironments.

The capability of *Manduca* big PTTH to fulfill additional physiological functions undoubtedly relies upon the neurohormone's binding to receptors on target cells. Thus, if PTTH exerts a local effect within the brain, CA, or subesophageal ganglion, the PTTH receptor should be localized in those sites. Currently nothing is known about the receptor for this neuropeptide that would enable its localization. We are attempting to localize the PTTH receptor and thus, the target cells for the peptide, through the generation of anti-idiotypic antibodies to the big PTTH MAb. This strategy has been successful in identifying vertebrate hormone and neuropeptide receptors (Couraud and Strosberg 1988; Farid 1989). The MAb to *Manduca* big PTTH is an excellent candidate for generating an anti-idiotypic antibody because of its high specificity and because it appears to bind at or near the neuropeptide's active site (O'Brien *et al.* 1988). Murine antiserum has been

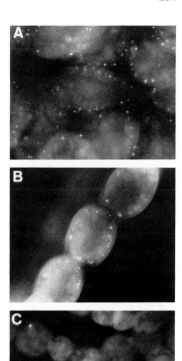

Figure 2. Photomicrographs of day 3 fifth larval instar prothoracic glands immunostained with the murine anti-idiotypic serum reveals punctate staining at the periphery of cells in the A) base of the gland and B) arm of the gland. C) Pre-immune serum does not stain the prothoracic gland cells.

obtained which blocks the enzyme linked immunosorbant assay (ELISA) for PTTH, suggesting that an anti-idiotypic antibody has been generated (Westbrook 1992). A fundamental test for an anti-idiotypic antibody is its ability to bind to the neuropeptide's target cells. In this case, the PG are being used as the target for the neuropeptide. Immunofluorescent staining of *Manduca* day 3 fifth larval instar PG with the anti-idiotypic serum revealed punctate staining around the periphery of each cell (Fig. 2). This staining pattern would be expected for a membrane-associated receptor. Staining was not obtained with either pre-immune serum or the anti-PTTH MAb.

Additional evidence supporting the generation of an anti-idiotypic antibody that binds the big PTTH receptor was provided by demonstrating that the immunostaining of the PG was dramatically reduced on day 4 of the fifth larval instar, which is when PTTH is released into the hemolymph to drive pupal commitment (Bollenbacher, 1988). This absence of staining may be due to saturation of the receptors by the released PTTH that would prevent binding of the anti-idiotypic antibody.

Studies are ongoing to determine whether this antibody does actually recognize the PTTH receptor. If so, the antibody may be employed to determine whether PTTH receptors are localized in regions of the nervous system where PTTH may exert local actions. The localization of neuropeptide and receptor will be critical in determining the multifunctional nature of PTTH.

Acknowledgments: The authors thank Drs. Geoff Haughton and Larry Arnold for their advice and assistance with the anti-idiotypic antibody generation. This work was supported by NIH grant #DK-31642 to WEB and a Sigma Xi Grant-in-Aid of Research to ALW.

REFERENCES

Agui, N. , Granger, N.A., Gilbert, L.I. and Bollenbacher, W.E., Cellular localization of the insect prothoracicotropic hormone: In vitro assay of a single neurosecretory cell, *Proc. Natl. Acad. Sci. USA*, 1979.

Bollenbacher, W.E., The interendocrine regulation of larval-pupal development in the tobacco hornworm, *Manduca sexta*: A model, *J. Insect Physiol.* 34, 941-947.

Bollenbacher, W.E. and Granger, N.A., Endocrinology of the prothoracicotropic hormone, in *Comprehensive Insect Physiology, Biochemistry, and Pharmacology*, Vol. 7, Kerkut, G.A. and Gilbert, L.I., Eds., Pergamon Press, New York, 1985, pp. 109-151.

Couraud, P.O. and Strosberg, A.D., Anti-receptor anti-idiotypic antibodies, in *Molecular Neuroanatomy*, VanLeeuwen, F.W., Buijs, R.M., Pool, C.W. and Pach, O., Eds., Elsevier, New York, pp. 191-203.

Farid, N.R., Antiidiotypic antibodies as probes for receptor structure and function, *Endocrine Rev.* 6, 1-23, 1989.

O'Brien, M.A., Katahira, E.J., Flanagan, T.R., Arnold, L.W., Haughton, G. and Bollenbacher, W.E., A monoclonal antibody to the insect prothoracicotropic hormone, *J. Neurosci.* 8, 3247-3257, 1988.

Scharrer, B., The neuropeptide saga, *Am. Zool.* 30, 887-895, 1990.

Truman, J.W. and Copenhaver, P.F., The larval eclosion hormone neurones in *Manduca sexta*: Identification of the brain-proctodeal neurosecretory system, *J. Exp. Biol.* 147, 457-470, 1989.

Westbrook, A.L., Development and function of the prothoracicotropic hormone neuroendocrine axis in the tobacco hornworm, *Manduca sexta*, Ph.D. dissertation, The University of North Carolina at Chapel Hill, 1992.

Westbrook, A.L. and Bollenbacher, W.E., The development of identified neurosecretory cells in the tobacco hornworm, *Manduca sexta, Devl. Biol.* 140, 291-299, 1990.

Westbrook, A.L., Haire, M.E., Kier, W.M. and Bollenbacher, W.E., Three-dimensional architecture of identified neurosecretory cells in an insect, *J. Morphol.* 208, 161-174, 1991.

Westbrook, A.L., Regan, S.A. and Bollenbacher, W.E., The developmental expression of prothoracicotropic hormone in the CNS of the tobacco hornworm, *Manduca sexta*, *J. Comp. Neurol.* 327, 1-16, 1993.

PHOTOPERIODIC INFLUENCES ON ECDYSIOTROPIN LEVELS IN THE HINDGUT OF PREPUPAE OF THE EUROPEAN CORN BORER, *OSTRINIA NUBILALIS*

Dale B. Gelman and Robert A. Bell

Insect Neurobiology and Hormone Laboratory
USDA, ARS, Beltsville, MD 20705, USA

Brain ecdysiotropins (PTTHs), because of their ability to stimulate ecdysteroid production by insect prothoracic glands (PTGs), play an important role in regulating insect molting and metamorphosis. Recently, new ecdysiotropins were discovered in the hindguts of the European corn borer and the gypsy moth, *Lymantria dispar* (Gelman *et al.*, 1991). The hindgut ecdysiotropin of *O. nubilalis* is a small heat-stable peptide whose estimated molecular weight is 500 to 1,500 daltons. It is located throughout the hindgut and is present in greatest concentration in those 5th instars that have undergone gut purge (OW prepupae). While both brain (PTTH) and hindgut ecdysiotropins utilize cyclic AMP as a second messenger, they are reported to activate different PTG receptors (Gelman *et al.*, 1993). Currently, studies are underway to determine the source and physiological role of the lepidopteran hindgut ecdysiotropins. In this paper, we report our results concerning the fluctuation of hindgut ecdysiotropin levels in *O. nubilalis* 5th instars as a function of time of day.

O. nubilalis larvae were reared as described in Gelman and Hayes (1982a) under a light-dark regimen of LD 16:8 and a temperature of 30 ± 1°C. To synchronize larvae, pharate 5th instars were transferred to 2-dram vials containing fresh medium. Hindguts were sampled around the clock at Lights On +5 h and +10 h and at Lights Off +2 h and +6 h. Dissected hindguts (proctodaea minus recta) were stored in polypropylene microcentrifuge tubes at -80°C. When needed, tissue was homogenized in ice cold Grace's medium, placed in a boiling water bath for 2.5 min and centrifuged at 16,000 x g and 4°C for 3 min. Supernatants were stored on ice. An *in vitro* PTG assay developed by Bollenbacher *et al.* (1975) and modified by Gelman et al. (1993) was utilized to measure levels of hindgut ecdysiotropin. Briefly, to allow ecdysteroid production to fall to low levels, *L. dispar* PTGs were preincubated in 25-μl of Grace's medium. After transfer to 25-μl of hindgut extract, the glands were incubated for an additional 2 h. To convert the 3-dehydroecdysone produced to RIA-detectable ecdysone, PTGs were removed from the incubation medium, and a 25-μl drop containing 0.5 μl of hemolymph from a diapause bound day-8 5th instar (contains ketoreductase) was added. After an additional 1 h of

0-8493-4591-X/94/$0.00 + $.50
© 1994 by CRC Press, Inc.

incubation, aliquots were taken and ecdysteroid content was determined by RIA (Bollenbacher *et al.*, 1975; Gelman *et al.*, 1993).

In prepupae that had experienced gut purge (OW), ecdysiotropic activity was significantly lower at Lights Off +6 than at any other time sampled (Fig. 1A). In day-3 5th instars, however, hindgut ecdysiotropin levels did not appear to vary with time of day (Fig. 1B). When 5th instars were exposed to 24 h of darkness beginning on day 4, ecdysiotropin levels in hindguts from OW prepupae were lower at both Lights Off +2 and Lights Off +6 than at the two times sampled during the photophase (Fig. 2A). Thus, decreased levels of ecdysiotropin were observed through most of the scotophase. For 5th instars exposed to 24 h of darkness beginning on day 2, ecdysiotropin levels in hindguts from OW prepupae were lowest at Lights On +5 and at Lights Off +2 (Fig. 2B). Therefore, exposure to 24 h of darkness for varying lengths of time alters the levels of ecdysiotropin in the *O. nubilalis* hindgut.

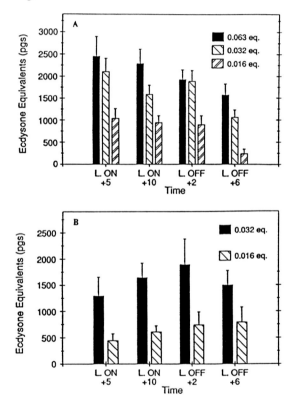

Figure 1. Ecdysiotropin levels in hindguts from <u>O. nubilalis</u> OW prepupae (A) or day 3 5th instars (B) as a function of time of day. Titers are expressed as ecdysone equivalents (measured by RIA) produced/PTG/2 h of incubation. One equivalent = 1 hindgut. Each bar represents the mean ±SE of at least eight separate determinations.

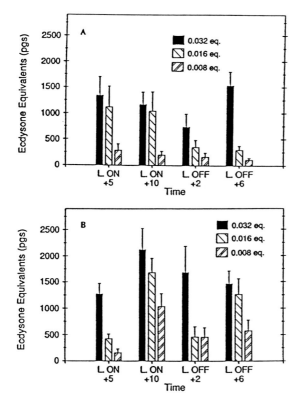

Figure 2. Ecdysiotropin levels in hindguts from <u>O</u>. <u>nubilalis</u> OW prepupae that had been exposed to 24 h of darkness beginning on day 4 (A) or on day 2 (B) of the 5th instar. Determination of ecdysiotropin levels as in Figure 1.

Similarly, the times of pharate pupal and pupal formation vary with time of day (Gelman, et al., 1982a). Pharate pupal formation occurs primarily during the scotophase, and pupal formation during the photophase. Placing 5th instars in 24 h of darkness beginning on day 4 resulted in a broadening of the curve which describes pupation. and what appears to be a 2-h phase shift (Fig. 3A), while exposure to 24 h of darkness beginning on day l, destroyed the bell-shaped curve completely (Fig. 3B). Percent pupation did not increase gradually with time, peak, and decrease as it did in the controls, but occurred throughout the 24-h day. Similar results (not shown) were observed for pharate pupal formation. In *O. nubilalis* prepupae, results from double ligation experiments (Gelman and Hayes, 1982b) revealed that PTTH release begins 3 to 4 h prior to the onset of darkness and ecdysone production/release just prior to Lights On. It is tempting to speculate that the observed reduction in levels of hindgut ecdysiotropin during the latter portion of the scotophase might be attributed to its release and in some way be associated with the increased hemolymph ecdysteroid levels observed at this time. Experiments designed to test this hypothesis are underway.

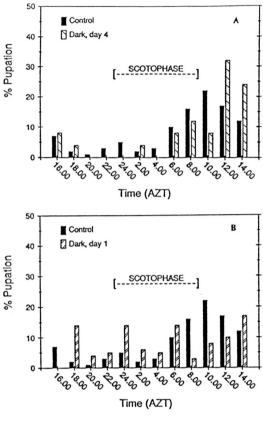

Figure 3. Percent pupation as a function of time of day. Control larvae were maintained at LD 16:8. Experimental larvae were exposed to 24 h of darkness beginning on day 4 (A) or on day 1 (B) of the 5th instar.

REFERENCES

Bollenbacher, W. E., Vedeckis W. V., Gilbert L. I. and O'Connor J. D., Ecdysone titers and prothoracic gland activity during the larval-pupal development of Manduca sexta: a model. *Dev. Biol.*, 44, 46, 1975.

Gelman D. B., Thyagaraja B. S., Kelly, T. J., Masler, E. P., Bell R. A. and Borkovec A. B., The insect gut: A new source of ecdysiotropic peptides. *Experientia*, 47, 77, 1991.

Gelman, D. B. and Hayes, D. K., Methods and markers for synchronizing maturation of fifth-stage larvae and pupae of the European corn borer, *Ostrinia nubilalis* (Lepidoptera: Pyralidae). *Ann. Ent. Soc. Am.*, 75, 485, 1982a.

Gelman D. B. and Hayes D. K., Critical periods for the brain and prothoracic glands of fifth instars of the European corn borer, *Ostrinia nubilalis* (Hubner) *Comp. Biochem. Physiol.*, 73A, 81, 1982b.

Gelman, D. B., Thyagaraja, B. S. and Bell, R. A., Ecdysiotropic Acitivity in the Lepidopteran Hindgut - An Update. *Insect Biochem. Molec. Biol.*, 23, 25, 1993.

Acknowledgements: We thank A. A. Khalidi and J. C. Harder, for technical assistance and Drs. R. S. Hakim and D. K. Hayes for critical readings of this manuscript.

EFFECT OF NONSTEROIDAL ECDYSONE MIMIC RH-5849 ON CIRCADIAN RHYTHMS OF LARVAL-PUPAL TRANSFORMATION IN *SPODOPTERA LITTORALIS* (BOISD.)

Maciej A. Pszczolkowski, Andras Szecsi
and Bronislaw Cymborowski

Department of Invertebrate Physiology, Warsaw University,
Xwirki i Wigury 93, 02-089 Warszawa, Poland.

INTRODUCTION, MATERIALS AND METHODS

The process of moulting in insects is initiated and directed by the hormone 20-hydroxyecdysone (20HE) (Riddiford and Truman 1978) which causes the secretion of a new cuticle. When the synthesis of new cuticle is completed, the old one is shed off during ecdysis. A neurosecretory peptide, the eclosion hormone (EH) triggers ecdysis by acting on the central nervous system to elicit the ecdysial motor pattern and other physiological changes associated with imaginal ecdysis. In *Lepidoptera* the eclosion hormone is regarded to be connected with all ecdyses during development (Truman et al. 1983).

The regulatory role of ecdysteroids in pupal and imaginal moulting of *Lepidoptera* has been examined by Truman et al. (1983), however circadian aspects of exogenic ecdysone action onto pupation rhythm has not been demonstrated so far. In *Spodoptera littoralis* the changes of ecdysteroid level in the last day of the last larval instar of are described as a typical pre-ecdysial pattern with a large peak about 20 h before pupation (Zimowska et al. 1985). Later on the level of ecdysteroids declines and the mean time of larval-pupal ecdysis is correlated with a comparatively low level of ecdysone. Ecdysis is gated and exhibits a clear circadian pattern of distribution. The question arises whether is it possible to influence the pupation rhythm (advance or delay the moulting) using an ecdysone agonist. This paper shows that in *S. littoralis* the nonsteroidal ecdysone agonist RH-5849 can delay or advance the time of pupation in a circadian manner. Possible mechanisms of these responses are discussed.

Larvae of *Spodoptera littoralis* (Boisd.), *Lepidoptera*, *Insecta*, were reared in LD 16:8, at 24 C on a semiartifical diet (Pszczolkowski et al. unpublished). They were separated from the stock colony as young penultimate (5th) instar larvae and kept in transparent plastic boxes until they reached the last (6th) instar. Only second gate larvae were used for experiments (see Zimowska et al. 1985 for details). To examine the circadian effect of RH-5849 the larvae were treated with this ecdysone agonist at different times before the expected moment of pupation. RH-5849 (kind gift of Dr. K. Wing, Roehm and Haas Co.), dissolved in acetone, was applied

0-8493-4591-X/94/$0.00 + $.50
© 1994 by CRC Press, Inc.

263

topically. Two kinds of controls were made: larvae treated with acetone only and untreated. Petri dishes with experimental and control larvae were examined every three hours until the larvae pupated. Median of the time of pupation of untreated controls has been taken as acrophase of pupation rhythm. The difference between this acrophase and the time of pupation of treated larvae has been taken as phase delay (when untreated controls pupated later) or as phase advance (when they pupated earlier). During scotophases all manipulations (applications of compounds and data acquisition) were performed in dim red light of luminance less than 1 lux. The middle of the penultimate scotophase has been taken as Arbitrary Zeitgeber Time zero (AZT 0).

RESULTS AND DISCUSSION

The peak of pupation rhythm of untreated controls occurs at AZT 30 and the median of the circadian distribution of moultings is equal to AZT 27. Larvae treated with 5 µg of RH-5849 during the penultimate dark phase delay their pupation by nearly 12 h. When treated 30 min after the last light switch on (AZT 4.5), they advance the time of pupation about 12 h (Fig. 1). Then there is no evident effect of RH-5849 application within the period between AZT 9 and AZT 21. The time of larval-pupal moulting delays when RH-5849 is applied at AZT 24, AZT 25.5 and AZT 27, when the rate of pupation is high. This advancing or delaying effect occurs in a dose dependent manner both at AZT 4.5 and at AZT 25.5 (Table 1).

Truman et al. (1983) showed that application of 20HE can delay both pupal and adult ecdysis in Manduca. For pupation the delay is dose dependent and varies between 8 and 20 h. In our study we found that nonsteroidal ecdysone agonist RH-5849 can both delay and advance pupation in *Spodoptera*. Truman and Morton (1990) reported that a similar delay can be obtained for *Manduca* larvae treated with ecdysone up to 8 h before the expected time of pupation. Thereafter the larvae were insensitive to ecdysone. In the case of *S. littoralis* the delay occurs when RH-5849 is applied to larvae during the penultimate scotophase and the last dark period (between 9 and 4.5 h before pupation). The second delayed effect is in accord with the data published by Truman et al. (1983) and Truman and Morton (1990). The delaying response after application of RH-5849 during the penultimate scotophase, the advance of pupation if RH-5849 was applied just after the last light switch on and the dead zone for RH-5849 treatment may be explained as evidence of a circadian system involving EH and ecdysteroids. We propose that the system may act as follows.

During the penultimate dark phase the level of ecdysteroids is low (Zimowska et al. 1985) but EH is ready to be released or a small amount of EH is released at the end of scotophase. Application of exogenic ecdysone agonist may temporally block an oscillating EH system and delay the moment of pupation. At the beginning of photophase the exogenic level of ecdysteroids increases rapidly (Zimowska et al. 1985) and the release of EH is impossible up to the normal time of ecdysis. If RH-5849 is then applied the endogenous titer of ecdysone declines due to a short feedback loop between ecdysteroid titer and ecdysteroid release. EH can then be released

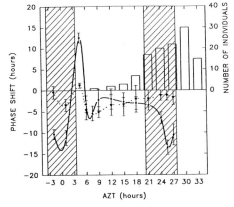

Figure 1. The phase shifting effect of nonsteroidal agonist RH-5849 applied onto Spodoptera littoralis larvae at various times of their development. The solid line (and the solid circles) depicts phase response for 5 µg of RH-5849, the dotted line (and the solid triangles) represents phase response for solvent. Each point refers to the mean ± SEM of 15-25 individuals, with the exception of the points corresponding with AZT 4.5 (43 individuals) and AZT 25.5 (29 larvae). Open bars in right upper corner depict circadian pattern of pupation (median is AZT 27). According to Zimowska et al. (1985) ecdysteroid titers in haemolymph of untreated larvae was nearly 15 ng/ml$*10^{-2}$ at AZT 0, 40 ng/ml$*10^{-2}$ at AZT 4.5 and 12 ng/ml$*10^{-2}$ at AZT 25.5. Shaded areas represent scotophase, unshaded-photophase. Other details in the text.

Table 1.

The dose dependent phase shift of pupation time in *Spodoptera littoralis* larvae treated with nonsteroidal ecdysone agonist RH-5849 at two different times of their development. Numbers in parentheses represent numbers of individuals.

dose of RH-5849	phase shift (h + SEM)	
	AZT 4.5	AZT 25.5
0.05	+1.38 + .8 (12)	-1.9 + 0.4 (17)
0.5	+2.9 + 0.1 (15)	-7.2 + 0.9 (16)
5	+12.1 + 1.2 (43)	-12.3 + 1.1 (29)
20	-3.0 + 1.1 (14)	-12.7 + 2.3 (13)
50	-6.8 + 0.7 (13)	-13.3 + 1.7 (12)
solvent	+1.2 + 0.6 (23)	-1.8 + 0.5 (21)

and pupation can occur earlier than normal. The second delaying effect of RH-5849 can be explained as a dose dependent delay of pupation (Table 1) due to inhibition of EH release, an effect similar to that described by Truman et al. (1983) for *Manduca* larvae treated with ecdysone.

This system is similar to that proposed by Truman (1984) for eclosion in *Manduca*. In his model two kinds of oscillators have been presented. In the first one, the gating system was photosensitive and EH release dependent. The second one was photo-insensitive and caused by changes in ecdysteroid level. The gating system has been presented as modulated by changes in ecdysteroid titers (Truman 1984).

The postulated short feedback loop between ecdysteroid level and ecdysteroid release can be supported by our results (Table 1). Low doses of RH-5849 can advance the pupation but high doses delay it. The postulated feedback may occur both at low and high doses of RH-5849 but higher doses can persist longer in the organism, combined with endogenous ecdysteroids and finally delay the pupation.

We realize that both the feedback and the two oscillator hypotheses should be confirmed by further experiments. We plan to scrutinize them by examining the effects of RH-5849 and ecdysteroid titers in *Spodoptera* under various lighting conditions.

This work has been supported in part by Grant KBN-14-501-GR-44.

REFERENCES

Riddiford, L., and Truman, J. W., Biochemistry of Insect Hormones and Insect Growth Regulators. In: *Biochemistry of Insects*, Academic Press, New York, San Francisco, London, pp 313-320, 1978.

Truman, J. W., Physiological aspects of the two oscillators that regulate the timing of eclosion in moths. In: *Photoperiodic Regulation of Insects and Molluscan Hormones*. Ciba Foundation Symp. Pitman, London, pp 221-239, 1984.

Truman, J. W., and Morton, D. B., The eclosion hormone system: an example of coordination of endocrine activity during the molting cycle of insects. In: *Progress in Comparative Endocrinology*, Wiley-Liss Inc., pp 300-308, 1990.

Truman, J. W., Roundtree, D. B., Reiss, S. E., and Schwartz, L. M., Ecdysteroids regulate the release and action of eclosion hormone in the tobacco hornworm, *Manduca sexta* (L.). *J. Insect Physiol.* 29, 895-900, 1983.

Truman, J. W., Taghert, P. H., Copenhaver, P. F., Tublitz, N. J., and Schwartz, L. M., Eclosion hormone may control all ecdysis in insects. *Nature*, Vol. 291, No. 5810, 70-71, 1981.

Zimowska, G., Handler, A. M., and Cymborowski, B., Cellular events in the prothoracic glands and ecdysteroid titres during the last-larval instar of *Spodoptera littoralis*. *J. Insect Physiol.* 31, 331-340, 1985.

DISRUPTION OF DIAPAUSE AND INDUCTION OF PRECOCIOUS DEVELOPMENT AND METAMORPHOSIS IN THE GYPSY MOTH WITH KK-42, A NOVEL ANTI-JH COMPOUND

Robert A. Bell, *Albert B. DeMilo and Thomas J. Kelly

Insect Neurobiology and Hormone Laboratory and
*Insect Chemical Ecology Laboratory, USDA, ARS,
Beltsville, MD 20705, USA.

Nearly 10 years have elapsed since the initial discovery that KK-42, a terpenoid imidazole compound, induced novel anti-juvenile hormone effects in the silkworm, *Bombyx mori* (Kuwano *et al.* 1983; Kuwana *et al. 1985).* Shortly thereafter, it was found that KK-42 not only suppressed secretion of juvenile hormone (JH) *in vivo* but also the secretion of ecdysteroid hormones from isolated prothoracic glands maintained *in vitro* (Yamashita *et al.* 1987; Akai and Mauchamp, 1989).   A subsequent discovery of considerable importance was the ability of KK-42 to induce termination of late embryonic diapause in the wild silkmoth, *Antheraea yamamai* (Enzhi *et al.* 1986; Suzuki *et al. 1990).* Although the effects of KK-42 on silkworm development has been well documented, there is a lack of evidence that KK-42 may exert such effects in other species of insects.   However, the results of several studies summarized here demonstrate that KK-42 is effective in disruption of the normal pathways of development and diapause in the gypsy moth, *Lymantria dispar.*

A gift of KK-42 (1 g) was generously provided by Dr. K. Suzuki, Iwate University, Morioka, Japan. An additional 10 g was synthesized at Beltsville according to previously described procedures (Kuwano *et al.* 1985). To test the effects on embryonic diapause, groups of eggs were de-haired by hand and immersed in various concentrations of KK-42 in acetone solution for 30-60 sec. Treated eggs were maintained at 25°C, 90-98% R.H and observed daily for yolk depletion and hatch. To determine effects of KK-42 on post-embryonic development, KK-42 was dissolved in acetone and incorporated into the larval diet at various concentrations. Observations were made on rate of growth, development, duration of larval instars and evidence of precocious developmental changes.

KK-42 was effective in disrupting diapause when applied at the proper dosage and age or stage of embryonic development. The most effective dose (8000 ppm), applied to 15-16 day old eggs, prevented diapause and induced 60-80% hatch within 1-2 weeks post-treatment (Table 1). Treated eggs that were less than 10 days old were not affected by KK-42 and entered diapause in a normal manner. Likewise, eggs that had already entered diapause (25-30 day-old) were not responsive to KK-42 and continued in a state of

0-8493-4591-X/94/$0.00 + $.50
© 1994 by CRC Press, Inc.

developmental arrest. This is in contrast to results obtained in *Antheraea yamamai* where KK-42 was effective in reactivating development in eggs after entry into diapause (Enzhi *et al.* 1986). After diapausing eggs were chilled for 2 months, KK-42 was effective in accelerating the rate of diapause termination. The mechanism by which KK-42 prevents or precociously terminates diapause in the gypsy moth is not known. However, in *Antheraea yamamai*, it is believed to interfere with an inhibitory center located in the mesothoracic part of the pharate embryonic larvae that is required for the maintenance of diapause (Suzuki *et al.* 1990). Thus, the proposed mechanism regulating late embryonic diapause differs from previously proposed mechanisms in post-embryonic larval/pupal diapause in that the brain neuroendocrine and prothoracic gland systems are not thought to be involved.

TABLE 1

Diapause Suppressing Effects of KK-42 in Relation to Age or Developmental Stage of *Lymantria dispar* Eggs

Age of eggs (days)	Morphogenetic stage	Hatch %
1	---------	0
5	segmentation of embryo	0
10	nervous endocrine systems and cuticle formation	20-40
15-16	mature, pharate larva	60-80
20	onset of diapause	30-40
25-30	intense diapause	0
31-60	diapause eggs chilled one month	0
90-120	diapause eggs chilled 2-3 months	75-90

Each treatment involved three reps (n = 60-70 eggs) 1 rep immersed in acetone solutions of KK-42 at 8000 ppm (30-40 sec); diapause eggs were chilled at 5-7°C, then incubated at 25°C for 24 h before treatment. Control groups (treated with acetone) showed no hatch except in 2-3 month chilled eggs in which hatch was delayed by 1-3 weeks beyond that of the KK-42 treated group.

TABLE 2

Developmental Effects Induced by KK-42 Incorporated in Artificial Diet and Fed to *Lymantria Dispar* Larvae.

Observed effect	Effective dose range (ppm)
Precocious metamorphosis	120-240
L_4 phenotype induced in L_3 larvae	80-240
Pupal antennae formed in 5th instar larvae	30-120
Increased duration and weight of larval instars 1-4	40-320

n = 10-20 larvae used/treatment; generally the larvae were fed KK-42 during the instar preceding the observed effect. Effective dose range refers to the minimum and maximum amount required to induce the lowest and highest response respectively.

We have also confirmed and extended the findings of earlier Japanese investigators by showing that KK-42 readily induces precocious metamorphosis when penultimate instars of the gypsy moth are fed on diet containing dosages of 120-240 ppm (Table 2). Preliminary data also showed that ecdysteroid hormone secretion was suppressed by KK-42 in a dose dependent manner in isolated gypsy moth prothoracic glands *in vitro*. There was also a dose-dependent increase in the weight and duration of larval instars 1-4, but not in the 5th (final) instar.

We also observed effects of KK-42 in the gypsy moth that are different from those previously reported in silkworms. These include the precocious formation of pupal antennae in newly molted 5th (final) instar larvae induced by relatively low doses (30-60 ppm) during the 4th instar. This malformation was lethal due to inhibition of larval feeding, thereby resulting in death from starvation. Another effect of KK-42 was the precocious appearance of 4th larval instar head capsule coloration in newly molted 3rd instar larvae. This finding suggests that juvenile hormone may play a role in the color changes of the head capsule that normally accompany progression from the 3rd to 4th larval instar. The potential for using KK-42 as a new anti-hormonal agent for insect control will require further study. It is evident, however, that the compound is useful as a probe to detect developmental events in embryonic and post-embryonic development that are mediated by juvenile and ecdysteroid hormones. The discovery that diapause can be circumvented with KK-42 should facilitate our efforts to determine the mechanism of diapause in the gypsy moth, and to further improve laboratory and mass rearing for biological control programs.

REFERENCES

Kuwano, E., Takeya, R. and Eto, M., Terpenoid imidazoles: New anti-juvenile hormones, *Agric. Biol. Chem.*, 47(4), 921, 1983.

Kuwano, E., Takeya, R. and Eto, M., Synthesis and anti-juvenile hormone activity of 1-substituted-5-[(E)-2, 6-dimethyl-1,5-heptadienyl] imidazoles, *Agric. Biol. Chem,* 49(2), 483, 1985.

Yamashita, O., Kadona-Okuda, K., Cain, E. and Eto, M., An imidazole compound as a potent anti-ecdysteroid in an insect, *Agric. Biol. Chem.* 510, 2295, 1987.

Akai, H. and Mauchamp, B., Suppressive effects of an imidazole derivative, KK-42, on JH levels in hemolymph of *Bombyx* larvae., *J. Seric. Sci. Jpn.*, 58(1), 73, 1989.

Enzhi, T., Suzuki, K., Cain, E., Abe, S. and Kurihara, M., Effect of anti-JH (KK-42) treatment on the breaking of the diapause of the eggs of the silkmoth *Antheraea yamamai, J. Seric. Sci. Jpn.,* 55(4), 305, 1986.

Suzuki, K., Minigawa, T., Kumagai, T., Naya, S., Endo, Y., Osanai, M. and Cain, E., Control mechanism of diapause of the pharate first-instar larvae of the silkmoth *Antheraea yamamai., J. Insect Physiol.* 36(11), 855, 1990.

NEUROENDOCRINE CONTROL OF DIAPAUSE IN MALES OF PYRRHOCORIS APTERUS

M. Hodková[1] and R.M. Wagner[2]

[1]Institute of Entomology, Czech Academy of Sciences, 370 05 České Budějovice, Czech Republic, [2]United States Department of Agriculture, Agricultural Research Service, Livestock Insects Laboratory, Beltsville, MD 20705-2350, USA

INTRODUCTION

Pyrrhocoris apterus has adult diapause controlled by photoperiod in both males and females (Hodek 1968, Ždárek 1970, Hodková et al. 1991). Photoperiodic signals are transmitted via the neuroendocrine system. The involution of ovaries in short-day females is due to inactivity of the corpora allata (CA) which are inhibited by the pars intercerebralis (PI) of brain. Under long-day conditions, the PI stimulates the CA (Hodková 1976).

In males of *P. apterus*, short-day conditions inhibit mating behaviour. This effect of short days is probably not due to inactivity of the CA because these glands are not necessary for mating activity in long-day males (Ždárek 1968).

The results presented here are concentrated on the role of the PI in regulation of mating behaviour in males of *P. apterus* under diapause inducing short days and diapause preventing long days.

MATERIAL AND METHODS

Pyrrhocoris apterus L. (Heteroptera) was reared from the egg stage at 25°C under either long-day (18L:6D) or short-day (12L:12D) conditions. Males and females were fed as individual pairs on linden seed and water in Petri dishes. Their mating activity was evaluated by (1) number of females ovipositing viable eggs, and (2) number of copulating pairs.

The PI or the CA were removed under insect saline through an incision to the neck membrane (Sláma 1964). In sham-operated controls, the neck membrane was incised but the PI or the CA were only touched.

RESULTS AND DISCUSSION

All females of long-day pairs oviposited viable eggs irrespective of whether males were PI-ectomized or sham-operated (Table 1). The average frequency of mating during 24 h was similar in intact and PI-ectomized males (Hodková, unpubl.). The results indicate that factors stimulating mating behaviour in

0-8493-4591-X/94/$0.00 + $.50

© 1994 by CRC Press, Inc.

long-day males are outside of the PI. In contrast, Pener et al. (1972) reported that a neurosecretory factor from the PI stimulates mating behaviour (Pener et al. 1972).

Two weeks after surgical treatment, short-day males were given a single, 2.5 h opportunity to copulate. Extirpation of the PI stimulated mating activity. While only 3.3% (n=60) sham-operated males mated, as much as 45% (n=53) of PI-ectomized males mated within this period.

Table 1

Effect of extirpation of pars intercerebralis in males on viability of eggs

	Days after surgical treatment	n	% females ovipositing viable eggs
Long-day conditions			
Extirpation of PI	21-27	9	100
Sham-operation	17-24	14	100
Short-day conditions			
Extirpation of PI	19-26	13	100
Sham-operation	20-34	18	28

Data from Hodkova (1990)

Short-day PI-ectomized and long-day intact males showed similar average frequency of mating during 24 h (Hodková, unpubl.). The difference between PI-ectomized and sham-operated short-day males was confirmed by comparison of the viability of eggs (Table 1). The results indicate that factors stimulating mating behaviour are suppressed by the PI in diapause males of *P. apterus*. The inhibitory effect of the PI on mating behaviour has not been found so far in other species.

The inhibition of mating behaviour by the PI does not seem to be due to an inhibition of the CA. Allatectomized short-day males transferred to long-day conditions were readily activated (Table 2). In contrast to *P. apterus*, the mating behaviour is assumed to be controlled by the CA in acridids with adult diapause (Pener 1992).

Table 2

Effect of allatectomy of males on viability of eggs

	Days after surgical treatment	n	% females ovipositing viable eggs
Extirpation of CA	5-12	10	90
	12-18	10	100
	22-30	10	90
Sham-operation	4-15	9	100
	15-21	10	100
	22-29	9	89

Males were transferred from short days to long days on the day of operation

While neurosecretory factors regulating activity of the CA have been identified as peptidic hormones, allatotropins and allatostatins (Gäde 1992), the chemical nature of cerebral factors controlling mating behaviour is not known. Our attempt at identification of these factors has been started by comparison of brain extracts from diapause and active males of *P. apterus*. HPLC tracings showed differences in three peaks representing peptides (Wagner and Hodková, this volume).

REFERENCES

Gäde, G., Insect neuropeptides, in *Advances in Regulation of Insect Reproduction*, Benettova B., Gelbic I. and Soldan T., Eds., Inst. Entomol., Czech Acad. Sci., 1992, pp. 67-80.

Hodek, I., Diapause in females of *Pyrrhocoris apterus* L. (Heteroptera), *Acta Entomol. Bohemoslov.*, 65, 422-435, 1968.

Hodkova, M., Nervous inhibition of corpora allata by photoperiod in *Pyrrhocoris apterus*, *Nature*, 263, 422-435, 1976.

Hodkova, M., Transmission des signaux photopériodiques chez *Pyrrhocoris apterus* L. (Heteroptera), in *Régulation des Cycles Saisonniers chez les Invertébrés*, Ferron P., Missonnier J. and Mauchamp B., Eds., INRA, Paris, 1990, pp. 209-212.

Hodkova, M., Zieglerova, J. and Hodek, I., Diapause in males of *Pyrrhocoris apterus* and its dependence on photoperiod and activity of females, *Zool. Jb. Syst.*, 118, 279-285, 1991.

Pener, M. P., Environmental cues, endocrine factors, and reproductive diapause in male insects, *Chronobiol. Int.*, 9, 102-113, 1992.

Pener, M. P., Girardie A. and Joly P., Neurosecretory and corpus allatum controlled effects on mating behavior and color change in adult *Locusta migratoria migratoides* males, *Gen. Comp. Endocrinol.*, 19, 494-508, 1972.

Slama, K., Hormonal control of respiratory metabolism during growth, reproduction, and diapause in female adults of *Pyrrhocoris apterus* L. (Hemiptera), *J. Insect Physiol.*, 10, 283-303, 1964.

Wagner, R. M. and Hodkova M., Photoperiodic control of mating behavior of *Pyrrhocoris apterus*, this volume.

Zdarek, J., Le comportement d'accouplement à la fin de la diapause imaginale et son contrôle hormonal dans le cas de la punaise *Pyrrhocoris apterus* L. (Pyrrhocoridae, Heteroptera), *Ann. Endocr. Paris*, 29, 703-707, 1971.

Zdarek, J., Mating behaviour in the bug, *Pyrrhocoris apterus* L. (Heteroptera): ontogeny and its environmental control, *Behaviour*, 37, 253-268, 1970.

Peptides Affecting Reproduction

TESTIS ECDYSIOTROPIN: SEQUENCE AND MODE OF ACTION

M.J. Loeb[1], R.M. Wagner[2], D.B. Gelman[1], J.P. Kochansky[1],
W.R. Lusby[1], C.W. Woods[3], and K. Ranga Rao[4]
[1] Insect Neurobiology and Hormone Lab, [2] Livestock Insects Lab,
U.S. Department of Agriculture, Beltsville, MD, USA; [3] Dcpt. of
Animal Sciences, Univ. of MD, College Park, MD, USA;
[4]Dept. of Biology, Univ. of West FL, Pensacola, FL, USA.

Testis ecdysiotropin (TE) can be extracted from brains of *Heliothis
virescens* or *Lymantria dispar* at   the same developmental times that
excised testes synthesize ecdysteroid *in vitro* (Loeb *et al.* 1984, 1987,
1988). Brain extracts stimulated RIA-detectable synthesis *de novo* in
early last instar testes of either species, and boosted already existing
synthetic activity in testes from pupae (Loeb *et al.* 1984, 1988); extracts
were interchangeable between species (Loeb *et al.* 1988). Activity was
bioassayed in early last instar testes *in vitro*; dose response to TE was a
sigmoidal curve (Loeb *et al.* 1988). TE appeared in peptidic fractions at
several molecular weight ranges, with high activity in the 2 KDa range
(Loeb *et al.* 1990).

In this study, approximately 13,000 brains from day 7 *L. dispar* pupae
were dissected and kept at -20° C in methanol: H_2O: acetic acid:
thiodiglycol (90: 9: 0.9: 0.1), containing an anti-protease cocktail (1:100)
(Evans 1987). Brains were sonicated in this fluid, and centrifuged at
16,000 g for 5 min; supernatants were combined and concentrated (Speed
Vac) to 500 μl at room temperature. The concentrate was diluted in 10%
acetonitrile (CH_3CN)/H_2O/0.1% trifluoroacetic acid (TFA), passed
through doubled C-18 Sep Pak cartridges (Waters), eluted with 80%
aqueous CH_3CN/0.1%TFA. The eluate was reconcentrated to 1000 μl.
Peptide standards, followed by four 250 μl aliquots of the TE preparation,
were subjected to High Pressure Size Exclusion chromatography
(HPSEC) through two I-125 (Waters) columns arranged in tandem,
preceded by a TSK SWXL progel guard column (Supelco Inc.) and eluted
with 40% CH_3CN/0.1%TFA, 1 ml/min; fractions of 1-2 KDa were
collected and bioassayed. All active material was pooled, reconcentrated,
and subjected to HPLC through a Vydac C-18 column (Fig 1). One of
the peaks (41.8 min), appeared to be clean and was TE bioactive (Fig 1).
Edman degradation strongly suggested a 21-mer with the sequence: Ile-
Ser-Asp-Phe-Asp-Glu-Tyr-Glu-Pro-Leu-Asn-Asp-Ala-Asp-Asn-Asn-Glu-
Val-Leu-Asp-Phe. However, it is also possible that a weak terminal signal
indicates that the native molecule is a 22-mer which ends in Leu.

The 21-mer peptide was synthesized as an amide on a Milligen/-
Biosearch 9600 synthesizer, using the Boc protocol on p-methylbenzhy-

0-8493-4591-X/94/$0.00 + $.50
© 1994 by CRC Press, Inc.

drylamine resin, with synthesis routines modified to allow for *in situ* neutralization during synthesis. It was cleaved from the resin with HF/anisole and purified by standard reverse phase HPLC. The sequence was confirmed by Edman degradation and MS. Molecular weight was 2472 Da. However, the HPLC retention time for the amide was 40.4 min, while the retention time for the acid form (prepared with carboxypeptidase y) was 42.8 min, both close to but not the same as the retention time of the native material (41.8 min). It is possible that native TE is actually the 22-mer. However, bioassays of the synthetic 21-mer amide were positive, with maximum steroidogenic effects at 10^{-10} and 10^{-15} M when bioassayed with early last instar *L. dispar* testes. It boosted ecdysteroid synthesis by testes from day 7 *L. dispar* pupae at 10^{-9} and 10^{-13} M (Fig. 2). These titers are well within a physiological range of activity.

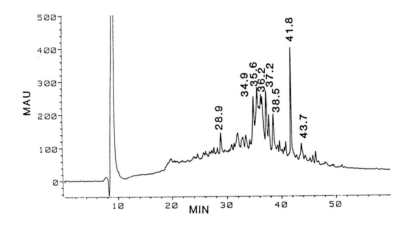

Figure 1: Approximately 1-2 kDa fraction from HPSEC subjected to HPLC on Vydac C-18; absorption at 210 mμ. The peak eluting at 41.8 min was analyzed.

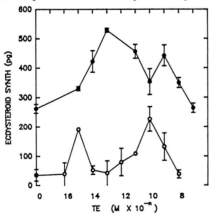

Figure 2: Effects of synthetic TE on ecdysteroidogenesis (larvae- open circles. pupae- closed circles). n = 4.

The molecular weight and sequence of TE differ from the prothoracic ecdysiotropins Bombyxin (Ishizaki *et al.* 1987) and big PTTH (Kataoka *et al.* 1987) from *B. mori*, as well as from a partial sequence of the egg development neurohormone, a mosquito ovarian ecdysiotropin (Masler, personal communication). It does not elicit RIA-detectable ecdysteroid or 3-dehydroecdysone production by *L. dispar* prothoracic glands. Therefore, TE may be a unique peptide which functions to regulate ecdysteroid secretion by testes.

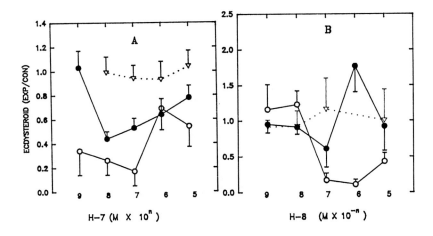

Figure 3: Phosphokinase inhibitor effects on ecdysteroidogenesis induced by synthetic TE (10^{-10} M) in larval testes (open circles), pupal testes (closed circles) and in the absence of TE in pupal testes (triangles). A: H-7, B: H-8, n = 8.

We showed that low Ca^{2+} titer (Loeb 1991) and diacylglycerol mediated the action of TE in testes, although cross talk with cAMP was probable since TE elicited production of cAMP; cAMP alone induced ecdysteroid synthesis by larval testes. However, cAMP in the presence of TE served to inhibit ecdysteroidogenesis (Loeb *et al.* 1993). In contrast, PTTH is mediated at the prothoracic glands by high Ca^{2+} titers and cAMP (Smith *et al.* 1984). The isoquinoline sulfonamide, H-7, acts to block the phosphorylation induced by phosphokinase C (PKC) (Hidaka *et al.* 1984); H-7 inhibited synthetic TE-induced (10^{-10} M) ecdysteroid synthesis in larval and pupal testes (Fig 3a), supporting previous evidence for mediation by the inositol phosphate system and PKC. H-8 blocks the phosphorylation induced by phosphokinase A (Hidaka *et al.* 1984); it also inhibited synthetic TE-induced larval synthesis but not TE-induced pupal synthesis (Fig 3b). Thus, cAMP may function in *de novo* TE-stimulated ecdysteroid synthesis, but not in TE-boosted or ongoing pupal synthesis. The ADP ribosylating agent, pertussis toxin (Heideman and Bourne 1990) inhibited ecdysteroidogenesis by larval (100% inhibition) and pupal (50% inhibition) testes exposed to synthetic TE at concentrations from 60 to 2 pg/µl, indicating that G_i protein (Heideman and Bourne 1990) is part of the TE-induced cascade. H-7, H-8, and pertussis toxin had no significant effect on ongoing synthesis in pupal testes, suggesting that ecdysteroidogenesis by testes stimulated *in vivo,* prior to excision, is constitutive.

REFERENCES

Evans, W.H., Organelles and membranes of animal cells, in *Biological Membranes. A practical approach*, Findlay, J.B. and Evans, W.H., Eds., IRL Press, Oxford. 1987. chap. 1.

Ishizaki, H., Mizoguchi, A., Hatta, M., Suzuki, A., Nagasawa, H., Kataoka, H., Isogai, A., Tamura, S., Fujino, M., and Kitada, C., Prothoracicotropic hormone (PTTH) of the silkmoth, *Bombyx mori*: 4K-PTTH. *Molec. Entomol.* 19, 119-128, 1987.

Kataoka, H., Nagasawa, H., Isogai, A., Tamura, S., Mizoguchi,A., Fujiwara, Y., Suzuki, C. Ishizaki, H. and Suzuki, A., Isolation and partial characterization of a prothoracicotropic hormone of the silkworm, *Bombyx mori.* Agric.Biol. Chem. 51, 1067-1076, 1987.

Hidaka, H., Inagaki, M., Kawamoto, S., and Sasaki, Y., Isoquinolinesulfonamides, novel and potent inhibitors of cyclic nucleotide dependent protein kinase and protein kinase C. *Biochemistry* 23, 5036-5041, 1984.

Heideman, W. and Bourne, H.R., Structure and function of G protein α chains, in *G Proteins,* Iyengar, I. and Birnbaumer, L., Eds. Academic Press, N.Y. 1990 pp 17-40.

Loeb, M.J., Brandt, E.P. and Birnbaum, M.J., Ecdysteroid production by testes of the tobacco budworm, *Heliothis virescens,* from last larval instar to adult, *J. Insect Physiol.* 30, 375-381, 1984.

Loeb, M.J., Brandt, E.P., Woods, C.W. and Borkovec, A.B., An ecdysiotropic factor from brains of *Heliothis virescens* induces testes to produce immunodetectable ecdysteroid *in vitro*, *J. Exp. Zool.* 243, 275-282, 1987.

Loeb, M.J., Brandt E.P., Woods,C.W. and Bell R.A., Secretion of ecdysteroid by sheaths of testes of the Gypsy moth, *Lymantria dispar,* and its regulation by testis ecdysiotropin. *J. Exp Zool.* 248, 94-100, 1988 .

Loeb, M.J., DeMilo, A.B., and Sheppard, C.A., Characterization of testis ecdysiotropin (TE) from brains of Lepidoptera, in *Insect Neurochemistry and Neurophysiology 1989*, Borkovec, A.B. and Masler, E.P., Humana Press, Clifton, N.J. 1990 pp 259-262.

Loeb M.J., Effect of calcium ions on ecdysteroid secretion by testes of *Heliothis virescens.* *Comp. Biochem. Physiol.* 100B:303-308, 1991.

Loeb M.J., Gelman, D.B. and Bell, R.A., Second messengers mediating the effects of testis ecdysiotropin in testes of the Gypsy Moth, *Lymantria dispar.* Arch. Insect Biochem. Physiol., 23, 13-28, 1993.

Smith, W.A., Gilbert, L.I., and Bollenbacher, W.E. The role of cyclic AMP in the regulation of ecdysone synthesis. *Molec. Cell Endocrinol.,* 37, 285-294, 1984.

Acknowledgements: Thanks to Melissa Champion for technical assistance, and to Irene Bedard for clerical assistance.

IDENTIFICATION OF A PEPTIDE FROM THE FEMALE REPRODUCTIVE SYSTEM OF THE HOUSE FLY, *MUSCA DOMESTICA*

R. M. Wagner[1*], J. P. Kochansky[2], A. Richeson[3], J. A. Hayes[3], J. C. Hill[4] and B. A. Fraser[4]

United States Department of Agriculture, Agricultural Research Service, [1]Livestock Insects Laboratory, and [2]Insect Neurobiology and Hormone Laboratory, Beltsville, MD 20705,

[3]Department of Animal Sciences, University of Maryland, College Park, MD 20809, and

[4]Food and Drug Administration, Center for Biological Evaluation and Research, Bethesda, MD 20892

INTRODUCTION

The regulation of reproductive behavior and function in insects is a complex series of events including development of reproductive tissues, initiation or inhibition of mating behavior, and those which aid in the fertilization process, maturation of oocytes, or oviposition. Several factors which regulate these processes have been identified, such as juvenile hormones, allatostatins and allatotropins, mating inhibition factors, oviduct stimulating factors, pheromone-biosynthesis activating hormones, egg-development hormones, enzymes and other compounds. In general, these factors have been isolated from the brain, *corpora cardiaca*, *corpora allata* or other tissues of the nervous system. To date, the only peptides isolated from reproductive tissues of insects have been mating inhibitory (Baumann *et al.* 1975; Chen *et al.* 1988; Chen and Balmer 1989) and oviduct stimulating factors (Paemen *et al.* 1991d; Paemen *et al.* 1991b) from male accessory glands and proctolin from oviducts (Holman and Cook 1985; Orchard and Lange 1987; Cook and Wagner 1990; Lange 1990) and oostatic hormone from ovaries (Borovsky *et al.* 1990) of female insects (although immunoreactivity to other peptides has been identified in insect oviducts; Lange *et al.* 1991; Sevala *et al.* 1992). We report here the first peptide factor isolated from the accessory reproductive glands of a female insect, a factor found in the house fly, *Musca domestica.*

0-8493-4591-X/94/$0.00 + $.50
© 1994 by CRC Press, Inc.

281

ISOLATION AND PURIFICATION

The factor, which we have named *Musca* AG peptide, was isolated from accessory glands of 100 3-5 day-old female house flies. The tissues were extracted into methanol/1 N HCl/acetonitrile/water (75:25:50:50, v/v/v/v) containing 0.05% (by volume) trifluoroacetic acid by homogenization for 30 sec, followed by cooling intervals of 5 min in crushed ice (Wagner *et al.*, 1993). Extracts were centrifuged through 0.45 um filter units, followed by removal of proteins using 10,000 NMWL centrifugal ultra-filtration units. Samples were concentrated under a stream of argon and diluted with 10% acetonitrile/0.1% TFA prior to HPLC purification. *Musca* AG peptide was separated from other components on a Vydac C-18 column (218TP54) using a one-hour gradient of 10 - 50% acetonitrile in water containing 0.1%TFA at a flow rate of 0.4 ml/min (Jaffe *et al.* 1989). Eluting peaks were monitored at 210 nm. The peptide, which eluted at 51.4 min (Fig 1A), was determined by amino acid analysis to contain: Leu(6), Asx(2), Ala(2), Pro(1), Ser(2), Thr(1), Gly(1).

STRUCTURAL IDENTIFICATION

An aliquot of this same fraction was subjected to automated Edman degradation with the resulting sequence: Leu-Leu-Asn-Ala-Leu-Pro-Leu-Asp-Ala-Leu-Ser-Ser-Leu-Thr-Gly. Plasma desorption mass spectrometry (Bio-Ion) confirmed the sequence, with an m/z of 1519 for the $[M + Na^+]^+$ ion and an m/z of 1501 for the $[M + Na^+ - H_2O]^+$ ion, which are consistent with a C-terminal amide residue. Chemical synthesis of this structure (**Leu-Leu-Asn-Ala-Leu-Pro-Leu-Asp-Ala-Leu-Ser-Ser-Leu-Thr-Gly-NH$_2$**) resulted in an identical retention time during HPLC analysis to that of the native compound (Fig. 1C).

DISCUSSION

Extracts of the oviducts of female house flies did not contain this material to any appreciable extent (Fig 1B), and there was little or no difference between levels of *Musca* AG peptide in accessory glands of mated and unmated flies (data not shown). The function of this peptide is unknown, although both the synthetic and natural products have some stimulatory activity on house fly oviduct contraction. We are presently examining other biochemical, physiological and behavioral assays to ascertain the role(s) of this peptide in housefly reproductive function.

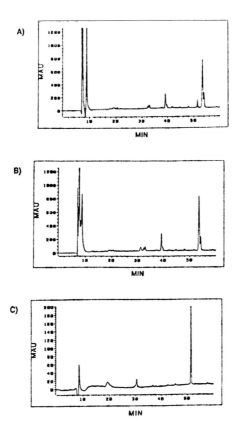

Figure 1. HPLC elution profiles of extracts of oviducts and accessory glands from female house flies on a Vydac C-18 column at 28°C and 0.4 ml/min. Solvent A = 0.1% TFA in water, solvent B = 60% acetonitrile/40% water (v/v) containing 0.1% TFA. Linear gradient elution from 16.7% solvent B to 83.3% solvent B in 60 min. A) extract of accessory glands from 67 unmated 3-5 day-old females, B) extract of oviducts from 100 unmated 3-5 day-old females, C) purification of synthetic peptide.

REFERENCES

Baumann, H., Wilson, K. J., Chen, P. S., and Humbel, R. E., The amino-acid sequence of a peptide (PS-1) from *Drosophila funegris:* a paragonial peptide from males which reduces the receptivity of the female,. *Eur. J. Biochem.* 52, 521-529, 1975.

Borovsky, D., Carlson, D. A., Briffin, P. R., Shabanowitz, J, and Hunt, D. F., Mosquito oostatic factor: a novel decapeptide modulating trypsin-like enyzme biosynthesis in midgut. *FASEB J.* 4, 3015-3020, 1990.

Chen, P. S., Stumm-Zollinger, E., Aigaki, T., Balmer, J., Bienz, M., and Bohlen, P., A male accessory gland peptide that regulates reproductive behavior of female *D. melanogaster*, *Cell* 54, 291-298, 1988.

Chen, P. S. and Balmer, J., Secretory proteins and sex peptides of the male accessory gland in *Drosophila sechellia*, *J. Insect Physiol.* 35, 759-764, 1989.

Cook, B. J. and Wagner, R. M., Prospective chemical regulators of female reproductive muscle function in the stable fly *Stomoxys calcitrans.*, in *Insect Neurochem. Neurophysiol.*, Masler, E., Kelly, T. and Borkovec, A., Eds, Humana Press, New York, 1990, pp 413-416.

Holman, G. M. and Cook, B. J., Proctolin, its presence in and action on the oviduct of an insect, *Comp. Biochem. Physiol. C* 80, 61-64, 1985.

Jaffe, H., Raina, A. K., Riley, C. T., Fraser, B. A., Nachman, R. J., Vogel, V. W., Zhang, Y., and Hayes, D. K., Primary structure of two neuropeptide hormones with adipokinetic and hypotrehalosemic activity isolated from the corpora cardiaca of horse flies (Diptera), *Proc. Nat. Acad. Sci., USA* 86, 8181-8164, 1989.

Lange, A. B., The presence of proctolin in the reproductive system of *Rhodnius prolixus*, *J. Insect Physiol.* 36, 345-351, 1990.

Lange, A. B., Orchard, I., and TeBrugge, V. A., Evidence for the involvement of a Schisto-FLRF-amide-like peptide in the neural control of the locust oviduct, *J. Comp. Physiol. A* 168, 383-391, 1991.

Orchard, I. and Lange, A. B., Cockroach oviducts: the presence and release of octopamine and proctolin, *J. Insect Physiol.* 33, 265-268, 1987.

Paemen, L., Tips, A., Schoofs, L., Proost, P., Vandamme, J. and DeLoof, A., Lom-AG-Myotropin I: a novel peptide from the male accessory glands of *Locusta migratoria, Peptides* 12, 7-10, 1991a 1991a.

Paemen, L., Schoofs, L., Proost, P., Decock, B. and DeLoof, A., Isolation, identification and synthesis of Lom-AG-Myotropin II, a novel peptide in the male accessory reproductive glands of *Locusta migratoria, Insect Biochem.* 21, 243-248, 1991b.

Sevala, V. L., Sevala, V. M., Davey, K. G. and Loughton, B. G., A FMRF-amide-like peptide is associated with the myotropic ovulation hormone in *Rhodnius prolixus, Arch. Insect Biochem. Physiol.* 20, 193-203, 1992.

Wagner, R. M., Woods, C. W., Hayes, J. A., Kochansky, J. P., Hill, J. C. and Fraser, B. A., Isolation and identification of a novel peptide from the accessory sex gland of the female house fly, *Musca domestica, Biochem. Biophys. Res. Commun.*, 194, 1336-1343, 1993.

ANTAGONISTS AND AGONISTS OF JH ACTION ON THE FOLLICLE CELLS OF *LOCUSTA MIGRATORIA*

[1]K. G. Davey, [1]V. L. Sevala, [2]G. D. Prestwich and [3]L. L. Jurd

[1]Department of Biology, York University, North York, ON, Canada, M3J 1P3; [2]Department of Chemistry, SUNY, Stony Brook, NY 11794, USA; [3]Plant Protection Research, USDA, Albany, CA 94710, USA.

BACKGROUND

Among the functions of juvenile hormone (JH) in reproduction in the female insect is the control of patency, a condition whereby large extracellular spaces develop between the cells of the follicular epithelium surrounding the vitellogenic oocyte. This allows the vitellogenin circulating in the hemolymph to have access to the oocyte surface, where receptor-mediated endocytosis incorporates the vitellogenin into the yolk granules of the oocyte. The development of patency has been shown in the blood feeding hemipteran *Rhodnius prolixus* to result from a shrinkage of the cells mediated by the activation of a JH-sensitive Na^+/K^+ ATPase. The activation of the Na^+/K^+ ATPase requires JH and is mediated via protein kinase C which phosphorylates a 100 kDa peptide. All of these actions of JH occur in the membrane of the follicle cell, and membrane preparations bind JH I in a specific and saturable manner (K_D 6.5 nM, B_{max} 1.9 pmol/mg). [Ilenchuk and Davey, 1987; Sevala and Davey, 1993]. More recently, the follicle cells of *Locusta migratoria* have been shown to react in a similar way to JH III. When follicle cells are exposed *in vitro* to JH III, the optical path difference as measured by interference microscopy increases [Davey et al., 1993]. The optical path difference (opd) is a measure of the dry mass of the cell, and changes in opd are inversely proportional to changes in the volume of the cell [Abu-Hakima and Davey, 1977]

A variety of compounds has been examined for their ability to act as agonists in causing the follicle cells to reduce in volume in *in vitro* tests. In *Rhodnius*, JH II and JH III are without effect: only JH I among the known functional JHs is effective [Ilenchuk and Davey, 1987]. In *Locusta*, JH III and, to a lesser degree, JH II, are capable of causing the cells to shrink, while JH I is ineffective. The bis-epoxide of JH III is neither an agonist nor an antagonist to JH III [Davey et al., 1993].

The action of JH I on the follicle cells of *Rhodnius* is antagonised by an antigonadotropin, a peptide which emanates from neurons in a series of

0-8493-4591-X/94/$0.00 + $.50
© 1994 by CRC Press, Inc.

abdominal neurosecretory organs [Davey, 1993]. The action of JH III on the follicle cells of *Locusta* is likewise antagonised by extracts prepared from the abdominal neuroscretory organs of *Rhodnius*, and the thoracic ganglia of *Locusta* contains similar activity [Davey et al., 1993].

NEW ANTAGONISTS

In seeking additional antagonists to the action of JH on follicle cells, we have identified two sorts of compounds, which, when present in the *in vitro* medium for locust follicle cells, inhibit the action of JH III in bringing about an increase in the opd. Certain sulphonamides, such as ethoxyzolamide, acetazolamide, or dansyl amide act as antagonists to JH III. When they are present in the medium bathing isolated locust follicle cells at concentrations greater than 10^{-5} M, the action of JH in increasing the opd is markedly reduced or abolished. These sulphonamides are inhibitors of carbonic anhydrase, and their action was predicted from their effects on ecdysis in the nematode *Haemonchus contortus*. In this nematode, ecdysis from the second to the third larval stage is stimulated by a host signal involving CO_2. JH and certain juvenoids at very high concentrations can mimic the action of CO_2, and ethoxyzolamide antagonises both CO_2 and JH [Davey et al., 1983].

A second compound which antagonises the action of JH III on isolated locust follicle cells *in vitro* is J2581, one of a series of compounds isolated from Panamanian hardwood [Jurd et al., 1979]. J2581 (5-ethoxy-6-(4-methoxyphenyl)methyl-1,3-benzodioxole) has been reported as interfering with the uptake of vitellogenin in *Drosophila*. It has a wide variety of effects [Ma, 1993]. In tests with the follicle cells from *Locusta*, when J2581 is present in the medium at concentrations of 10^{-5} M, the effect of JH III in causing an increase in the opd is all but abolished.

PRELIMINARY BINDING STUDIES

Using the photoaffinity ligand for JH III, EFDA [Koeppe et al., 1984], preparations of membranes from follicle cells of *Locusta* have been shown to contain a peptide of 35 kDa which binds the ligand and from which the binding is displaced by competition with JH III. Similar preparations do not bind EBDA, the photoaffinity label for JH I, but do bind EHDA, the label for JH II. These binding characteristics of the photoaffinity labels thus reflect the biological activity of the analogous JHs. In *Rhodnius*, using labelled JH I as the ligand, earlier studies had shown that neither JH II nor JH III exhibited competition for the binding site [Ilenchuk and Davey, 1987].

While it has not yet been possible to conduct full competition assays, incubating preparations of membranes from locust follicle cells in the presence of 10^{-4} M J2581 reduces the binding of EFDA when the membranes are subsequently extracted and run on SDS-PAGE. Densitometry of the gels shown that J2581 reduces the amount of EFDA bound to about 50% of the value in the absence of the antagonist. Similarly, ethoxyzolamide at 10^{-4} M reduces the binding of EFDA to about 55% of the control value.

IMPLICATIONS

These preliminary observations strongly suggest that J2581 and ethoxyzolamide owe their inhibitory action to an ability to interact with the binding site for JH on the follicle cell membrane. Taken together with the observations related to the specificity of the binding by JH, whereby JH I, but not JH II or JH III, interacts with the site in *Rhodnius*, while JH III and JH II, but not JH I appear able to bind to the site in *Locusta*, some tentative conclusions are possible about the nature of the interaction. First, the difference between the hormones in the series JH I, II, and III has to do with the nature of the groups attached to carbons 7 and 11. The position of these carbons is thus of importance in defining the binding interaction. Second, the fact that J2581 and the sulphonamides share in common a benzene ring, coupled with the fact that many compounds which possess JH activity also possess a benzene ring [Zaoral and Slama, 1970; Niwa et al., 1990] strongly suggests that the conformation adopted by the hormones is one which resembles one or more open rings. Modelling studies are currently in progress.

REFERENCES

Abu-Hakima, R. and Davey, K. G., The action of juvenile hormone on the follicle cells of *Rhodnius prolixus*: the importance of volume changes, *J. Exp. Biol.*, 69, 33, 1977.

Davey K. G., Hormonal integration of egg production in *Rhodnius prolixus. Amer. Zool.* In press, 1993.

Davey, K. G., Rogers, W. P. and Sommerville, R. I., The effect of a mimic of insect juvenile hormone, an inhibitor of carbonic anhydrase, nor-adrenaline and iodine on changes in the optical path difference of the excetory cells accompanying exsheathment in *Hemonchus contortus*, *Internat. J. Parasit.*, 12, 509, 1983

Davey, K. G., Sevala, V. M. and Gordon, D. R. B., The action of juvenile hormone and antigonadotropin on the follicle cells of *Locusta migratoria*, *Invert. Reprod. Devel.*, in press, 1993.

Ilenchuk, T. T. and Davey, K. G., Effects of various compounds on Na/K ATPase activity, JH I binding capacity and patency response in the follicles of *Rhodnius prolixus*, *Insect Biochem.*, 17(), 1085, 1987.

Jurd, L., Fye, R. L. and Morgan, J., New types of insect chemosterilants: Benzylphenols and benzyl-1,3-benzodioxole derivatives as additives to house fly diet, *J. Agric. Food Chem.*, 28, 1007, 1979.

Koeppe, J. K., Kovalick, G. E. and Prestwick, G. D., Photoaffinity labelling of juvenile hormone-binding proteins in *Leucophaea maderae*, in *Synthesis, Metabolism and Mode of Action of Invertebrate Hormones*, Hoffmann, J. and Porchet, M., Eds., Springer-Verlag, Berlin, 1984, 438.

Ma, M., The effects of benzodioxole on *Drosophila* reproductive physiology, in *Host Regulated Developmental Mechanisms in Vector Arthropods, Third Symposium*, Borovsky, D. and Van Handel, E., Eds., University of Florida-IFAS, Vero Beach, FL, 1993, 133.

Niwa, A., Iwamura, H., Nagagawa, Y. and Fujita, T., Development of N,O-disubstituted hydroxylamines and N,N-disubstituted amines as insect juvenile hormone mimetics and the role of the nitrogenous function for activity, *J. Agric. Food Chem.*, 38, 514, 1990.

Sevala, V. L. and Davey, K. G., Juvenile hormone dependent phosphorylation of 100 kDa polypeptide is mediated by protein kinase C in the follicle cells of *Rhodnius prolixus, Invert. Reprod. Devel.*, in press, 1993.

Zaoral, M. and Slama, K., Peptides with juvenile hormone activity, *Science*, 170, 92, 1970.

Localisation, purification and partial characterisation of a male accessory gland factor by neuropeptide-specific monoclonal antibody MAC-18 in Leptinotarsa decemlineata,

Hans M. Smid and Hugo Schooneveld.

Wageningen Agricultural University, Department of Entomology, Binnenhaven 7, 6709 PD Wageningen, The Netherlands

The insect's male accessory gland function varies widely between different species (For review, see Chen, 1984). In many species, these glands produce the proteins required for the formation of the spermatophore. In others, they produce female-specific peptide hormones which are transferred via the semen to the female, were they can have different functions, such as oviposition stimulation and reduction of the female's receptivity. This is to assure that the female starts oviposition only after mating, and that the deposition of unfertilized eggs is prevented.

The female Colorado potato beetle, either mated or unmated, starts oviposition 5-6 days after emergence. The mated female's fecundity then stays at a high level, while the unmated female's fecundity decreases to close to zero. (Dortland, 1979) A once mated female continues to produce on average 70-80 fertilized eggs per day for several months. We wish to know how the event of copulation results in stimulation of oviposition. The present work concentrates more specifically on the possible role of the male accessory glands.

The male Colorado potato beetle has one pair of tubular accessory glands. The epithelium consists of a single layer of cylindric cells, all having a similar ultrastructure (Smid *et al.* 1992). In contrast to this seemingly simple organization, the gland's lumen contains at least 50 different proteins, as determined with SDS-PAGE (Peferoen and De Loof, 1984). The Colorado potato beetle does not use a spermatophore for sperm transfer, so another function of these proteins was to be sought.

We used immunological methods to investigate the presence of possible neuropeptide-like substances in the accessory gland. The rationale of this approach is that we have earlier produced a bank of 18 different monoclonal antibodies directed against different, yet unknown antigens, most of them located in the secretory granules of the peptidergic neurones in the brains the Colorado potato beetle Schooneveld *et al.*, 1989; Schooneveld and Smid, 1990). One of these

0-8493-4591-X/94/$0.00 + $.50
© 1994 by CRC Press, Inc.

antibodies, encoded MAC-18, immuno-labelled about 100 cells in the epithelium of the accessory glands of 7 day-starved males (Smid and Schooneveld, 1992). Immuno-electron microscopy revealed that MAC-18 recognized the secretory granules of these gland cells. These granules contained a characteristic fibril-like material, which was also present in large clusters in the lumen of the gland. We concluded that the MAC-18 positive cells released their immuno-reactive products into the lumen; it should be transferred to the female during copulation.

The coincidence of the immuno-labelling with MAC-18 of the accessory gland cells and some peptidergic neurones, made us consider the possibility that the antigen in the accessory glands is a neuropeptide-like substance. We therefore started a purification program to isolate this compound from the accessory glands. An extract in 80% methanol of 100 pairs of accessory glands taken from normally fed, reproducing beetles was fractionated with RP-HPLC, on a C_4 column with water acetonitrile gradient containing 0.1% TFA and detection at 220 nm. Fractions were lyophilized and aliquotes were subjected to SDS-PAGE followed by electro-blotting and immuno-detection with MAC-18. One HPLC fraction showed immuno-reactivity, two sharp bands both of approximately 7 kD. appeared on the immunoblot. This fraction was subjected to a second HPLC run for final purification, with the same column, but a different water-acetonitrile gradient. Separate peaks were collected and again tested for immuno-reactivity with MAC-18 as described above. It was found that both immunoreactive bands were eluted together in one single HPLC peak. This peak, containing two apparently very similar substances, was subjected to amino-acid analysis.

The N-terminal amino-acid sequence of could be determined up to position 40. A single signal was obtained despite of the two bands detected with SDS-PAGE, except for position 7, were two residues were detected. Unfortunately, there were no suitable sites detected for protease digestion, necessary for complete sequence analysis. The sequence was rather characteristic in that it was composed of a series of imperfect hexa-repeats. The amino-acid sequence was compared with the EMBL/Genbank database using the T-FASTA program (Pearson and Lipman, 1988). This homology search revealed 45% identity and 63% similarity with chicken prion-protein (Harris *et al.*, 1991). No similarity was found with any known insect neuropeptide.

The chicken prion-protein is a brain-factor which stimulates acetyl-choline receptor formation (Harris *et al.*, 1991). Its working mechanism is not fully understood. The term prion refers to its sequence identity with mammalian prion-proteins. These mammalian prions are infectious particles, causing spongiform encephalopathies, a fatal brain disease occurring in different species, including man (Chesebro, 1990). Interestingly, the MAC-18 positive fibril-like structures in both the secretory granules of the gland cells and in the lumen of the accessory gland, are very similar to these fibril-like prion-protein complexes. The working mechanism of prions is not understood, thus the structural homology of the MAC-18 accessory gland factor with prion proteins does not give us a key towards a physiological function in our species.

Future experiments will be carried out to reveal more structural and functional information about this MAC-18 antigen, isolated from the male accessory glands from the Colorado potato beetle. We will investigate its possible role in the oviposition stimulation after its transfer to the female. The monoclonal antibody MAC-18 will be an important tool during these studies.

REFERENCES.

Chesebro, B., Spongiform encephalopathies: The transmissible agents, in *Virology*, 2nd Edition, Fields, B. N., Ed., Raven Press, New York, 1990.

Chen, P. S., The functional morphology and biochemistry of insect male accessory glands and their secretions. *Ann. Rev. Ent.* 29, 233-255, 1984.

Dortland, J. F., The hormonal control of vitellogenin synthesis in the fat body of the female Colorado potato beetle, *Gen. comp. Endocrinol.* 38, 332-344, 1979.

Harris, D. A., Falls, D. L., Johnson, F. A. and Fischbach, G. D., A prion-like protein from chicken brain copurifies with an acetylcholine receptor-inducing activity, *Proc. Natl. acad. Sci. USA*, 88, 7664-7668, 1991.

Pearson, W. and Lipman, D. J., Improved tools for biological sequence comparison, *Proc. Natl. Acad. Sci. USA*, 85, 2444-2448, 1988.

Schooneveld, H. and Smid, H. M., Monoclonal antibodies obtained through insect brain homogenates: Tools for cell-specific neuroanatomical studies in the Colorado potato beetle, Entomol. exp. appl. 51, 283-296, 1989.

Schooneveld, H., Smid, H. M., Boonekamp, P. and Van Haeften, T., Peptidergic neurons in the Colorado potato beetle, *Leptinotarsa decemlineata* (Say), identified immunocytochemically by monoclonal antibodies against homogenates of neuroendocrine tissue. *Cell Tissue Res.* 257, 29-39, 1989.

Peferoen, M. and De Loof, A., Intraglandular and extraglandular synthesis of proteins secreted by the accessory reproductive glands of the Colorado potato beetle, *Leptinotarsa decemlineata, Insect Biochem.*, 14, 407-416, 1984.

Smid, H. M. and Schooneveld, H., Male accessory sex glands contain a new class of exocrine peptidergic cells in *Leptinotarsa decemlineata* (Say), identified with the neuropeptide specific monoclonal antibody MAC-18, *Inv. Reprod. Dev.* 21, 141-148, 1992.

Smid, H. M., Schooneveld, H. and Van Ham, I., Immunological evidence that the male accessory glands in the Colorado potato beetle produce regulatory peptides with possible post-copulatory effects in the female. *Proc. Exper. & Appl. Entomol., N.E.V. Amsterdam,* 3, 153-157, 1992.

Biological Activity of Four Synthetic Domains
of the Locust Ovary Maturating Parsin

Josiane Girardie[1], Olivier Richard[1], Serge Geoffre[2],
Gilles Précigoux[2], Adrien Girardie[1] and
Michel Hospital[2]

Laboratoire de Neuroendocrinologie/URA CNRS 1138[1],
Laboratoire de Cristallographie et Physique
Cristalline/URA CNRS 144[2], Université Bordeaux I,
33405, Talence Cedex, France

I. INTRODUCTION

A neurohormone inducing precocious oocyte maturation was previously isolated from the nervous part of corpora cardiaca of the African locust (Girardie et al. 1992). This molecule, denoted as Lom OMP for Ovary Maturating Parsin of Locusta migratoria, is an original unique polypeptide chain of 65 residues (Girardie et al. 1991). The circular dichroism spectra of the Lom OMP has not revealed any significant periodic secondary structure. Its hydropathy profile has indicated three hydrophilic domains located at both ends (Y1-D7; A49-V65) and in the central part (A28-D40) of the sequence.

Because these hydrophilic domains probably trigger the biological activity of the Lom OMP, four synthetic peptides belonging to the three hydrophilic domains were chemically synthesized and tested for their activity on oocyte maturation.

II. MATERIAL AND METHODS

A. CHEMICAL PEPTIDE SYNTHESIS

Four peptides were synthesized: [1]YYEAPPD[7], [3]EAPPD[7], [28]ASWPHQQRRQALD[40] and [49]AADAQFQDEEEDGGRRV[65].

Chemical synthesis of the peptides was performed automatically on an Applied Biosystem synthesizer model 431A, according to the general stepwise Merrifield's solid phase method (Barany and Merrifield 1980). Functional amino acid side chains were protected with the usual hydrogen fluoride (HF) labile type groups. Elongation was done on Boc-amino acyl PAM Resins (0.3-0.6 meq/g) according to the following operational cycle protocol: deblocking with 50% trifluoroacetic acid in dichloromethane (DCM), washing with DCM, neutralization with 10% diisopropylethylamine (DIEA)/DCM (5 min), washing with DCM then coupling with preformed 1-hydroxybenzotriazole esters in N-methylpyrrolidone during 40 min (Tam 1987). Dimethylsulfoxyde and DIEA were added to increase the coupling efficiencies (Tonolio et al. 1985). After coupling of the final amino acid

0-8493-4591-X/94/$0.00 + $.50
© 1994 by CRC Press, Inc.

residue, cleavage of the crude peptides from the solid support was achieved by stirring for 1 hr at 0°C in liquid HF (10 ml/g resin) with anisole (1 ml) and ethanedithiol (0.1 ml) as scavengers.

The crude products were purified by gel filtration on Sephadex G25 media (Pharmacia Sweden)using 10% acetic acid as eluant. Final purification was carried out by semi-preparative RP-HPLC on a C18 Spherisorb ODS 2.3 μm column in water-acetonitrile 0.05% TFA. Characterization of the final products was done by amino acid analysis realized on analytical HPLC following hydrolysis of the purified peptides by 6N HCL treatment in evacuated tubes (Pico-Tag system from Waters) for 24-48 h at 110°C.

B. BIOLOGICAL ANALYSIS

The gonadotropic effect of the four synthetic peptides was analyzed according to the procedure previously used for the extracted Lom OMP (Girardie et al. 1992). Five experimental series (a series for each peptide and for comparison a series with the extracted Lom OMP) of 12 to 14 locust females were realized. Thirty nmoles of each peptide or 2 corpora cardiaca equivalents of extracted Lom OMP (about 20 pmoles) dissolved in 5 μl of water were injected every day into adult females from day 1 to day 10 or 11. Five μl of water were injected into 13 locust females (control series) using the previous protocole. The gonadotropic effect was evaluated in each female by the mean length of ten basal oocytes measured at day 11 or 12 and in each series by the mean of these mean-lengths. Statistical differences between experimental and control series were estimated by student's t-test.

Dose-response curves were established for peptides found to stimulate ovarian maturation. Five experimental series of 11 to 13 adult females were realized for each gonadotropic peptide injected at five increasing doses: 0.25, 0.5, 1, 2 or 3 nmoles. Females were killed at day 11. Because of the time needed to achieve these dose-response curves, two control series of 12 and 13 females were necessary. The gonadotropic effect of each dose of synthetic peptides was determined as previously.

III. RESULTS

The four peptides were successfully synthesized and purified at homogenity.

The twelve females injected with the synthetic peptides E3-D7 or A28-D40 had at respectively day 12 and 11 oocyte lengths of respectively 3.513±0.548 mm and 2.052±0.249 mm exactly comparable to those of controls injected in the same conditions: 3.144±0.621 mm (11 locusts at day 12); 2.149±0.339 mm (12 locusts at day 11). On the contrary, the synthetic peptides Y1-D7 or A49-V65 injected into respectively 12 and 13 locusts significantly accelerated the oocyte growth. The means of the oocyte mean-lengths obtained at day 11 with each both peptides were not significantly different although slightly higher (5.152±0.396 mm) following injections of Y1-D7 than following injections of A49-V65 (4.773±0.488 mm). The both oocyte mean-lengths were slightly lower but not significantly different from the oocyte mean-lengths of 14 females injected with extractable Lom OMP (5.703±0.240 mm).

The gonadotropic effect of the synthetic peptide Y1-D7 was significant as soon as following injections of 0.5 nmol (3.506±0.516 mm; 2.061±0.280 mm for controls) whereas only injections of 1 nmole were efficient for the synthetic peptide A49-V65 (2.753±0.271 mm). Then, gonadotopic effect of both peptides raised with increasing doses and respectively reached a comparable level following injections of 2 (4.313±0.537 mm; 4.464±0.478 mm) or 3 (5.274±0.385mm; 4.976±0.470mm) nmoles.

Females injected with 3 nmoles of each synthetic peptide had oocyte mean-lengths comparable to those of females injected with 30 nmoles of each synthetic peptide.

IV. CONCLUSION-DISCUSSION

Four peptides (Y1-D7, E3-D7, A28-D40 and A49-V65) corresponding to the three hydrophilic domains of the Lom OMP, a new neurohormone of 65 amino acids which stimulates oocyte maturation in the African locust, were synthesized. Only the N-terminal and the C-terminal synthetic peptides stimulated oocyte maturation as the Lom OMP did. The smallest molar amounts of both synthetic peptides needed for the maximal expression of the biological activity were 150 times more important than the efficient amount of the entire molecule itself. These results are in agreement with literature reports concerning biological activity of synthetic fragments of vertebrate hormones.

The N-terminal peptide deprived of its two beginning tyrosines did not stimulate the oocyte growth indicating that these amino acids are necessary for the total expression of the biological activity of the Lom OMP.

Sequence of the two active synthetic peptides are very different in their amino acids composition. So, the Lom OMP could have either two receptors on two distinct target tissues which contribute to ovarian maturation (fat body and ovary), or only one target tissue but with two receptors or one receptor with two sites of binding. Binding studies require large amounts of Lom OMP and for this reason we are working to obtain the entire Lom OMP by chemical synthesis.

REFERENCES

Barany, G. and Merrifield, R. B., *Analysis, Synthesis, Biology,* in The Peptides, Vol. 2, Gross, E. and Meienhofer, J., Eds., Academic Press, New York, 1980, pp. 1-284.

Girardie, J., Richard, O., Huet, J. C., Nespoulous, C., Van Dorsselaer, A. and Pernollet, J. C., Physical characterization and sequence identification of the ovary maturating parsin, a new neurohormone purified from the nervous corpora cardiaca of the African locust (*Locusta migratoria*), *Eur. J. Biochem.* 202, 1121-1126, 1991.

Girardie, J., Richard, O. and Girardie, A., Time-dependent variations in the activity of a novel ovary maturating neurohormone from the nervous corpora cardiaca during oogenesis in the locust, *Locusta migratoria* migratorioides, *J. Insect Physiol.* 38, 215-221, 1992.

Tam, J. P., Synthesis of biologically active transforming growth factorae, *Int. J. Peptide Protein Res.* 29, 421-431, 1987.

Toniolo, C., Bonora, G., Moretto, V. and Bodansky, M., Self-association and solubility of peptides. Solvent titration study of peptides related to the C-terminal decapeptide sequence of porcine secretine, in *Peptides: Structure and Function*, Beber, C., Hruby, V. J. and Koppel, K., Eds., Pierce Chemical Co., Rockford, IL, 1985, pp. 419-422.

Acknowledgements - This work was supported by grants from the Conseil Général d'Aquitaine.

PHOTOPERIODIC CONTROL OF MATING BEHAVIOR OF *PYRRHOCORIS APTERUS*

R. M. Wagner[1*] and M. Hodkova[2]

[1]United States Department of Agriculture,
Agricultural Research Service, Livestock Insects Laboratory,
Beltsville, MD 20705-2350, USA, and

[2]Czech Academy of Sciences, Institute of Entomology,
Ceske Budejovice, 370 05 Czech Republic

INTRODUCTION

It has been shown that photoperiod affects sexual activity of the linden bug, *Pyrrhocoris apterus*, including growth of reproductive organs, receptivity and mating behavior of males and females. Short day photoperiod (12L:12D) induces adult diapause, which is characterized by deterioration of reproductive organs and inhibition of mating behavior, while long day photoperiod (18L:6D) stimulates mating and development of reproductive organs (Slama 1964a, 1964b, Hodek 1968, Zdarek 1970, Hodkova *et al.* 1991). This photoperiodic control appears to be mediated by the neuroendocrine system, as removal of *corpora allata (CA)* from diapausing males causes a temporary stimulation of mating behavior, while extirpation of the *pars intercerebralis (PI)* results in permanent stimulation (Hodkova 1990). In non-diapausing males, however, removal of the *PI* or *CA* has no effect on the active sexual status. We therefore chose to examine the effects of brain extracts of diapausing males on mating behavior when injected into active males, and extracts of active males on behavior of diapausing males.

MATING STUDIES

Insects were raised from the egg stage under either a long-day (18L:6D) or short-day (12L:12D) regimen. One male and 1-2 active (non-diapausing) female adults were placed into individual petri plates containing linden seed and water. Just prior to the light cycle, male adults were anaesthetized by immersion in water for 20 min, then injected with saline or the test compound and mating was

0-8493-4591-X/94/$0.00 + $.50
© 1994 by CRC Press, Inc.

observed for the duration of the light cycle. Initial extracts for injection were prepared by homogenization of the brain/*CC/CA* complexes of either diapausing or active (non-diapausing) males in methanol. After filtration and evaporation, the residue was resuspended in saline (0.9% NaCl) and used for injection.

Active males and females that were not separated the day prior to mating experiments exhibited a photoperiodic pattern of mating behavior, with a major peaks of sexual activity at 6 and 14 hrs after lights on during long-day conditions, with the lowest frequency of mating at 9 hr. If males and females were separated on the day previous to the experiment, mating behavior was exhibited within one hr after lights on and continued at a high level until 10 hr, when there was a sudden drop in mating activity. Anaesthesia had no effect on these results.

Active males which were separated from females and injected with 0.5 eq of brain extract from diapausing males exhibited a maximum peak of activity at 6 hrs after lights on and a minimum of activity at 10 hrs, in contrast with the control insects, which mated continuously. Injection of 0.5 eq of brain extract from active males into diapausing males resulted in mating of a few insects after 8 hr. In contrast, uninjected diapausing males did not mate at all, and active control males mated continuously during this period.

HPLC OF EXTRACTS

Extracts of brain/*CA/CC* complexes were prepared from either 25 diapausing male insects or 25 active male insects by homogenization in methanol/1 N HCl/acetonitrile/water containing 0.05% (by volume) trifluoroacetic acid. Extracts were centrifuged through 0.45 *u*m filter units, followed by removal of proteins using 10,000 NMWL centrifugal ultra-filtration units (Wagner *et al.* 1993). 1993). Samples were concentrated under a stream of argon and diluted with 10% acetonitrile/0.1% TFA prior to HPLC purification. Peptides and other components of *P. apterus* brain/*CA/CC* were separated on a Vydac C-18 column using a linear gradient as per Jaffe *et al.* (1989). Eluting peaks were monitored at 210 nm.

Elution profiles for the two sets of extracts were similar, with the exception of the areas of the peaks at retention times of 13 and 22 min and the presence of a peak at 28 min in the active (non-diapausing) extract (Fig 1). These fractions will be tested in the future for their effect on mating behavior of males of opposite photoperiod.

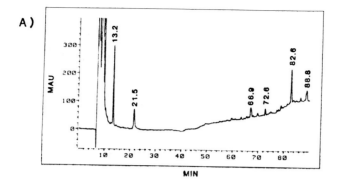

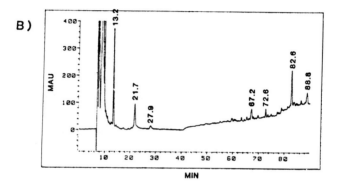

Figure 1. HPLC elution profiles of extracts of brain/CC/CA complexes of male linden bugs on a Vydac C-18 column at 28°C and 0.4 ml/ min. Solvent A = 0.1% TFA in water, solvent B = acetonitrile/water (60:40, v/v) containing 0.1% TFA. Linear gradient elution from 16.7% solvent B to 83.3% solvent B in 60 min. A) diapausing males, B) active males.

DISCUSSION

Results of mating studies with brain extracts agree with those of micro-surgical intervention, suggesting that both inhibitory and stimulatory compounds are present in the nervous system of *P. apterus*. Diapausing males which were injected with brain extracts of active males exhibited a slight stimulation in mating activity after 8 hr, while active males separated from females and injected with brain extracts of diapausing males exhibited a change in mating behavior which paralleled that of active males which had been reared with adult females. The HPLC profiles of the extracts for active and diapausing male brain extracts were similar, but there was one compound present in the extract of active males that was not present in that of diapausing males. This information will be used in isolation of compounds from *P. apterus* which are responsible for these differences in mating behavior.

REFERENCES

Jaffe, H., Raina, A. K., Fraser, Riley, C. T., Fraser, B. A., Nachman, R. J., Vogel, V. W., Zhang, Y., and Hayes, D. K., Primary structure of two neuropeptide hormones with adipokinetic and hypotrehalosemic activity isolated from the corpora cardiaca of horse flies (Diptera), *Proc. Nat. Acad. Sci., USA* 86, 8181-8164, 1989.

Hodek, I. Diapause in females of *Pyrrhocoris apterus* L. (Heteroptera), *Acta ent. bohemoslov.*, 65, 422-435, 1968.

Hodkova, M. Transmission des signaux photoperiodiques chez *Pyrrhocoris apterus* L. (Heteroptera), in *Regulation des cycles saisonniers chez les intertebres*, INRA, Eds, 1990, pp. 209-212.

Hodkova, M., Zieglerova, J., and Hodek, I., Diapause in males of *Pyrrhocoris apterus* L. (Heteroptera) and its dependence on photoperiod and the activity of females. *Zool. Jb. Syst.*, 118, 279-285, 1991.

Slama, K., Hormonal control of respiratory metabolism during growth, reproduction and diapause in female adults of *Pyrrhocoris apterus* L. (Hemiptera), *J. Insect Physiol.*, 9, 283-303, 1964a.

Slama, K., Hormonal control of respiratory metabolism during growth, reproduction and diapause in male adults of *Pyrrhocoris apterus* L. (Hemiptera), *Biol. Bull.*, 127, 499-510, 1964b.

Wagner, R. M., Woods, C. W., Hayes, J. A., Kochansky, J. P., Hill, J. C. and Fraser, B. A., Isolation and identification of a novel peptide from the accessory sex gland of the female house fly, *Musca domestica*, *Biochem. Biophys. Res. Commun.*, 194, 1336-1343, 1993.

Zdarek, J., Mating behavior in the bug, *Pyrrhocoris apterus* L. (Heteroptera): Ontogeny and its environmental control, *Behavior*, 37, 253-268, 1970.

New Neuropeptides

STIMULATION OF SILK GLAND BY A NEUROPEPTIDE
OF BRAIN ORIGIN IN *Galleria mellonella*

Joanna Michalik, Ewa Szolajska

Institute of Biochemistry and Biophysics
Polish Academy of Sciences
Rakowiecka 36, 02-532 Warsaw, Poland

The brain of *Galleria mellonella* is the source of the neuropeptide activating silk gland in the penultimate(VI) and last (VII) instar larvae (Michalik and Szolajska 1992, Kodrik and Sehnal 1991) *in vitro*. Also the influence of juvenile hormone (Sehnal et al. 1983) and ecdysteroid (Sehnal and Michalik 1984) applied *in situ* during the last instar has been shown. The constant presence of the brain factor positively controlling silk gland activity continues the efficient processes of silk protein synthesis on the level of transcription and translation in the last larval instar until the degenerating phase i.e. shortly before larva-pupa transformation. The response of the silk glands to the factor stimulating transcription, measured as [^{3}H] uridine incorporation into total RNA during 3 hr incubation *in vitro*, can be observed when the glands are dissected from isolated abdomens (larvae ligated behind prothorax) at day 1 and 3 of the last instar, in debrained larvae at day 3 of the last instar, and at three to five days after surgery. Also last instar larvae, starved for four days starting at day 3 (VII/$_3$), were the source of the glands responding to stimulation, [Table 1].

Table 1

Donors of silk glands from last instar larvae and their response to brain factor during *in vitro* incubation

Larvae: VII/$_1$, VII/$_2$, VII/$_3$, VII/$_4$	-
Isolated abdomens: VII/$_1$, VII/$_3$	+ +
Isolated abdomens: VII/$_6$	-
VII/$_3$ debrained	+ +
VII/$_3$ starved	+ +

In silk glands isolated from larvae controlled by the brain *in situ*, no stimulation *in vitro* by brain or brain extract could be detected. It can be concluded that the absence of the neurohormone in isolated abdomens or debrained larvae caused a decrease in the level of transcription and translation which could be increased by more than twice in the presence of crude brain extract (0.1 brain equivalent). Three days of starvation resulted in a decrease in neuropeptide synthesis, so its absence or a very low concentration caused

0-8493-4591-X/94/$0.00 + $.50
© 1994 by CRC Press, Inc.

the silk glands to respond to stimulation. Interestingly, in a late stage (VII_6) of the last larval instar, when the synthetic activity sharply drops, the silk gland tissue becomes insensitive to brain influence.

We have checked the activity of the silk gland stimulating factor in different developmental stages as well as in different parts of the nervous system.

Table 2

The presence of the factor in crude extracts of different neurohemal tissues

Brain: $VI/_3$, $VII/_1$, $VII/_3$	+ +
Brain $VII/_6$	-
Brain - MNSC (median neurosecretory cells)	+ +
MNSC (median neurosecretory cells)	-
Brain $VII/_3$ starved	+
Brain $VII/_3$ ($18°C$)	+
CC/CA (corpora cardiaca/corpora allata complex)	-
Brain *B.mori*	+ +

The results presented in Table 2 show that the neuropeptide is present in brain tissue in penultimate and the first part of the last larval instar during high silk gland activity and massive silk protein synthesis. In late stages of the last instar, when the silk gland ends its function as a silk protein producer, the stimulating factor is no longer present in brain tissue. The brains of larvae kept for one week in the suboptimal temperature of $18°C$ (the treatment which results in a high juvenile hormone level and permanent larval stage for several months (Smietanko et al. 1989)) synthesizes the neuropeptide in much lower quantity. Also, starvation caused restricted synthesis and lower concentration of neurohormone in the *Galleria* brain. The brain is the only source of the peptide. Neither the CC/CA complex nor the suboesophageal ganglion contains it. The presence of the neuropeptide was not detected in median neurosecretory cells. Most probably it is synthesized in the area of the brain containing lateral cells. The silk glands of *Galleria* are also stimulated by crude brain extract from last larval instar *Bombyx mori* with efficiency similar to *Galleria* brain [Table 2]. It seems that the character and structure of the factor is not unique to one species, but is probably similar or common to several spinning Lepidoptera. The presence of the factor in the brain of *Lymantria dispar* versus *Spodoptera litoralis* has to be checked.

The hormonal status of the silk gland is crucial for maintaining its proper actvity. It has been shown that a juvenile hormone analogue applied to larvae in preparatory (day 1-2), and acceleratory (day 3-4) phases of the last instar suppressed both nucleic acid and protein synthesis (Sehnal et al. 1983, Sehnal and Michalik 1984). Twenty-hydroxyecdysone (20HE) applied in the reggression phase (VII/$_6$) accellerated silk gland degeneration.

We examined the action of neurohormone in the presence of juvenile hormone II (JH-II) and 20HE *in vitro*.

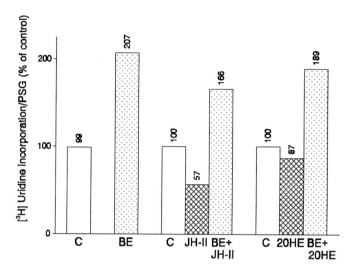

Figure 1. The influence of crude brain extract on transcription in posterior silk gland (PSG) during 3 hrs incubation *in vitro* in the presence and absence of JH-II (10⁻⁸M) and 20-HE (10⁻⁶M). Experimental conditions as in (Michalik and Szolaska 1982).

As shown in Fig. 1, total RNA synthesis in posterior silk gland incubated *in vitro* with crude brain extract of *Galleria* is stimulated more than two times (C, C+BE). The presence of JH-II in the incubation medium at a concentration of 10⁻⁸M caused as much as 43% inhibition in [³H] uridine incorporation into total RNA. However this effect was supressed by 0.2 brain equivalents of neuropeptide in the incubation medium (C, C+JH, C+JH+BE). The presence of 10⁻⁶M 20 HE slightly diminished [³H] uridine incoporation into total RNA (87% of the control) but the presence of both neuropeptide and 20 HE resulted in stimulation of RNA synthesis approximately two times more than in the presence of neuropeptide alone (C, C+20-HE, C+20-HE+BE).

REFERENCES

Michalik, J. and Szolajska E., Brain factor from *Galleria mellonella* (Lepidoptera) stimulating silk gland activity, *Experientia* 48, 762, 1992.

Kodrik, D. and Sehnal, F., Neurohormonal stimulation of posterior silk gland in *Galleria mellonella*, in *Wild silkmoth*, Akai, H. and Kiuchi, M., Eds., 1991, chap. 43.

Sehnal, F., Janda, V. and Nemec, V., Composition, synthetic and cytolitic activities of *Galleria mellonella* silk glands during the last-larval instar under the action of juvenile hormone, *J.Insect Physiol.* 29, 237-248, 1983.

Sehnal, F. and Michalik, J., Control of activity and regression of the silk glands in the last-larval instar of *Galleria mellonella, J. Insect Physiol.* 30, 119-126, 1984.

Smietanko, A., Wisniewski, J. R. and Cymborowski, B., Effect of low rearing temperature on development of *Galleria mellonella* larvae and their sensitivity to juvenilizing treatment, *Comp. Biochem. Physiol.* 92A, 163-169, 1989.

PURIFICATION OF NEUROPEPTIDE SERICOTROPIN FROM THE BRAINS OF *GALLERIA MELLONELLA*

Dalibor Kodrík[1], František Sehnal[1], Rainer Keller[2], Siegward Burdzik[2] and A. Krishna Kumaran[3]

[1]Entomological Institute, Academy of Sciences, 370 05 České Budějovice, Czech Republic, [2]Institute of Zoophysiology, The University, D-5300 Bonn, Germany, [3]Department of Biology, Marquette University, Milwaukee, WI 53233, U.S.A.

I. INTRODUCTION

Lepidopteran silk glands are modified salivary glands, which are adapted for the production of huge quantities of fibrous proteins. Development and function of the silk glands are under hormonal control (Kodrík and Sehnal 1993). Each molt-inducing surge of ecdysteroids causes their functional regression, from which they recover as long as the juvenile hormone is present (Akai 1965). Juvenile hormone absence in the last larval instar is a prerequisite for the rise of silk production prior to cocoon spinning, but the glands afterwards histolyze when the surge of ecdysteroids induces pupal moult (Fukuda 1942). Silk production in the last larval instar is enhanced in response to a brain factor (Sehnal and Michalik 1984).

The brain factor in question was identified in *Galleria mellonella* as a 5-10 kD peptide, and was called sericotropin (Kodrík and Sehnal 1991, Michalik et al. 1992). It is released from the brain both under *in vivo* and *in vitro* conditions, and seems to stimulate silk production by elevating RNA synthesis in the silk glands. In this report we describe purification of sericotropin by FPLC and HPLC, and present partial sequence of its N-terminus.

II. MATERIAL AND METHODS

Maintenance of the culture of the waxmoth, *Galleria mellonella*, ligation of the larvae, dissection of the brain and silk glands, cultivation of the silk glands *in vitro*, and performance of sericotropin bioassay were described in our previous paper (Kodrík and Sehnal 1991). Biological activity of the extracts was measured as percentile increase of ^{3}H-uridine incorporation into the silk gland RNA during a 1 hr incubation *in vitro*.

In total, 15,000 brains from the last instar larvae 72 - 84 hr after ecdysis were processed in several batches, each of 1,000 - 3,000 brains. The crude extraction of sericotropin was done according to our previously published protocol by heating aqueous brain homogenate to 60°C for 5 min and collecting the supernatant (Kodrík and Sehnal 1991). Subsequent purification included three steps: (1) FPLC (Pharmacia), using Superose ™12 column and phosphate buffer. (2) Active fractions from FPLC (between 5-30 kDa)

0-8493-4591-X/94/$0.00 + $.50
© 1994 by CRC Press, Inc.

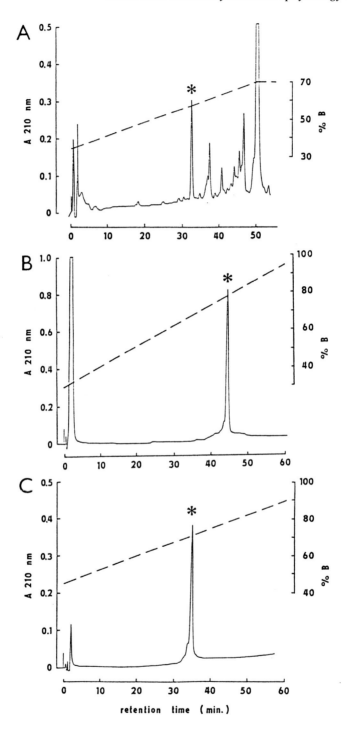

were desalted and concentrated on the C_{18} Sep-Pak column (Waters) which was washed with a mixture of 60% CH_3CN, 0.1% TFA (= B) and 0.11% TFA (= A). (3) Active eluent was taken for further three-step RP HPLC purification on Waters equipment:

a) μBondapak ^{TM}Phenyl column, 2 mm x 30 cm, gradient 35 - 70% B in A.

b) C_{18} μBondapak column, 2 mm x 30 cm, gradient 30 - 95% B.

c) The same type of column, gradient 45 - 90% B.

Apparently pure sericotropin was lyophilized and sent for sequencing to the protein/nucleic acid shared facility, Cancer Center of the Medical College of Wisconsin, Milwaukee. Sequencing of the aminoterminus was done with Edman degradation on an automatic equipment (Applied Biosystems 477A Protein Sequencer).

III. RESULTS

Brain homogenate in a buffer caused in average a 37% rise of RNA synthesis in posterior silk glands, cultured *in vitro* for 1 hr. The heating of aqueous brain extract caused precipitation of a large volume of material without a detectable loss of biological activity. The supernatant was lyophilised and, after re-solvation in water, injected onto the Superose ^{TM}12 column. Fractions between 5-30 kDa, which contained almost 100% of the initial activity, were pooled and concentrated on the C_{18} Sep-Pak column. More than 20% of the activity was recovered when the column was washed with an elution mixture consisting of 70% B (see Material and Methods). The eluent was loaded onto the μBondapak ^{TM}Phenyl column and chromatographed under a linear increase of B from 35 to 70% in 50 min. Fractions were collected manually and their aliquots were tested in the bioassay. Nearly all activity was associated with the fraction that is marked * in Fig. 1A. An aliquot of this fraction, which corresponded to 10 brains in the initial homogenate, stimulated RNA synthesis in the posterior part of silk glands to 176% of the control value. We estimated that about 82% of the initial activity had been lost when this purification step was completed.

The active fraction was concentrated on SpeedVac centrifuge, and the concentrate injected onto the C_{18} μBondapak column. Using linear gradient from 30 to 95% B, we were able to elute nearly all activity in a single and clearly defined fraction (Fig. 1B). This fraction passed unchanged through another C_{18} μBondapak column which was washed with a linear gradient from 45 to 90% B (Fig. 1C). The biological activity remained virtually unchanged during both steps of RP HPLC purification. An aliquot corresponding to 10 brains stimulated RNA synthesis in posterior silk glands by 68%.

Figure 1. (Opposite page) RP HPLC of sericotropin (*): **A.** Separation of the μBondapak ^{TM}Phenyl column, 1,500 brain equiv.; **B.** C_{18} μBondapak column, 3,000 brain equiv.; **C.** C_{18} μBondapak column, 2,700 brain equiv.

About 6,000 brain extracts were purified in 1991, using the above described procedure, and another 9,000 in 1992. In both years, 100 pmol of apparently pure sericotropin was obtained from about 3,400 brains. Independent sequencing of sericotropin samples 1991 and 1992 revealed slightly different amino acid compositions of the N-terminus:

1. Leu-Thr-Asp-Glu-Gln-Lys-Glu-Lys-Leu-Lys-Lys-His-Arg-Ser-Glu.
2. Leu-Thr-Asp-Glu-Gln-Lys-Lys-Glu-Leu-Leu-Lys.

IV. DISCUSSION

Present data confirm that the larval brain of *G. mellonella* contains a protein, presumably a neurohormone, which stimulates RNA synthesis in the silk glands (Sehnal and Michalik 1984). Sequence data indicate that there are two forms of this peptide but we do not know if they are derived from different alleles of a single gene or from separate genes of a gene family. A molecular biological approach has recently been undertaken in our laboratory to resolve this question and to elucidate the full structure of sericotropin. It should be mentioned that existence of a large gene family has been demonstrated in the case of bombyxin (Iwami et al. 1990).

Stimulation of RNA synthesis by neurohormone has been proposed in a few cases but seems to be rather exceptional (Kodrík and Sehnal 1991). Our data show that sericotropin causes this effect in the silk glands at concentrations around 0.1 ppm. Judging from the N-terminal sequence, sericotropin is not related to any of the known insect neurohormones.

V. REFERENCES

Akai, H., *Bull. Sericult. Exp. Stn.*, 19, 375-484, 1965.
Fukuda, S., *Zool. Mag.*, 54, 11-13, 1942.
Iwami, M., Adachi, T., Kondo, H., Kawakami, A., Suzuki, Y., Nagasawa, H., Suzuki, A. and Ishizaki, H., *Insect Biochem.*, 20, 295-303, 1990.
Kodrík, D. and Sehnal, F., in *Wild Silkmoths '89 '90*, Akai, H. and Kiuchi, M., Eds., Int. Soc. for Wild Silkmoths, pp. 43-50, 1991.
Kodrík, D. and Sehnal, F., *Int. J. Insect Morphol. Embryol.*, 22, 1993
Michalik, J., Szolajska, E. and Lassota, Z., *Experientia*, 48, 762-765, 1992.
Sehnal, F. and Michalik, J., *J. Insect. Physiol.*, 30, 119-126, 1984.

PURIFICATION AND PARTIAL CHARACTERIZATION OF NEUROPEPTIDE F-LIKE MATERIAL FROM THE NEUROENDOCRINE SYSTEM OF AN INSECT

Peter Verhaert[1], Aaron Maule[2], Chris Shaw[2], David Halton[2], Lars Thim[3], and Arnold De Loof[1]

[1]Zoological Institute of the University, Naamsestraat 59, B-3000 Leuven (Belgium)
[2]Comparative Neuroendocrinology Research Group, Schools of Clinical Medicine and Biology, & Biochemistry, The Queen's University of Belfast (Northern Ireland)
[3]Novo Nordisk A/S, Bagsvaerd (Denmark)

INTRODUCTION

The vertebrate peptide transmitters neuropeptide Y (NPY), pancreatic polypeptide (PP), and some intestinal peptides such as peptide YY (PYY) are homologous 36-amino acid (AA) peptides. They all share a characteristic amidated C-terminal aromatic amino acid (Tyr or Phe) and, particularly in the C-terminal half of the molecule, other essential residues (for refs. see McKay *et al.* 1990, Rajpara *et al.* 1992, Marks *et al.* 1993; Table 1a). This results in the adoption by these peptides of a relatively rigid conformation in aquaeous solution, known as the "PP-fold". Accordingly, these peptides are grouped in one big peptide class designated the NPY superfamily of peptides.

Since already more than 15 years immunochemical data are accumulating which indicate that also invertebrates possess these types of

Table 1
Primary structure of members of the NPY superfamily

a. some representative vertebrate members

Human PP	APLEPVYPGDNATPEQMAQYAADLRRYINMLTRPRYNH₂
Crow PP	APAQPAYPGDDAPVEDLLRFYNDLQQYLNVVTRPRYNH₂
Alligator PP	TPLQPKYPGDGAPVEDLIQFYDDLQQYLNVVTRPRFNH₂
Common Frog PP	APSEPHHPGDQATQDQLAQYYSDLYQYITFVTRPRFNH₂
Salmon PYY (= PP)	YPPKPENPGEDAPPEELAKYYTALRHYINLITRQRYNH₂
Rat PYY	YPAKPEAPGEDASPEELSRYYASLRHYLNLVTRQRYNH₂
Frog NPY	YPSKPDNPGEDAPAEDMAKYYSALRHYINLITRQRYNH₂
Human NPY	YPSKPDNPGEDAPAEDMARYYSALRHYINLITRQRYNH₂

b. invertebrate members "NPFs"

Moniezia	PDKDFIVNPSDLVLDNKAALRDYLRQINEYFAIIGRPRFNH₂
Artioposthia	KVVHLRPRSSFSSEDEYQIYLRNVSKYIQLYGRPRFNH₂
Helix	STQMLSPPERPREFRHPNELRQYLKELNEYYAIMGRTRFNH₂
Aplysia	DNSEMLAPPPRPEEFTSAQQLRQYLAALNEYYSIMGRPRFNH₂

Double underlined amino acids (AAs, single letter codes) are common to all typical vertebrate NPY-type peptides known to date. The ultimate AA, a well as AA #17 counting back from the C-terminus (underlined once) are always aromatic residues (either Tyr or Phe).

0-8493-4591-X/94/$0.00 + $.50
© 1994 by CRC Press, Inc.

peptide   messengers.   Both   immunohistochemistry   (IHC)   and
radioimmunoassay   (RIA),   particularly   with   C-terminally   specific
antibodies,   demonstrate   the   extensive   occurrence   of   NPY-superfamily
immunoreactivity (IR) in the nervous systems of all invertebrate phyla
examined (see refs. in Rajpara *et al.* 1992). Nevertheless, following the
establishment that the PP antibodies employed crossreacted with $FMRFNH_2$
(one of the first invertebrate regulatory peptides to be identified) and the
subsequent isolation of various invertebrate "FaRPs" ($FMRFNH_2$ related
peptides, i.e., short -$RFNH_2$ terminating peptides; see Walker 1992), the
assumption that these "typical invertebrate" peptides, and not "genuine
NPY-type" molecules, would be responsible for the observed PP/NPY-IR
got generally accepted.

However,   the   recent   identification   of   some   -$RFNH_2$ ending
invertebrate peptides with considerably more structural features in common
with NPY (the so-called "neuropeptide F (NPF)s"; Maule *et al.* 1991,
Curry *et al.* 1992, Leung *et al.* 1992, Rajpara *et al.* 1992; Table 1b) as
well as the discovery of a NPY-type receptor from an insect (Li *et al.*
1992), strongly argument in favour of the original idea that also
invertebrate members of the NPY superfamily of regulatory peptides exist.

This   paper   briefly   summarizes   our   immunochemical   and
physicochemical data which demonstrate the presence of insect NPY/NPF-
like material. Focus will be on data obtained in the American cockroach,
Periplaneta americana, although our studies in various other species
yielded similar results.

EXPERIMENTS & RESULTS

A. Immunochemistry

1. RIA. An antibody highly specific to invertebrate NPF was employed.
Previous radioimmunometrical studies have clearly demonstrated the
essential epitope of the antibody to be -$RXRFNH_2$, reflected in its lack of
crossreactivity with $FMRFNH_2$ or mammalian PP (Marks *et al.* 1992).
Using this antibody 0.5 nmol NPF-IR could be detected in a crude extract
(EtOH:0.7 M HCl; 3:1) of 100 *Periplaneta* brains, i.e. 5   pmol/brain.

2. IHC. NPF-like material was detected in various types of neuronal cell
bodies and their processes within the entire central nervous system of the

Table 2
Results of immuno-absorption tests in cockroach and fly CNS

absorbed with antibody	NPF 1-39	NPF 30-39	FMRFNH$_2$	PP 31-36
NPF FMRFNH$_2$ PP	- + -	- + -	+ + - +	+ + +- -

Legend: -, staining disappeared; +-, staining significantly reduced; +, partial disappearance
    of staining; + + no effect.

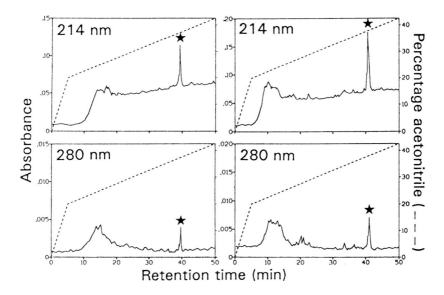

Figure 1. Absorbance profiles (214 & 280 nm) of *Periplaneta* NPF peaks 1 and 2 (left and right panel respectively) after final purification step (Vydac C8 analytical reverse-phase column eluted at 1ml/min). Asterisks indicate fraction with NPF-immunoreactivity. The absorbance at 280 nm may be caused by the presence of one or more Tyr-residues, a typical feature of NPY/NPFs (see Table 1).

cockroach, including brain and all ventral ganglia, as well as in the neuroendocrine retrocerebral complex and in (mostly open) endocrine-type epithelial cells lining the midgut (a more detailed description will be given elsewhere). A series of absorption tests (Table 2) fully confirmed the serum specificity and selectivity data as determined above in RIA: differential (cross-)absorption experiments revealed that on the one hand, the FMRFNH$_2$ antiserum used -although not crossreacting with mammalian PP (-RPRYNH$_2$)- does recognize invertebrate NPF, whereas, on the other hand, the NPF antibodies do not show any affinity towards *F*MRFNH$_2$ at all. This, in addition to double immunolabellings which showed a considerable overlap of NPF- and FMRFNH$_2$-immunostaining patterns, indicates the presence of a more NPF-like substance in *P. americana*.

B. Physicochemistry

1. Chromatography. One thousand individually dissected cockroach brains were extracted in acidified ethanol, dried under N$_2$ and reconstituted in 2M acetic acid. Gel permeation on Sephadex G-50 (where the cockroach NPF-like peptide eluted very closely after the elution position of the NPF(*M. expansa*) standard) was followed by several reverse phase HPLC runs (C18 semi-preparative Partisil ODS3 followed by analytical C8 columns Supelcosil LC308 and Vydac). Each time the columns were eluted with a linear acetonitrile gradient containing 0.1 % TFA, and the NPF-IR was determined by RIA. Final purification achieved on the Vydac C8 column resolved two intensely NPF-immunoreactive peaks (a third one was only weakly immunopositive). Both NPF-peaks were rechromatographed (Figure 1), treated with pyroglutamase to remove potential pGlu aminoterminal blocking and rechromatographed again. No shift in retention time was observed after this treatment, indicating the absence of a N-terminal pGlu.

2. *Mass Spectrometry*. One tenth of each purified NPF-peak was subjected to plasma desorption mass spectrometry (BioIon 20K). In the case of the second (most abundant) peak a single peptide was resolved with an approximate molecular mass of 3196 Da. Unfortunately, the remainder of the fractions did not appear sufficient to obtain unequivocal results from automated gas-phase Edman degradation (Applied Biosystems 470A).

DISCUSSION & CONCLUSION

Our results indicate that the peptides under consideration are different from the presently identified insect FaRPs, particularly from those of *Periplaneta americana* sequenced to date (Veenstra 1989). Presently the purification of *P. americana* NPF from dissected nervous systems is repeated (midguts are also being processed) and the isolation of NPF from more than 20.000 fleshfly heads underway. Our data indicate that these efforts may lead to the first identification of a genuine member of the NPY-superfamily of peptides from insects (and arthropods in general).

REFERENCES

Curry, W.J., Shaw, C., Johnston, C.F., Thim, L. and Buchanan, K.D. Neuropeptide F: primary structure from the turbellarian, *Artioposthia triangulata*, *Comp. Biochem. Physiol.* 101C, 269-274, 1992

Leung, P.S., Shaw, C., Maule, A.G., Thim, L., Johnston, C.F. and Irvine G.B. The primary structure of neuropeptide F (NPF) from the garden snail, *Helix aspersa, Regul. Pep.* 41, 71-81, 1992

Li, X.-J., Wu, Y.-N., North, A. and Forte M. Cloning, functional expression, and developmental regulation of a neuropeptide Y receptor from *Drosophila melanogaster, J. Biol. Chem.* 267, 9-12, 1992

Marks, N.J., Maule, A.G., Halton, D.W., Shaw, C., Johnston, C.F. and Moore S. Distribution of neuropeptide F immunoreactivity in the caryophyllidean tapeworm, *Caryophyllaeus laticeps*, from the intestine of the bream, *Abramis brama, Regul. Pep.* 39, 267, 1992

Marks, N.J., Shaw, C., Halton, D.W., Maule, A., Curry, J., Verhaert, P. and Thim L. Crow pancreatic polypeptide: a novel avian variant, *Abstract Volume XII Int. Congress on Comp. Endocrinol.* A-97, 1993

Maule, A.G., Shaw, C., Halton, D.W., Thim, L., Johnston, C.F., Fairweather, I. and Buchanan, K. Neuropeptide F: a novel parasitic flatworm regulatory peptide from *Moniezia expansa* (Cestoda: Cyclophyllidea), *Parasitol.* 102, 309-316, 1991

McKay, D.M., Shaw, C., Thim, L., Johnston, C.F., Halton, D.W., Fairweather, I. and Buchanan, K.D. The complete primary structure of pancreatic polypeptide from the European common frog, *Rana temporaria, Regul. Pep.* 31, 187-198, 1990

Rajpara, S.M., Garcia, P.D., Roberts, R., Eliassen, J.C., Owens, D.F., Maltby D., Myers, R.M. and Mayeri, E. Idenitification and molecular cloning of a neuropeptide Y homolog that produces prolonged inhibition in *Aplysia* neurons, *Neuron* 9, 505-513, 1992

Veenstra, J.A. Isolation and structure of two gastrin/CCK-like neuroeptides from the American cockroach homologous to the leucosulfakinins, *Neuropeptides* 14, 145-149, 1989

Walker, R.J. Neuroactive peptides with an RFamide or Famide carboxyl terminal, *Comp. Biochem. Physiol.* 102C, 213-222, 1992

The authors thank the British Council for supporting the Leuven-Belfast collaboration. P.V. is senior research associate of the Belgian N.F.W.O..

IDENTIFICATION AND CHARACTERIZATION OF PROTEINS IMMUNOLOGICALLY SIMILAR TO CELLULAR RETINOIC ACID AND CELLULAR RETINOL BINDING PROTEINS (CRABP AND CRBP) IN THE CNS OF *MANDUCA SEXTA* AND *DROSOPHILA MELANOGASTER*

David P. Muehleisen, Rosemary S. Gray, Anne L. Westbrook, Zeng F. Wang, Frank Chytil[1], David E. Ong[1], and Walter E. Bollenbacher

Department of Biology, University of North Carolina at Chapel Hill, Chapel Hill, NC 27599-3280, USA
[1]Department of Biochemistry and Medicine, Vanderbilt University School of Medicine, Nashville, TN 37232-0164, USA

Partial amino acid sequence data of the big prothoracicotropic hormone (PTTH) from *Manduca sexta* revealed the protein had sequence similarity with the vertebrate cellular retinoic acid and retinol binding proteins (CRABP and CRBP, respectively) (Muehleisen *et al.* 1993). These results led us to consider the possible existence of CRABP and CRBP in *Manduca*. Using antibodies to CRABP and CRBP, immunoreactive cells to the proteins were identified in the brains and ganglia of *Manduca* and *Drosophila melanogaster*.

Both CRBP I & II and CRABP I & II are intra-cellular binding proteins that appear to have key roles in the action of vitamin A, i.e., retinol (R), and its derivative retinoic acid (RA) (Ruberte *et al.* 1993). Both retinoid binding proteins are members of the superfamily of cytoplasmic hydrophobic ligand binding proteins (Takahashi *et al.* 1982). Members of the superfamily are low molecular weight proteins (14-18 kDa) occurring in the cell's cytoplasm and binding hydrophobic ligands (e.g., fatty acids, sterols, and retinoids). Other family members include the peripheral nerve myelin P2 protein, bovine mammary derived growth inhibitor, and tissue specific fatty acid-binding proteins (FABP). FABPs have recently been identified in the insects *Schistocerca gregaria* (Haunerland and Chisholm 1990) and *Manduca* (Smith *et al.* 1992).

In vertebrates, CRBP and CRABP appear to be involved in an array of physiological processes involving development (Maden *et al.* 1989), reproduction (Porter *et al.* 1985), photoreception (Saari and Bunt-Milam 1984), and homeostasis (Blomhoff *et al.* 1990). CRBP I and II specifically bind all trans-R, and in the visual system CRBP I transports and stores R for conversion to retinal for use in photoreception (Bok *et al.* 1984; Saari and Bunt-Milam 1984). The CRBPs are also involved in intracellular transport of R to the nucleus for reasons that are presently unclear (Takase *et al.* 1979). In contrast, CRABP I and II specifically bind all trans-RA, and they are thought to transport RA to the nucleus where, presumably, RA regulates genes that direct morphogenesis and reproduction (Takase *et al.* 1986). Together, CRBP and CRABP may influence development by regulating local RA levels (Dolle *et al.* 1990). A non-overlapping spatial expression of the two binding proteins could establish a RA gradient that directs cell differentiation and development.

0-8493-4591-X/94/$0.00 + $.50
© 1994 by CRC Press, Inc.

315

This investigation of CRABP and CRBP's possible presence in *Manduca* and *Drosophila* (Canton S strain) was conducted with anti-human CRBP I (MacDonald *et al.* 1990) and ligand affinity purified anti-rat CRABP I (Porter *et al.* 1985) antibodies. A tissue whole mount immunostaining protocol was used (O'Brien *et al.* 1988; Westbrook *et al.* 1991).

In *Manduca* late embryo and larval brains, two sets of previously unidentified protocerebral neurons express a CRABP-like phenotype. Staining of these neurons persisted into the pupal stage, but staining was not evident in the adult. One neuron group, near the lateral neurosecretory cell group III (L-NSC III), consists of two bilaterally paired dorsolateral neurons (CRABP-A), and the other group, near the brain midline, is a single bilaterally paired neuron (CRABP-B). Both the somata and processes of the neurons immunostain. Neuronal collaterals form 4 discrete dendritic fields with varicosities indicative of release sites (see Westbrook *et al.* 1991). These findings suggest the CRABP-like phenotype and/or other peptides produced by the neurons are released locally in the brain.

In *Drosophila* larvae and puparia, paired cerebral CRABP-immunoreactive neurons appear to extend collaterals into the ventral nervous system (VNS). These cells may be homologous to the *Manduca* CRABP-A neurons. CRABP-immunoreactive neurons are also evident in the *Drosophila* VNS. Their somata are arrayed segmentally with the thoracic neuromeres having 2 bilaterally paired neurons, and the abdominal neuromeres having single bilaterally paired neurons.

CRBP immunoreactive neurons appear to be present in *Manduca* and *Drosophila* as well. In *Manduca* the medial neurosecretory cell group IIa$_2$ (M-NSC IIa$_2$) and the smaller medial neurosecretory cell group IIa$_1$ (M-NSC IIa$_1$) immunostain. Here the entire neurons, i.e., somata and collaterals, appear to stain, including terminals within the corpora cardiaca (IIa$_1$) and corpora allata (IIa$_2$). At present it appears the *Drosophila* CNS has CRBP immunoreactive neurons, as well. While the immunostaining data for *Manduca* and *Drosophila* suggest CRBP and CRABP are present, it must be emphasized that these immunostaining data are preliminary and do not confirm the presence of the proteins.

Using bovine CRABP (Nilsson *et al.* 1988) and rat CRBP (Sherman *et al.* 1987) cDNAs as probes, the genomes of *Manduca* and *Drosophila* were examined by Southern blots under moderately stringent conditions. Both CRABP- and CRBP-like genes appear to be present in the *Manduca* and *Drosophila* genomes.

Finally, using primers based on two highly conserved regions of CRBP, we have amplified by polymerase chain reaction a 1200 bp region of *Drosophila* genomic DNA (Fig. 1). The product is currently being sequenced to determine if a CRBP-like gene has been amplified.

If CRBP and CRABP-like proteins are present in *Manduca* and *Drosophila*, their functions could be similar to those suggested for the vertebrate proteins (Chytil and Ong 1987; Blomhoff *et al.* 1990, 1991). CRBP could have a role

bp

Figure 1. *Drosophila* genomic DNA was amplified by PCR using primers corresponding to the vertebrate CRBP. The product was run on a 3% agarose gel, and one 1200 bp product was observed, suggesting a CRBP-like gene is present in *Drosophila.*

2000 —
1000 —
700 —
500 —
400 —
300 —
200 —

100 —

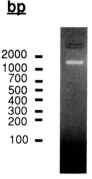

in photoreception, possibly in photoperiodic and circadian clock control of insect postembryonic development (see Bollenbacher and Granger 1985). This could involve extraretinal photoreceptors like those in the *Manduca* brain (Bowen *et al.* 1984). These extraretinal photoreceptors appear to regulate the photoperiodic clock control of PTTH release, which determines if and when *Manduca* develops into a adult. In such a process CRBP might transport retinol to the photoreceptor sites (Saari and Bunt-Milam 1984). Also, CRABP and possibly CRBP as well, could play a role in the morphogenesis of the insect CNS, which could occur during either embryogenesis or postembryonic development.

While these data and suggested roles for *Manduca* and *Drosophila* CRBP and CRABP are currently speculative, they raise the intriguing possibility that the proteins are present in invertebrates, warranting a more rigorous investigation of their presence. If the proteins are shown to be in *Manduca* and *Drosophila*, these animals promise to be useful models in understanding the evolution of CRBP and CRABP and basic aspects of their function in the control of neuron development and extraretinal light perception.

Acknowledgments: The authors thank Ms. Susan Whitfield for assistance with the illustration. This work was supported by a grant to WEB from the National Institutes of Health (NIH) (DK-31642) and a NIH Postdoctoral training Fellowship to DPM (NS-07166).

REFERENCES

Blomhoff, R., Green, M. H., Berg, T. and Norum, K. R., Transport and storage of vitamin A, *Science* 250, 339-404, 1990.

Blomhoff, R., Green, M. H., Berg, T. and Norum, K. R., Vitamin A metabolism: new perspectives on absorption, transport and storage, *Physiol. Rev.* 71, 952-990, 1991.

protein in the rat retina. *Invest. Opthal. & Vis. Sci.* 25, 877-883, 1984.

Bollenbacher, W. E. and Granger, N. A., Endocrinology of the Prothoracicotropic Hormone, in *Comprehensive Insect Physiology, Biochemistry, and Pharmacology*, vol 7, Kerkut, G. A. and Gilbert, L. I., Eds., Pergamon Press, New York, 1985, pp. 109-151.

Bowen, M. F., Saunders, D. S., Bollenbacher, W. E. and Gilbert, L. I., *In vitro* reprogramming of the photoperiodic clock in an insect brain-retrocerebral complex, *Proc. Natl. Acad. Sci. USA* 81, 5881-5884, 1984.

Chytil, F., and Ong, D. E., Intracellular vitamin A-binding proteins, *Ann. Rev. Nutr.* 7, 321-335, 1987.

Dolle, P., Ruberte, E., Leroy, P., Morriss-Kay, G. and Chambon, P., Retinoic acid receptors and cellular retinoid binding proteins I. A systemic study of their differential pattern of transcription during mouse organogenesis, *Develop.* 110, 1133-1155, 1990.

Haunerland, N. H. and Chisholm, J. M., Fatty acid binding protein in flight muscle of the locust *Schistocerca gregaria*, *Biochim. Biophys. Acta* 1047, 233-238, 1990.

MacDonald, P. N., Bok, D. and Ong, D. E., Localization of cellular retinol-binding protein and retinol-binding protein in cells comprising the blood-brain barrier of rat and human, *Proc. Natl. Acad. Sci. USA* 87, 4265-4269, 1990.

Maden, M., Ong, D. E., Summerbell, D. and Chytil, F., The role of retinoid-binding proteins in the generation of pattern in the developing limb, the regenerating limb and the nervous system, *Devel (Suppl.)* 107, 109-119, 1989.

Muehleisen, D. P., Gray, R. S., Katahira, E. J., Thomas, M. K. and Bollenbacher, W. E., Immunoaffinity purification of the neuropeptide prothoracicotropic hormone from *Manduca sexta*, *Peptides* 14, 531-541, 1993.

Nilsson, M. H. L., Spurr, N. K., Saksena, P., Busch, C., Nordlinder, H., Peterson, P. A., Rask, L. and Sundelin, J., Isolation and characterization of a cDNA clone corresponding to bovine cellular retinoic-acid-binding protein and chromosomal localization of the corresponding human gene, *Eur. J. Biochem.* 173, 45-51, 1988.

O'Brien, M. A., Katahira, E. J., Flanagan, T. R., Arnold, L. W., Haughton, G. and Bollenbacher, W. E., A monoclonal antibody to the insect prothoracicotropic hormone, *J. Neurosci.* 8, 3247-3257, 1988.

Porter, S. B., Ong, D. E., Chytil, F. and Orgebin-Crist, M-C., Localization of cellular retinol-binding protein and cellular retinoic acid-binding protein in the rat testis and epididymis, *J. Andro.* 6, 197-212, 1985.

Ruberte, E., Friederich, V., Chambon, P. and Morriss-Kay, G., Retinoic acid receptors and cellular retinoid binding proteins III. Their differential transcript distribution during mouse nervous system development, *Develop.* 118, 267-282, 1993.

Saari, J. C. and Bunt-Milam, A. H., Retinoids and cells of the retina, in *The Retinoids*, vol 2, Sporn, M. B., Roberts, A. B. and Goodman, D. S., Eds., Academic Press, Orlando, 1984, pp.1-15.

Sherman, D. R., Lloyd, R. S. and Chytil, F., Rat cellular retinol-binding protein: cDNA sequence and rapid retinol-dependent accumulation of mRNA, *Proc. Natl. Acad. Sci. USA* 84, 3209-3213, 1987.

Smith, A. F., Tsuchida, K., Hanneman, E., Suzuki, T. C. and Wells, M. A., Isolation, characterization, and cDNA sequence of two fatty acid-binding proteins from the midgut of *Manduca sexta* larvae, *J. Biol. Chem.* 267, 380-384, 1992.

Takahashi, K., Odani, S. and Ono, T., A close structural relationship of rat liver z-protein to cellular retinoid binding proteins and peripheral nerve myelin P2 protein. *Biochem. Biophys. Res. Commun.* 106, 1099-1105, 1982.

Takase, S., Ong, D. E. and Chytil, F., Cellular retinol-binding protein allows specific interaction of retinol with the nucleus *in vitro*, *Proc. Natl. Acad. Sci. USA* 76, 2204-2208, 1979.

Takase, S., Ong, D. E. and Chytil, F., Transfer of retinoic acid from its complex with cellular retinoic acid-binding protein to the nuclei. *Arch. Biochem. Biophys.* 247, 328-334, 1986.

Westbrook, A. L., Haire, M. E., Kier, W. M. and Bollenbacher, W. E., Three-dimensional architecture of identified cerebral neurosecretory cells in an insect, *J. Morph.* 208, 161-174, 1991.

PSEUDOPEPTIDES AND A NON-PEPTIDE THAT MIMIC THE BIOLOGICAL ACTIVITY OF THE MYOSUPPRESSIN INSECT NEUROPEPTIDE FAMILY

Ronald J. Nachman[1], Daisuke Yamamoto[2],
G. Mark Holman[1] and Ross C. Beier[1]

[1]Veterinary Entomology Research Unit, FAPRL, USDA,
College Station, TX 77845, USA;
[2]Mitsubishi-Kasei Life Sciences Institute, 11 Tokyo, Japan

Because insect neuropeptides regulate critical processes in insects, several strategies for the development of future pest insect management agents have been advocated.[1,2,3] However, insect neuropeptides in and of themselves hold little promise as traditional insect control agents because of their susceptibility to enzymatic degradation in the target insect, lability under environmental conditions, and inability to pass through the hydrophobic insect cuticle. The removal of the peptide nature (i.e., the constituent peptide bonds) of insect neuropeptides leading to pseudopeptide and non-peptide analogs represents a strategy that could overcome these limitations.

In this paper, we discuss pseudopeptide mimetic analogs and a non-peptide that mimic the biological activity of the myosuppressin insect neuropeptide family. The first member of this family was isolated from the cockroach on the basis of its ability to inhibit contractions of the isolated cockroach hindgut. Subsequently, examples of this peptide family have been found in a range of insect species.

MYOSUPPRESSINS

The myosuppressins are members of the FLRFamide peptide family, inhibit the spontaneous contractions of the isolated cockroach hindgut, and share the common C-terminal heptapeptide Asp-His-Val-Phe-Leu-Arg-Phe-NH$_2$. They have been isolated from the cockroach *Leucophaea maderae*, locusts *Schistocerca gregaria* and *Locusta migratoria*, and flies *Neobellieria bullata* and *Drosophila melanogaster*.[4] Leucomyosuppressin (LMS) inhibits evoked transmitter release at the neuromuscular junction of the mealworm *Tenebrio molitor*.[5] A related FLRFamide peptide, *Manduca*FLRFamide was isolated from *M. sexta* and has been found to increase contraction of the dorsal longitudinal flight muscles, indicative of a role in sustaining or promoting flight behavior patterns.[6]

0-8493-4591-X/94/$0.00 + $.50
© 1994 by CRC Press, Inc.

While the C-terminal pentapeptide represents the actual active-core sequence, it is markedly less potent than the parent leucomyosuppressin peptide. Therefore, development of pseudopeptide analogs of the inhibitory myosuppressin neuropeptide family has focused on the C-terminal heptapeptide sequence (Asp-His-Val-Phe-Leu-Arg-Phe-NH$_2$), the smallest fragment capable of approaching the myoinhibitory potency of the parent peptide. In order to retain the carboxylic acid side chain of Asp, important to full myosuppressin activity,[4] diacids of varying lengths were utilized to replace amino acid residues in the N-terminal region of the heptapeptide fragment. A pseudohexapeptide analog with the structure *Suc*-His-Val-Phe-Leu-Arg-Phe-NH$_2$ (Suc = succinoyl: HO$_2$C(CH$_2$)$_2$C(O)-) (Figure 1) was found to retain myoinhibitory activity on the hindgut preparation at a threshold concentration of 0.15 nM, 2-orders of magnitude more potent than the myosuppressin C-terminal hexapeptide (threshold concentration [TC]: 17 nM) and at least as active as the myosuppressin heptapeptide fragment (TC: 0.21 nM). In this pseudohexapeptide analog, succinic acid replaces the Asp of the myosuppressin C-terminal hepapeptide fragment. The full potency demonstrated by the pseudohexapeptide analog suggests that the N-terminal amino group has little influence on the interaction between the myosuppressin hepapeptide and its receptor site. A pseudotetrapeptide myosuppressin analog was also synthesized in which sebacic acid replaced the N-terminal residue block Asp-His-Val (Figure 1). The

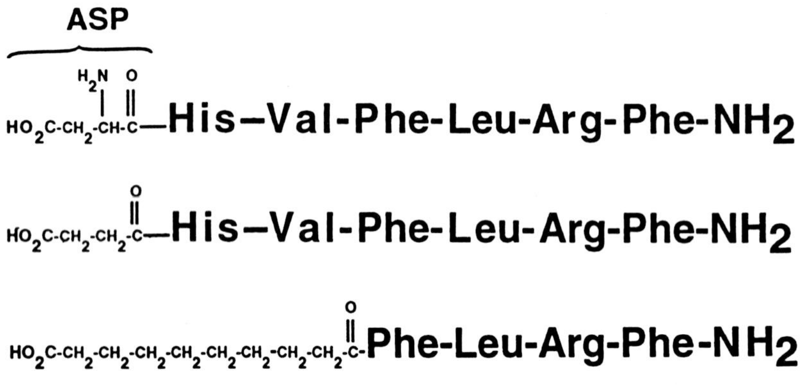

Figure 1. Acyl pseudopeptide analogs of the myosuppressin C-terminal heptapeptide fragment (top) include a pseudohexapeptide (middle) and a pseudotetrapeptide (bottom) that replace amino acids with succinnic and sebacic acid, respectively. The analogs retain both a carboxyl side chain and the myoinhibitory activity of the myosuppressins.[7]

pseudotetrapeptide analog *Sba*-Phe-Leu-Arg-Phe-NH$_2$ (Sba = sebacoyl: HO$_2$C(CH$_2$)$_8$C(O)-) retained myoinhibitory activity on the hindgut bioassay at a threshold of 2.7 µM, only about 75-fold less potent than the myosuppressin C-terminal pentapeptide fragment (TC: 35 nM).[7] The myosuppressin C-terminal tetrapeptide fragment (i.e., FLRFamide) is inactive in the isolated cockroach hindgut bioassay.[4] Therefore, the myosuppressin active core has been reduced from a pentapeptide to a pseudotetrapeptide.

The non-peptide benzethonium chloride (Bztc) shares several chemical features with the C-terminal pentapeptide myosuppressin sequence. Figure 2 reveals that the non-peptide has two separate zones with a phenyl ring analogous to the side chain phenyl ring of the two Phe residues of the myosuppressin pentapeptide. Zones with branched-chain and basic character in Bztc are analogous to the branched-chain Val and basic Arg residues, respectively, of the myosuppressins. On both the isolated

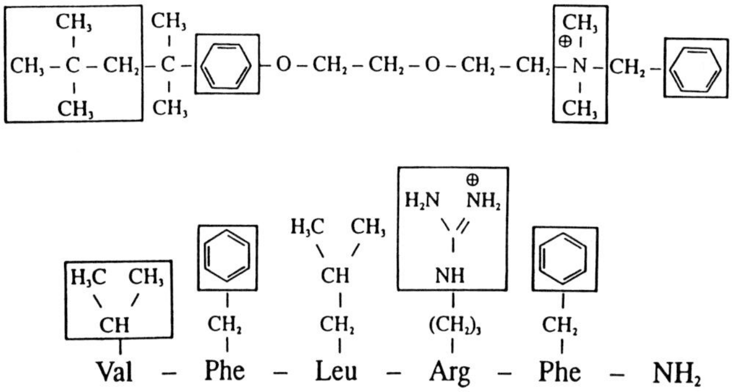

Figure 2. The non-peptide benzethonium chloride (top) shares several structural features with the myosuppressin C-terminal pentapeptide fragment (bottom).

cockroach hindgut preparation and the mealworm neuromuscular junction, Bztc mimics the inhibitory activity of leucomyosuppressin (LMS). On the hindgut, Bztc reversibly inhibits spontaneous contractions at a threshold concentration of 60 μM, about 3-orders of magnitude less potent than the myosuppressin C-terminal pentapeptide (Nachman, unpublished results). At the mealworm neuromuscular junction, Bztc induces reversible suppression of the evoked EPSPs at an approx. ID_{50} of 100 μM (Figure 3), or only about 20-fold less potent than the decapeptide LMS. Like LMS, both the suppression of neuromuscular junction EPSPs and inhibition of hindgut contraction evoked by Bztc is blocked by nordihydroguiaretic acid, an inhibitor of lipoxygenase. The non-peptide exhibited some side effects as it requires a higher stimulus intensity to fire in its presence and shows evidence of additional postsynaptic effects (Yamamoto and Nachman, unpublished data). A definitive answer as to whether Bztc operates via the

LMS receptor or another mechanism awaits the development of a myosuppressin receptor-binding assay.

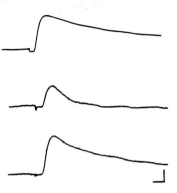

Figure 3. Amplitude of evoked extracellular EPSPs at mealworm neuromuscular junction[5] before (top), during perfusion of 100 μM Bztc (middle), and after its removal (bottom) (Yamamoto and Nachman, unpublished results). Bztc is only about 20-fold less potent than leucomyosuppressin (LMS).

REFERENCES

Nachman, R. J., Holman, G. M. and Haddon, W. F., Leads for insect neuropeptide mimetic development, *Arch. Insect Biochem. Physiol.*, 22, 181-197, 1993.

Keeley, L. L. and Hayes, T. K., Speculations on biotechnology applications for neuroendocrine research, *Insect Biochem.*, 17, 639-651, 1987.

Menn, J. J., New research horizons in insect control, *Pestic. Sci.*, 10, 372-377, 1985.

Nachman, R. J., Holman, G. M., Hayes, T. K. and Beier, R. C., Structure-activity relationships for inhibitory insect myosuppressins; contrast with the stimulatory sulfakinins, *Peptides, in press.*

Yamamoto, D., Ishikawa, S., Holman, G. M. and Nachman, R. J., Leucomyosuppressin, a novel insect neuropeptide, inhibits evoked transmitter release at the mealworm neuromuscular junction, *Neurosci. Lett.*, 95, 137-142, 1988.

Kingan, T. G., Teplow, D. B., Phillips, J. M., Riehm, J. P., Rao, K., Hildebrand, J. G., Homberg, U., Kammer, A. E., Jardine, I., Griffin, P. R. and Hunt, D. F., A new peptide in the FMRFamide family isolated from the CNS of the hawkmoth, Manduca sexta. *Peptides*, 11, 849-856, 1990.

Nachman, R. J., Tilley, J. W., Hayes, T. K., Holman, G. M. and Beier, R. C., Pseudopeptide mimetic analogs of insect neuropeptides, in *Natural and Derived Pest Management Agents*, Hedin, P., Hollingsworth, R. and Menn, J. J., Eds., American Chemical Society, *in press.*

Neuropeptide Metabolism

NEUROPEPTIDE BIOSYNTHESIS: BIOCHEMICAL, CELLULAR, AND BIOPHYSICAL STUDIES

Richard C. Rayne and Michael O'Shea

Sussex Centre for Neuroscience, University of Sussex, Brighton U.K.

Peptides are the most numerous and structurally diverse class of regulatory or signalling molecules in nervous and endocrine systems. In insects as in other animals, neuropeptides (transmitters, modulators and hormones) are involved in the regulation of a wide variety of essential physiological processes. While many aspects of the neurobiology of peptidergic systems in insects have been studied (for example, peptide action, release and anatomical localisation), few direct studies on biosynthesis exist. Over the past several years, however, the *corpora cardiaca* (CC) of the locust (*Schistocerca gregaria*) have been established as a model system to study neuropeptide biosynthesis (O'Shea and Rayne, 1992). Indeed, the locust CC glandular lobes are a neuropeptide "factory" much like the ELH-producing bag cells of *Aplysia* or the insulin-producing ß-cells (Sossin *et al.*, 1989), and provide an excellent model to study neuropeptide biosynthesis in an insect system.

As for other neuropeptides, biosynthesis of AKHs involves the enzymatic processing of precursor polypeptides. Owing to the cellular homogeneity of the CC and the ease with which the intact gland can be maintained alive *in vitro*, we have learned a great deal about the steps in AKH biosynthesis. Results of classical pulse-chase type experiments, protein chemistry, and cDNA cloning, have allowed us to establish a detailed model for AKH biosynthesis (O'Shea and Rayne, 1992). We are currently pursuing several lines of research directed toward defining further the steps in post-translational processing of the AKH precursors. Our approach is interdisciplinary and makes use of biochemical, cell biological, and biophysical techniques and is outlined below.

Biochemical Approaches

Exploiting the small size of the AKH I precursor (P1), we have produced a complete synthetic prohormone and used it as a substrate for assaying processing enzymes extracted from CC *in vitro*. We have established the existence in the CC of a specific endoproteolytic action which cleaves C-terminal to Arg^{13} at the Gly^{11}-Lys^{12}-Arg^{13} processing sites. The first products of *in vitro* processing, therefore, are AKH I extended by Gly^{11} -Lys^{12} -Arg^{13} and the dimer APRP 1. Other potential cleavage sites in the

0-8493-4591-X/94/$0.00 + $.50

© 1994 by CRC Press, Inc.

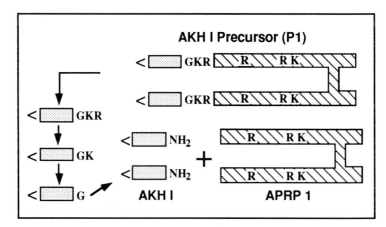

Figure 1. Reconstitution of AKH I precursor processing *in vitro*. The 10 amino acid sequence of AKH I is represented by the dot-filled rectangles; the symbol "<" represents residue number 1, pyroglutamic acid. Striped areas represent identical 28 amino acid residue chains which are linked by a single disulfide bond. G=glycine, K=lysine, R=arginine. See the text for further details.

prohormone, a dibasic site (Arg^{34}-Lys^{35}) and a monobasic site (Arg^{27}), are not cleaved *in vivo* or by CC extracts *in vitro*. This indicates therefore that the activity we observe in our *in vitro* assay may represent the *bona fide* AKH prohormone processing endopeptidase. Further processing of AKH-Gly-Lys-Arg in this system occurs sequentially: carboxypeptidase(s) trim the basic residues producing AKH-Gly-Lys, followed by AKH-Gly. Production of the amidated bioactive product, AKH-NH$_2$ (AKH I), from the glycine-extended peptide is a two step process requiring the cofactors ascorbate and Cu^{2+} (*cf.* Eipper *et al.*, 1992). This sequence of events is summarised in Fig. 1 (Rayne and O'Shea, 1991; 1993 submitted to *J. Exp. Biol.*).

We are particularly interested in understanding how processing enzymes "choose" the appropriate cleavage sites in neuropeptide precursors (recall that only one of three potential cleavage sites in the AKH I precursor is used *in vivo* and *in vitro*). It has been suggested that the specificity of enzyme activity to appropriate dibasic sites in prohormones is determined by secondary structure (e.g. ß-turns and Ω-loops) in the vicinity of appropriate processing sites (Geisow and Smyth, 1980; Rholam *et al.*, 1986). We have shown in pulse-chase experiments that replacement of lysine by its analog, thialysine, at the processing site in the AKH I precursor prevents cleavage of the prohormone by processing endopeptidases (not illustrated). Control experiments indicated that processing endopeptidases remain competent to cleave AKH prohormones in thialysine-treated *corpora cardiaca*, indicating that the observed defect in processing is due to incorporation of thialysine into proAKH I and not into the co-synthesized processing enzymes (Rayne and O'Shea, submitted to *Eur. J. Biochem.*). A molecular model of the AKH I precursor (discussed below) indicates that thialysine substitution causes this defect by disrupting a secondary structural motif at the processing site.

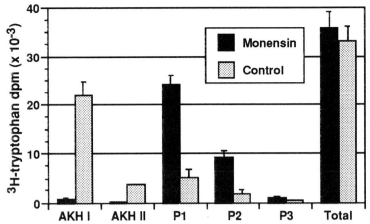

Figure 2. Monensin blocks processing of AKH precursors. Whole CC from adult locusts were maintained *in vitro* as described in Hekimi and O'Shea (1989). Control CC were "pulsed" for 1.5 h in the presence of ^{3}H-tryptophan then transferred to medium containing cycloheximide to block protein synthesis during a 12 hour chase. Treated CC were pulsed in the same way as control CC, but the chase medium contained 10 μM monensin. CC were homogenised and the extracts fractionated by RP-HPLC. In control incubations, most of the radiolabel incorporated into AKH precursors (P1, P2, and P3) is transferred to the products of processing, AKH I and AKH II. Addition of monensin to the chase medium prevents processing so that radiolabelled precursors are not converted to products. Data are expressed as mean ± S.D. for 3 individual CC.

Cell Biological Approaches

In addition to understanding the biochemical mechanisms of post-translational processing, we would like to identify the intracellular compartments in which these events take place. In neuroendocrine cells, prohormones are targeted to the regulated secretory pathway. Thus, neuropeptide precursors pass through the endoplasmic reticulum, the Golgi compartments, and are then packaged into secretory granules in which the products of processing are stored prior to release from the cell (Sossin *et al.*, 1990).

The ionophore monensin has been shown to block processing of prohormones in a number of systems (reviewed in Mollenhauer *et al.*, 1990); typically, the trans-Golgi is the site of the monensin-induced block. In pulse-chase experiments on intact CC maintained *in vitro*, we have shown that by including monensin in the chase medium, AKH precursor (P1, P2, and P3) processing is entirely blocked (Fig. 2). We can infer from this result that enzymatic cleavage of AKH precursors takes place in a post-trans-Golgi compartment, most likely the secretory granules.

Biophysical Approaches

Neuropeptides are produced by specific and limited proteolysis of precursor polypeptides. Although pairs of basic amino acids (Lys or Arg) often precede cleavage sites, such sequences are not invariably processing signals. For example, in each chain of the AKH I precursor, there are two dibasic sites, only one of which is recognised during processing *in vivo*. It has been suggested that various structural motifs (e.g. ß-turns, Ω-loops) which are associated with these dibasic sequences define actual cleavage sites (Rholam *et al.*, 1986; Bek and Berry, 1990; Paolillo *et al.*, 1992).

Using circular dichroism (CD) spectroscopy (Rayne, Drake, and O'Shea, in preparation) and ^{1}H 2D NMR (Horne, Doak, Rayne, O'Shea, and Campbell, in preparation) we are studying the solution structure of a complete, synthetic AKH I precursor. CD and NMR studies indicate that the unused processing site is "buried" within a region of α-helix, whereas the used processing site is "exposed". Application of a recently published structure prediction algorithm (Bek and Berry, 1990) indicates that a 7-residue Ω-loop (Pro^6 to Lys^{12}) resides adjacent to the used dibasic processing site in the AKH I precursor. We have used a molecular graphics system to construct a model of the AKH I precursor based on results obtained from CD, NMR and structure prediction algorithms. The model indicates that Lys^{12} is important in stabilizing the predicted Ω-loop and shows that "mutating" this residue to thialysine (see above) disrupts the Ω-loop (Rayne and O'Shea, submitted to *Eur. J. Biochem.*). The Ω-loop motif may therefore be important in directing processing endopeptidases to the correct cleavage site in the AKH I precursor.

Acknowledgements
    We thank Dr. A. Drake of Birkbeck College, University of London for collaboration on CD spectroscopy experiments. NMR spectroscopy is being carried out in collaboration with workers at the Oxford Centre for Molecular Sciences, Oxford University by Dr. T. Horne and Mr. D. Doak in the laboratory of Prof. I. Campbell.

REFERENCES

1. O'Shea, M. and Rayne, R.C., Adipokinetic hormones: Cell and molecular biology, *Experientia* 48, 431-438, 1992.
2. Sossin, W.S., Fischer, J.M. and Scheller R.H., Cellular and molecular biology of neuropeptide processing and packaging, *Neuron* 2, 1407-1417, 1989.
3. Eipper, B.A., Stoffers, D.A. and Mains, R.E., The biosynthesis of neuropeptides: peptide α-amidation, *Ann. Rev. Neurosci.* 15, 57-85, 1983.
4. Rayne, R.C. and O'Shea, M., Complete processing of AKH precursor reconstituted *in vitro*, presented at 21st Ann. Meet. Soc. for Neurosci., 1991.
5. Geisow, M. J. and Smyth, D. G., Proteolysis in prohormones and pro-proteins, in *The Enzymology of Post-Translational Modification of Proteins*, Vol. 1, Freedman, R. B. and Hawkins, H. C., Eds., Academic Press, London, 1980, pp. 259-287.
6. Rholam, M., Nicolas, P. and Cohen, P., Precursors for peptide hormones share common secondary structures forming features at the proteolytic processing sites, *FEBS Lett.* 207, 1-6, 1986.
7. Mollenhauer, H.H., Morré, D.J. and Rowe, L.D., Alteration of intracellular traffic by monensin: mechanism, specificity, and relationship to toxicity, *Biochim. Biophys. Acta* 1031, 25-246, 1990.
8. Bek, E. and Berry, R., Prohormonal cleavage sites are associated with Ω-loops, *Biochemistry* 29, 178-183, 1990.
9. Paolillo, L., Simonetti, M., Brakch, N., D'Auria, G., Saviano, M., Dettin, M., Rholam, M., Scatturin, A., DiBello, C. and Cohen, P., Evidence for the presence of secondary structure at the dibasic processing site of a prohormone: the pro-oxytocin model, *EMBO J.* 11, 2399-2405, 1992.
10. Hekimi, S. and O'Shea, M., Biosynthesis of adipokinetic hormones (AKHs): Further characterization of precursors and identification of novel products of processing, *J. Neurosci.* 9(3), 996-1003, 1989.

Metabolic Enzymes of Bioactive Neuropeptides in the Gypsy Moth, *Lymantria dispar*

Edward P. Masler and R. M. Wagner

Insect Neurobiology and Hormone Laboratory, Plant Sciences Institute and Livestock Insects Laboratory, Livestock and Poultry Sciences Institute, Agricultural Research Service, United States Department of Agriculture, Beltsville, Maryland, 20705-2350 U.S.A.

INTRODUCTION

Neuropeptides control essentially all critical life processes in insects. Their importance has led to a focus on neuropeptides as leads to novel insect control agents (Masler *et al.* 1993). This is especially relevant because insect control persists as a perplexing and costly agricultural problem. Environmental concerns and world economic pressures demand that exploration and development of new pest control agents be given significant attention. Discovery of such agents can be facilitated by the study of the biogenesis, mode of action and degradation of neuropeptides. These are complex processes which offer a number of biochemical targets at which to aim the development of new controls. These targets are either specific enzymes, or amino acid sequences which are recognized by the enzymes (Masler *et al.* 1993). The biosynthesis of neuropeptides involves the processing of peptide precursors by specific enzymes and, after binding to the receptor, neuropeptides are catabolized by specific proteolytic enzymes. Control of the attenuation of neuropeptide signals by these catabolic enzymes offers possibilities for the development of novel control agents. In order to exploit the importance of catabolic proteases, they must be identified and characterized. Membrane-bound proteases have been described in reports on vertebrate systems. These reports suggest that the membrane associated proteases may be most important for modulating neuropeptide activities, especially when the neuropeptide is bound to the receptor (Turner *et al.*, 1989; Turner, 1990). Initial studies on the presence and nature of peptide catabolic enzymes in insects (Starratt and Steele, 1984; Quistad *et al.*, 1984; Skinner *et al.*, 1987), indicated the presence of soluble proteolytic activities. More recently, membrane-bound proteolytic activities have been described in insects. Nearly all of this work has been done on the locust, *Shistocerca gregaria*, using locust adipokinetic hormone as substrate (Isaac 1987, 1988; Rayne and O'Shea 1992). Endopeptidase and aminopeptidase activities

0-8493-4591-X/94/$0.00 + $.50
© 1994 by CRC Press, Inc.

were found in tissue membrane preparations. Isaac and Priestly (1990) reported similar activities in *Drosophila melanogaster* head preparations. We have recently discovered peptidase activities in neural membrane preparations from the gypsy moth, *Lymantria dispar*. This is the first observation of such activity in the Lepidoptera.

METHODS

Lymantria dispar used in our laboratory were maintained according to methods described by Masler *et al.* (1991). Eggs were held at 5-7° C for 180 days to break diapause and then transferred to a rearing chamber and held at 25° C, 50-60 percent relative humidity and a 16hL:8hD photoperiod. Larvae hatched 2-3 days later. The larval-pupal molts and adult eclosion were monitored on a daily basis in order to stage animals to within +/- 12 hours. Tissues used for enzyme preparation were dissected from animals anesthetized by carbon dioxide and collected in 1.5mL polypropylene tubes on dry ice. Tissue was homogenized in 50mM Tris-HCl buffer, pH 7.3 at approximately 1 tissue equivalent per 50uL, on ice, and a 30,000xG pellet suspended in 100mM Tris-HCl, pH 7.3 was used as the source of enzyme for assay. The assay procedure is modified from Isaac (1988). Peptide substrate was prepared in 100mM Tris-HCl, pH 7.3, and incubated with enzyme. Following incubation, the mixture was fractionated on a reverse phase HPLC column (C-18 functionality) using a linear gradient of 10 to 60 percent acetonitrile in 0.1 percent aqueous trifluoroacetic acid. Elution was monitored at 214nm, the absorbing peaks were collected and subjected to amino acid sequencing to determine identity of the digestion products. The relative amounts of each digestion component were estimated by peak area integration and used to monitor the progress of the protease reactions. Specific protease inhibitors were prepared in incubation buffer and pre-incubated with enzyme preparations prior to addition of the substrate.

RESULTS and DISCUSSION

Crude membrane preparations degraded AKH over time as illustrated in Figure 1. Kinetics indicate that digestion, monitored by the decrease in the size of the AKH chromatographic peak over time, was slow during the first 2 hours of incubation. Digestion was most rapid between hours 2 and 4, and continued through 8 hours. Accompanying the degradation of AKH was the appearance of product peaks. Two of the most prominent were the tripeptide pGlu-Leu-Asn, which we term Pk1A, and the heptapeptide Phe-Thr-Pro-Asn-Trp-Gly-Thr, termed Pk5A. Other peaks appear to arise from the catabolism of Pk1A and Pk5A. Cleavage of the Asn_3-Phe_4 bond is inhibited by the endopeptidase inhibitor phosphoramidon. In comparing column B (phosphoramidon present) with column A (control, no inhibitor) in Figure 2, note that while the relative amount of AKH present after 8 hours of incubation is greater in B, just the opposite is observed with Pk5A and Pk1A. Obviously, inhibition of the degradation of AKH should have a negative effect on the production of peaks 1A and 5A. In the presence of amastatin (Fig.2, column C), the results are somewhat different. The inhibitor had essentially no effect upon degradation of AKH, and Pk1A was relatively unaffected. In contrast, Pk5A accumulates to a much higher degree when amastatin is present than in either the control or phosphoramidon treatments. Taken together, these data suggest that the initial cleavage is by a phosphoramidon-sensitive endopeptidase, yielding Pk1A and Pk5A. Subsequent action by an aminopeptidase causes degradation of Pk5A.

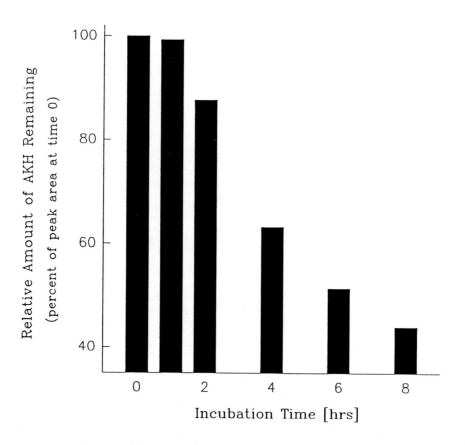

Figure 1. Degradation kinetics of AKH in the presence of *L. dispar* neural membrane preparations. Peptide and tissue preparations were incubated as described and terminated at the times indicated. The preparations were fractionated on reverse phase HPLC and the relative amount of AKH remaining in each incubation was estimated by peak integration and comparison with the time-0 control.

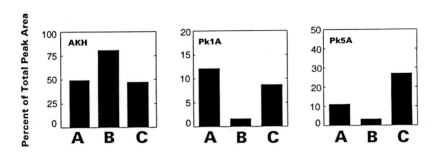

Figure 2. Effect of protease inhibitors on the fate of AKH. Incubations of AKH in the presence of *L. dispar* neural membrane preparations (A, control), condition A plus phosphoramidon (B) or condition A plus amastatin (C) were prepared as described. Incubations were done for 6 hr, the mixtures were fractionated by reverse phase HPLC, and the relative amounts of AKH, Pk1A and Pk5A under conditions A, B or C were estimated by peak integration.

REFERENCES

Isaac R. E. *Biochem. J.* 245, 365-370, 1987.

Isaac R. E. *Biochem. J.* 255, 843-847 1988.

Isaac R. E. and Priestly R. M., in *Insect Neurochemistry and Neurophysiology*. Borkovec A.B., Masler, E. P., Eds. Humana Press, Clifton, NJ. 1990, 267-270.

Masler E. P., Bell R. A., Thyagaraja B. S., Kelly T. J. and Borkovec A. B., *J. Comp. Physiol. B.* 161, 37-41, 1991.

Masler E. P., Kelly T. J. and Menn J. J. *Archives Insect Biochem. Physiol.* 22, 87-112, 1993.

Quistad G. B., Adams M. E., Scarborough R. M., Carney R. L. and Schooley D. A., *Life Sciences* 34, 569-572, 1984.

Rayne R. C. and O'Shea M., *Insect Biochem. Mol. Biol.*, 22, 25-34, 1992.

Skinner W. S., Quistad G. B., Adams M. E. and Schooley D. A., *Insect Biochem.* 17, 433-437, 1987.

Starratt A. N. and Steel R. W. *Insect Biochem.* 14, 97-102, 1984.

Turner A.J., Hooper N.M. and Kenny A.J., in *Neuropeptides: A Methodology*. Fink G., Harmar, A. J, Eds. Wiley and Sons, Chichester, Great Britain, 1989, 189-223.

Turner, A. J., in *The Biology and Medicine of Signal Transduction*. Nishizuka Y., et al., Eds. Raven Press, NY., 1990, 467-471.

Receptor Studies

GABA RECEPTOR SUBTYPES IN DORSAL UNPAIRED MEDIAN (DUM) NEURONS OF THE INSECT CENTRAL NERVOUS SYSTEM.

Véronique DUBREIL, Bruno LAPIED and Bernard HUE

Laboratoire de Neurophysiologie, URA CNRS 611, Université d'Angers, F-49045 ANGERS Cedex, FRANCE.

GABA is the main inhibitory neurotransmitter in the insect central nervous system (see for review Sattelle 1990). Until now, *in situ* electropharmacological studies of GABA receptors have been mainly performed on adult neurons such as unidentified dorsal median cells bodies (Kerkut *et al.*, 1969 ; Pitman and Kerkut 1970 ; Roberts *et al.*, 1981), cercal afferent-giant interneuron synapses (Callec 1974 ; Hue 1991), and fast coxal depressor motoneuron (Wafford and Sattelle 1986) of the cockroach. However, little is known about the localization and subtypes of GABA receptors on neurosecretory cells of the dorsal midline of insect ganglia, called Dorsal Unpaired Median (DUM) neurons (Hoyle *et al.*, 1974). In this work, we have obtained a better knowledge on GABA receptors of these cells located in the cockroach terminal abdominal ganglion (TAG) and particularly, the localization of these receptors.

All experiments were carried out *in situ*, on desheathed TAG of the cockroach *Periplaneta americana*, continuously superfused with saline (in mM : NaCl 200 ; KCl 3.1 ; $CaCl_2$ 5 ; $MgCl_2$ 4 ; HEPES 10 ; sucrose 50 ; pH 7.4). DUM neurons have a typical morphology : from their somata lying on the dorsal midline of the TAG, emerges a short primary neurite bifurcating into two lateral ramified branches running into the left and the right sides of the ganglia (Watson, 1984). Considering this particular anatomy and in order to point out the localization of GABA receptors, GABA pressure microejection (0.05 M ; 100 ms, except otherwise stated) has been performed onto the soma or lateral neurites. The response was always recorded using a microelectrode impaled in the soma.

0-8493-4591-X/94/$0.00 + $.50
© 1994 by CRC Press, Inc.

GABA microapplication onto the soma or onto the lateral neuritic arborization produced hyperpolarizing responses which obviously differ in the shape (see Fig. 1). Furthermore, it has been demonstrated that both GABA responses were sensitive to bath-applied picrotoxin (PTX) (10^{-4} M) suggesting that soma membranes as well as neuritic membranes possess GABA-dependent PTX-sensitive receptors.

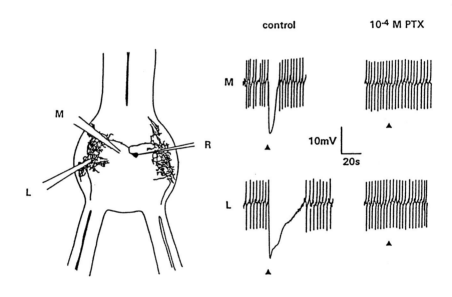

Figure 1. Dependence of GABA responses on the application site. (A) Experimental arrangement used to locate the GABA receptors on DUM cells, with the different positions of the GABA pipette compared with the morphology of the impaled DUM neuron. (B) GABA median ejection (M) induces a fast hyperpolarization, while GABA lateral ejection (L) induces a multiphasic hyperpolarization. Both responses are PTX-sensitive. R = recording electrode.

In order to clarify the ionic nature of GABA response, further experiments have been performed on soma membrane. Decrease (25%) in membrane resistance during GABA response has been recorded by passing hyperpolarizing current pulses through the cell membrane. This led to the conclusion that membrane conductance is increased by GABA-receptor activation.

Shift of the equilibrium potential for chloride ions (Δ E_{Cl^-} = +35.7 mV), by replacing half of NaCl by equimolar CH_3COONa in the saline, produced a decrease of the GABA-induced response. Moreover, in most of the cells studied, the GABA responses reversed close to E_{Cl^-} ($\approx$ -85 mV). These latter results well agree with the proposal that GABA receptors are coupled to chloride ions channels as vertebrate $GABA_A$ receptors. The specific $GABA_A$ antagonist bicuculline did not produce any effect on the GABA hyperpolarization.

Under PTX treatment (10^{-4} M up to 10^{-3} M), higher doses of GABA application onto the soma produced slower hyperpolarizations. These unexpected PTX-resistant GABA responses have been reduced in presence of low chloride saline (120 mM) indicating a chloride dependency of the response. Furthermore, it should be noted that high potassium saline (30 mM) also reduced this GABA response. According to Jensen *et al.* (1993), one may speculate that increasing external K^+ concentration indirectly induces a positive shift of E_{Cl^-} by inhibiting a K^+/Cl^- cotransport mechanism which extrudes Cl^-, which could explain GABA response depression.

In conclusion, these results demonstrate that DUM neurons have GABA receptors located on the soma and on lateral neurites. Furthermore, at least two kinds of GABA receptors subtypes have been detected on the soma membrane : one is PTX-sensitive and chloride-dependent while the other is PTX-resistant but chloride-dependent too.

REFERENCES

Callec, J.J., Synaptic transmission in the central nervous system of insect, in *Insect Neurobiology*, Treherne, J.E., Ed., North-Holland-American Elsevier, New York, 1974.
Hoyle, G., Dagan, D., Moberly, B.,and Colquhoun, W., *J. exp. Zool.*, 187, 159-165, 1974.
Hue, B., *Archiv. Insect Biochem. Physiol.*, 18, 147-157, 1991.
Jensen, M.S., Cherubini, E. and Yaari, Y., *J. Neurophysiol.*, 69, 764-771, 1993.

Kerkut, G.A., Pitman, R.M. and Walker, R.J., *Comp. Biochem. Physiol.*, 31, 611-633, 1969.

Pitman, R.M. and Kerkut G.A., *Comp. Gen. Pharmacol.*, 1, 221-230, 1970.

Roberts, C.J., Krogsgaard-Larsen, P. and Walker, R.J., *Comp. Biochem. Physiol.*, 69C, 7-11, 1981.

Sattelle, D.B., *Adv. Insect Physiol.*, 22, 1-113, 1990.

Wafford, K.A. and Sattelle, D.B., *Neurosci. Lett.*, 63, 135-140, 1986.

Watson, A.H.D., *J. Neurocytol.*, 13, 303-327, 1984.

DISTINCT TYPES OF CHOLINERGIC RECEPTORS ACTIVATED BY ARECOLINE ON *IN SITU* AND ISOLATED DUM NEURONS

Florence TRIBUT and Bruno LAPIED

Laboratoire de Neurophysiologie, URA CNRS 611, Université d'Angers, F-49045 ANGERS cedex, FRANCE

It is well established that axotomy and deafferentation procedures could produce some modifications in the electrophysiological properties of cell body membrane of invertebrate neurons (see for review Titmus and Faber 1990). These modifications observed in various preparations show differences in intensity, time course and domain of membrane involved. However, little is known concerning a possible qualitative post-axotomy evolution of the pharmacological properties of invertebrate cholinergic receptors. Recently, it has been shown that, among invertebrate neurons, some of them, called Dorsal Unpaired Median (DUM) neurons respond to axotomy and deafferentation treatment by a rapid increase in the density of sodium current involving a neosynthesis of sodium channels (Tribut *et al.*, 1991). To check whether this experimental procedure (i.e., axotomy and deafferentation) could produce some qualitative changes in the pharmacological properties of the diversity of functional cholinergic receptors present in these DUM neurons, the effect of a cholinergic ligand, arecoline, has been studied on both *in situ* and isolated DUM neurons.

All experiments were performed on isolated adult DUM neurons obtained after dissociation of the terminal abdominal ganglia (TAG) of the nerve cord of the adult male cockroaches, *Periplaneta americana* as previously described (Lapied *et al.*, 1989). The normal saline contained (in mM) : NaCl, 200 ; KCl, 3.1 ; CaCl$_2$, 5 ; MgCl$_2$, 4 ; HEPES buffer, 10 ; pH was ajusted to 7.4 with NaOH. The somata of DUM neurons were penetrated with conventional intracellular microelectrode (tip resistance was 40-50 MΩ when filled with 1 M potassium acetate - 1 M KCl solution). The cholinergic agonist, arecoline (ARE, 10^{-1} M) was directly applied by pneumatic pressure ejection (15 psi) onto the cell body as previously described (Lapied *et al.*, 1990). The cholinergic antagonists tested were dissolved in saline and applied through a gravity system perfusion. Experiments were carried out at room temperature (20^0 C). Results were expressed as mean $\pm$ S.E.M..

0-8493-4591-X/94/$0.00 + $.50
© 1994 by CRC Press, Inc.

339

Pressure ejection of ARE onto *in situ* DUM neurons (resting membrane potential -53 $\pm$ 1 mV, n = 20) induced a specific biphasic depolarization composed of a fast initial component followed by the development of a slow second phase (Fig 1A). Using both specific nicotinic and muscarinic antagonists, it has been possible to well pharmacologically discriminated this biphasic response. The fast depolarization was sensitive to nicotinic antagonists (mecamylamine 10^{-5} M, d-tubocurarine 5.10^{-5} M) but resistant to α-bungarotoxin, whereas the slow phase was inhibited by muscarinic antagonists (scopolamine 5.10^{-6} M, quinuclidinil benzilate (QNB) 5.10^{-6} M, atropine 10^{-5} M and pirenzepine 10^{-5} M). A similar ARE-induced biphasic depolarization was observed on isolated DUM neurons (resting membrane potential -52,4 $\pm$ 1,2 mV, n = 20) maintained in short-term culture (Fig 1). Interestingly, although the time course and the dose-dependence of the ARE-induced response were comparable in *in situ* and isolated DUM neurons using the same experimental conditions, it should be noted that both maximum amplitude and duration of the response were reduced from 6.7 $\pm$ 0.5 mV (n=15) to 3.8 $\pm$ 0.2 mV (n=15), and from 2.4 $\pm$ 0.5 min (n=15) to 1.1 $\pm$ 0.1 min (n=15) respectively, in isolated cells compared to *in situ* DUM neurons. This decrease in the maximum amplitude and the duration of the response in axotomized DUM neurons, could be attributed to a possible decrease in the sensitivity of DUM neuron cholinergic receptors after axotomy and deafferentation, as previously described in other invertebrate preparations (e.g., Bigiani and Pellegrino 1990).

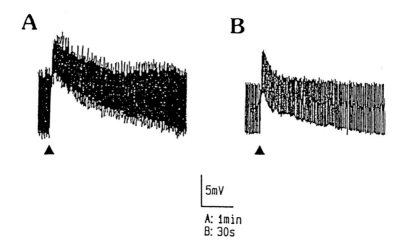

Figure 1. Micropressure ejection of ARE (▲ , 100 ms) onto (A) *in situ* and (B) isolated DUM neuron.

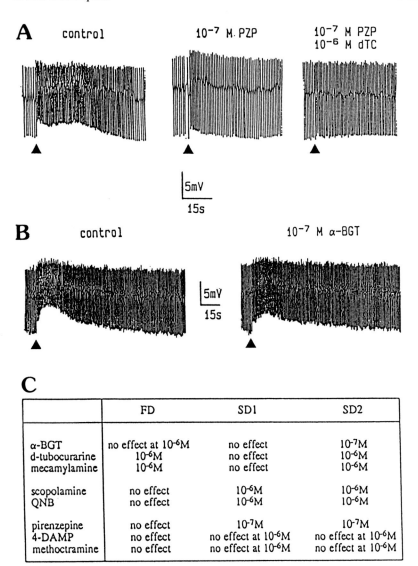

Figure 2. Effects of specific muscarinic and nicotinic antagonists on the response elicited by ARE (▲, 100 ms) on spontaneously active isolated DUM neuron. (A) Pretreatment with the selective M1 muscarinic antagonist pirenzepine (PZP) completely abolished the slow depolarization. Addition of d-tubocurarine (d-TC) blocked the residual component. (B) Pretreament with α-bungarotoxin (α-BGT) induced a slight reduction of the amplitude of the ARE-induced response. (C) The values given in the table represent the concentrations which produce 100% of inhibition ; FD : fast depolarization ; SD : slow depolarization.

In order to determine whether the pharmacological profile of these cholinergic receptors involved in this ARE-induced response could be affected by axotomy and deafferentation treatment, a same range of cholinergic antagonists used for *in situ* pharmacological studies has been tested. Fig. 2A shows that bath application of the selective M1 muscarinic antagonist, pirenzepine (PZP, 10^{-7} M) completely abolished the slow component of the ARE-induced response. The fast depolarization was blocked when d-tubocurarine (d-TC, 10^{-6} M) is added to the perfusion solution. Effects of the different cholinergic antagonists on the response elicited by ARE on isolated DUM neurons were summarized in the table (Fig. 2C). As shown in Fig 2B, interesting results have been obtained with α-BGT, the nicotinic antagonist well known to only blocked "mixte" (i.e., nicotinic/muscarinic) component in DUM neurons (Lapied *et al.*, 1990). α-BGT (10^{-7}M) reduced the amplitude of the slow depolarization of the ARE-induced response without any effect on the fast component and unmasked a resistant α-BGT slow depolarizing phase blocked by muscarinic antagonists. These results demonstrate that ARE produce a complex response composed of *i)* a fast nicotinic component resistant to α-BGT (FD in the table) *ii)* a slow muscarinic component mediated by a M1-like muscarinic receptor subtype (SD1) and *iii)* a "mixte" (nicotinic/muscarinic) component (SD2).

REFERENCES

1. **Bigiani A. and Pellegrino, M.**, *J. exp. Biol.*, 151, 423-434, 1990.
2. **Lapied, B., Malecot, C.O. and Pelhate, M.**, *J. exp. Biol.*, 144, 535-549, 1989.
3. **Lapied, B., Le Corronc, H. and Hue, B.**, *Brain Res.*, 533, 132-136, 1990.
4. **Titmus, M.J. and Faber, D.S.**, *Progress in Neurobiol.*, 35, 1-51, 1990.
5. **Tribut, F., Lapied, B., Duval, A. and Pelhate, M.**, *Pflügers Arch.*, 419, 665-667, 1991.

JHIII REGULATES PHONOTAXIS IN CRICKETS BY CONTROLLING EXPRESSION OF NICOTINIC RECEPTORS IN AUDITORY INTERNEURONS

J. Stout, J. Hao, G. Atkins, O. Stiedl, J. Ramseier, P. Coburn,
V. Hayes, J. Henley, and P. Kim

Biology Dept., Andrews University, Berrien Springs, MI, 49103, USA

Female crickets recognize and respond phonotactically to the calling song of a conspecific male. For *Acheta domesticus* this response begins 2-4 days following the imaginal molt as a result of declining behavioral thresholds to the male's call. The rate of juvenile hormone III (JHIII) synthesis increases on day 2 and, following a peak on day 4, is variable but generally declines until day 12 after which it remains variably lower. Removal of the corpora allata from responsive females does not eliminate phonotaxis (Loher et al. 1992), but does increase phonotactic threshold. Application of JHIII to allatectomized females restores phonotactic sensitivity within 6 hr (Stout et al. 1992).

L1, an ascending, prothoracic auditory interneuron of female *A. domesticus*, encodes the temporal structure of the male's calling song and is tuned to its carrier frequencies (4 - 6 kHz). The response of a single L1 is both necessary for phonotaxis by females to model calling songs (65 dB or lower) and sufficient to cause phonotaxis (Atkins et al. 1992). In 1-2 day old females both behavioral and L1 thresholds are 75 dB or higher. Application of JHIII to 1-day-old females caused their threshold for phonotaxis and the L1 neuron to drop from means of about 80 to 50-60 dB (typical of older females) within 12 hr (Stout et al. 1992). Neither age nor JHIII influence the auditory thresholds of the presently described prothoracic auditory neurons other than L1 (Stout et al. 1992). Since L1's response is necessary for phonotaxis by the female for calling songs (less intense than 65-70 dB, Atkins et al. 1992), hormonal regulation of the female's behavioral threshold could be solely the result of JHIII's control of L1's sensitivity.

An mRNA like that for locust α-L1 nicotinic receptors (Marshall et al. 1990) is produced by the L1 neuron (Stout et al. 1992). L1's in 1-2 day old females with high thresholds express less α-L1-like nicotinic receptor mRNA than do L1's in older females with low thresholds (Fig. 1 A&B). Addition of JHIII to 1-day-old females with high behavioral and L1 thresholds (Fig 1 C), increases the expression of this mRNA in L1's soma, while lowering both L1 and behavioral thresholds. Allatectomy of older females with low behavioral thresholds results in high behavioral and L1 thresholds and reduced expression of α-L1-like nicotinic receptor mRNA (Fig. 1 D). Increases in the expression of this receptor should increase the response of the L1 neuron to excitatory input. Thus, regulation of phonotactic sensitivity is, at least in part, through JHIII regulated changes in the expression of nicotinic receptors in the L1 neuron.

0-8493-4591-X/94/$0.00 + $.50
© 1994 by CRC Press, Inc.

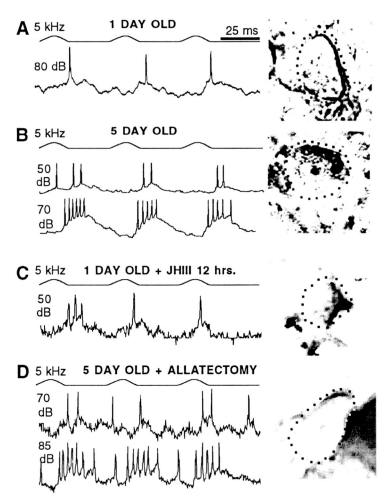

Figure 1. Recordings and *in situ* hybridization (of the soma which is outlined) from L1 auditory interneurons from females of the indicated ages and treatments. In C JHIII was topically applied and 12 hours later behavior was tested, followed immediately by recording . In D the female was allatectomized and tested behaviorally for 3 days during which the phonotactic threshold increased 20 dB. The first trace of each recording indicates the stimulus syllables.

SP is the most important temporal parameter of the male's calling song for female phonotaxis (Stout et al. 1992). Young virgin females (4-7 days) respond selectively to models of calling songs having SPs of 50-70 ms. As virgin females become older they become less selective for SP so that by 14 days of age they respond to all calling song SPs tested. Old females (28 days) were as selective for SP as young females 2-4 days following application of JHIII (Walikonis et al. 1991).

L3, an ascending prothoracic auditory interneuron, is tuned to both 5 and 16 kHz, is involved in the female's phonotaxis to the calling song, and selectively decrements in its spiking response to successive syllables of calling songs with different SPs (decrement is maximal to the male's modal SP of 50 ms). This SP-selective response is a neural correlate of SP-selective phonotaxis. Decrement, which is maximal (50%) around day 4, diminishes with age (20%) in 3-4-week-olds (Atkins et al. 1992) paralleling the changing phonotactic selectiveness. Expression of α-L1-like nicotinic receptor mRNA in the somata of L3 auditory neurons increases to a maximum in 4-5 day-old females and decreases in older females (Stout et al. 1992 and Fig. 2 A&B), correlating rather well with age related changes in both JHIII biosynthesis and the degree of decrement in the spiking response of L3 neurons, while allatectomy of 4-5 day-olds reduces both decrement and mRNA expression (Fig. 2C). In 24-28 day-old females, SP selective phonotaxis, L3 decrement and expression of the α-L1 like nicotinic receptor mRNA were maximal following JHIII application (Fig. 2 D) reaching levels typical for 4-5 day-olds.

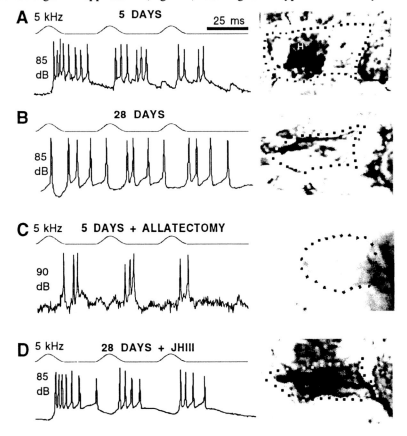

Figure 2. Recordings and *in situ* hybridizations (from the soma) of L3 neurons from females of different ages receiving the indicated treatments. Conditions for JHIII treatment and allatectomy were as described for Figure 1.

The change in L3's decrement with age is not paralleled by a change in threshold, but rather results from a change in the spiking activity to the first syllable. The response to the third syllable remains constant despite the degree of decrement. Therefore, these changes in spiking might involve regulation of excitatory and/or inhibitory inputs (Stout et al. 1992). The response of the hormone treated L3 was greater in response to the first syllable of the model call and then decremented down to a level of response to the third syllable that was typical of both young and old females (Fig. 2 D). These results suggest that JHIII may increase the excitatory and/or reduce the inhibitory inputs into the L3 neuron. The increased expression of α-L1-like nicotinic receptor mRNA induced in L3 by JHIII may provide a basis that is sufficient to explain L3's increased response to excitatory input as part of the regulation of its syllable period specific processing of input calling song information. Possibly JHIII also influences the expression of GABA or other inhibitory neurotransmitter receptors in this neuron as well.

JHIII which influences both the threshold of the female cricket's phonotactic response to, and her selectivity for the SP of the calling song, also influences the threshold of L1's response to, and L3's selective response for the SP of the calling song. The effect of JHIII on both interneurons increases their excitatory input. JHIII's influence on cricket phonotaxis provides a model for hormonal regulation of rather complex behavior by genetic regulation of the response properties of identified auditory neurons.

(For methods involved with: a. behavioral testing and hormonal treatments of females see Walikonis et al. (1991) and Stout et al. (1991); b. intracellular recording, staining and morphology see Atkins et al. (1992); c. *in situ* hybridization procedures, quantification of hybridization, sequence of locust α-L1 nicotinic ACh receptor gene see Stout et al. (1992), Emson (1993), Marshall et al. (1990) respectively.)

Acknowledgements

This research was supported by National Science Foundation grants BNS 88-19817 and IBN 92-22127.

Literature Cited

Atkins, G., Henley, J., Handysides, R., Stout, J., Evaluation of the behavioral roles of ascending auditory interneurons in calling song phonotaxis by the female cricket (*Acheta domesticus*). J. Comp. Physiol. 170, 363-372, 1992.

Emson, P., *In situ* hybridization as a methodological tool for the neuroscientist. Trends in Neurosciences 16, 9-16, 1993.

Loher, W., Weber, T., Rembold, H., Huber, F., Persistence of phonotaxis in females of four species of crickets following allatectomy. J. Comp. Physiol. 171, 325-341, 1992.

Marshall, J., Buckingham, D., Shingai, R., Lunt, G., Goosey, M., Darlison, M., Sattelle, D., Bernard, E. Sequence and functional expression of a single alpha subunit of an insect nicotinic acetyl choline receptor. EMBO J. 13, 4391-4398, 1990.

Stout, J., Atkins, G., Zacharias, D., Regulation of cricket phonotaxis through hormonal control of the threshold of an identified auditory neuron. J. Comp. Physiol. **169**, 765-772, 1991.

Stout, J., Hayes, V., Zacharias, D., Henley, J., Stumpner, A., Hao, J., Atkins, G., Juvenile hormone controls phonotactic responsiveness of female crickets by genetic regulation of the response properties of identified auditory interneurons, in *Insect Juvenile Hormone Research*, Mauchamp, B., Couillaud, F. and Baer, J., Eds. INRA, Paris, 1992, 265-283.

Walikonis, R., Schoun, D., Zacharias, D., Henley, J., Coburn, P., Stout, J. (1991) Attractiveness of the male *Acheta domesticus* calling song to females. III. The relation of age-correlated changes in syllable period recognition and phonotactic threshold to juvenile hormone III biosynthesis. J. Comp. Physiol. 169, 751-764.

INVESTIGATION OF THE BINDING OF

[3]H-TYRAMINE AND ^{3}H-OCTOPAMINE IN THE INSECT (*Locusta migratoria*) BRAIN

L. Hiripi[1], S. Juhos[1] and R. G. H. Downer[2]

[1] Balaton Limnological Research Institute of the Hungarian Academy of Sciences, H-8237
Tihany, Hungary

[2] Department of Biology, University of Waterloo, Waterloo, Ontario, Canada N2L 3G1

The biogenic amines are important regulatory compounds in the insect body and produce a variety of physiological effects. The phenylethylamine tyramine and octopamine are present in a significant concentration in the insect nervous system and their role as a neurotransmitters, neurohormons and neuromodulators are suggested (Downer, in this volume).

The actions of monoamines are mediated through their receptors. A large number of drugs also produce their effects by interacting with these receptors. In this paper we summarize our recent results concerning the identification of the tyramine and octopamine receptors in the locust brain.

The locust brain were homogenized in ice cold 50mM Tris-HCl buffer pH 7.4 and centrifuged three times at 50,000 g for 15 min. Membrane corresponding to 20mg wet tissue were incubated with ^{3}H-tyramine or ^{3}H-octopamine in the a total volume of 2 ml. Nonspecific binding was defined as the concentration of bound ligand in the presence 10 μM of tyramine or octopamine respectively. Following equilibrium (20 min. at 30° C) bound ligand was separated from free by vacuum filtration over Whatman GF/B glass fibre filters, which were then rinsed with 3X5ml ice-cold buffer. Radioactivity bound to the glass fibre filters was determined by liquid-scintillation spectrometry. The inhibition of specific binding of the radioligands by competing ligands was analyzed graphically to estimate the IC_{50}, using a non-linear least-squares fitting programme.

0-8493-4591-X/94/$0.00 + $.50
© 1994 by CRC Press, Inc.

Employing the assay method described above, specific binding observed with 1 nM of ^{3}H-tyramine and ^{3}H-octopamine. Nonspecific binding was less than 20% of total binding. The specific binding of both ligands reached equilibrium after 20 minutes and was reversible. The specific binding of both tyramine and octopamine was saturable and the Scatchard analysis indicates a single class of sites. The K_d and the B_{max} values for tyramine were 6.63 nM and 31.7 fmol/mg tissue, while for octopamine were 5.65 nM and 15 fmol/mg tissue respectively. The affinities of a series of compounds for tyramine and octopamine binding sites are listed in Table 1.

Both the kinetic and pharmacological results suggest that there are separate tyramine and octopamine receptors in the locust brain. In vertebrate brain the binding of ^{3}H-tyramine was also demonstrated, however tyramine and octopamine are equipotent displacers of tyramine binding suggesting that the tyramine binding associates with the dopamine receptor (Vaccari 1986). In locust brain however dopamine and octopamine have low affinities to tyramine binding. The ability of both tyramine and octopamine to inhibit their own binding is similar. However tyramine has higher affinity towards the octopamine receptor than octopamine has towards tyramine receptor. Epinephrine and norepinephrine have higher affinity to octopamine than tyramine receptor, whereas dopamine has higher affinity to tyramine than octopamine receptor. The pharmacological characterization shows that the alpha adrenergic naphazoline, tramazoline were particularly active competitors for octopamine but not for tyramine binding. Clonidine is less active, and it has higher affinity to tyramine than octopamine receptors, while synephrine and demethylchlordimeform have higher affinity to octopamine receptor. The meta isomer of octopamine has 100 times lower affinity to octopamine receptor than the para isomer.

The most portent antagonists of octopamine binding were mianserin, prazosin whereas the strongest inhibitors of tyramine binding were WB-4101, BOL and LSD. Yohimbine has 100 times higher affinity to the tyramine than octopamine receptors, , however it has low affinity for both tyramine and octopamine binding and it is not a useful ligand with which to characterize these receptors.

Mianserin has high affinity to octopamine receptor, however ^{3}H-mianserin binds in a partially irreversible fashion to locust brain tissue and therefore is not suitable for characterization of the octopamine receptor.

Prazosin has also high affinity to octopamine receptor. ^{3}H-prazosin binds with high affinity to the locust brain tissue (K_d=13.2 nM; B_{max}= 36.4 fmol/mg tissue), however only the prazosin and mianserin inhibit the ^{3}H-prazosin binding at low concentration (IC_{50}= 4 nM and 50 nM respectively). Naphazoline, octopamine,, synephrine, norepinephrine, clonidine, epinephrine, LSD, BOL, gramine, cyproheptadine, inhibit the prazosin binding at a concentration range of 10-100 μM.

Table 1. Affinity ratios of tyramine and octopamine binding sites for
drugs in locust brain

Affinity ratios relative to corresponding tyramine		Affinity ratios relative to corresponding octopamine	
Tyramine	1.0	p-Octopamine	1.0
p-Octopamine	0.02	Naphazoline	6.6
Apomorphine	0.018	Tramazoline	3.0
Synephrine	0.014	DCDM	1.25
Clonidine	0.011	Synephrine	0.98
DCDM	0.009	Tyramine	0.066
Dopamine	0.005	Epinephrine	0.014
Naphazoline	0.0011	Clonidine	0.008
Epinephrine	0.0005	m-Octopamine	0.009
Norepinephrine	0.0002	Apomorphine	0.0033
		Norepinephrine	0.0020
		Dopamine	0.0003
WB-4101	0.1	Mianserin	0.95
BOL	0.05	Prazosin	0.66
LSD	0.03	Phentolamine	0.099
Methysergide	0.017	LSD	0.099
Chlorpromazine	0.012	BOL	0.096
Butaclamol	0.010	Tetrabenazin	0.038
Yohimbine	0.0099	WB-4101	0.017
Mianserin	0.0096	Gramine	0.015
Phentolamine	0.0050	Cyproheptadine	0.012
Cyproheptadine	0.0010	Phenoxybenzamine	0.0099
Gramine	0.0005	Metoclopramide	0.0036
Prazosin	0.0003	Chlorpromazine	0.0033
		Yohimbine	0.0001

The present results indicate no similarities with the OA_1 receptor type
and some similarities, but also several major differences, to OA_2 type
classified by Evans (1981) in the locust extensor tibiae muscle.
The pharmacological properties of the octopamine receptor shows
similariries with the octopamine receptors identified in the nervous
tissues of other insects (Downer, 1988; Konings, et al., 1989; Roeder and
Gewecke, 1990)

Acknowledgement.
This study was supported by grants from Hungary (OTKA, No. 2477 and
5422) and the Natural Science and Engineering Research Council of
Canada.

REFERENCES

Downer, R.G.H., Octopamine-and dopamine-receptors and cyclic AMP production in insect. In: Molecular basis of drug and pesticide action, G.G.Lunt ed., Elsevier(Biomedical Division), Amsterdam 1988, pp.255-265.

Evans,P.D., Multiple receptor types for octopamine in the locust, *J.Physiol.*, 318, 99-122,1981.

Konings, P.N.M., Vullings, H.G.B., van Gemert, W.M.J.B., DeLeeuw, R., Diederen, J.H.B. and Jansen W.F., Octopamine-binding sites in the brain of Locusta migratoria, *J. Insect Physiol.*, 35, 519-524, 1989.

Roeder,T.A. and Gewecke,M., Octopamine receptors in locust nervous tissue. *Biochem. Pharmacol.*, 39, 1793-1797,1990.

Vaccari, A., High affinity binding of [3H]-tyramine in the central nervous system, Br. *J.Pharmacol.*, 89,15-25,1986.

AGONIST SPECIFIC COUPLING OF A CLONED *Drosophila* OCTOPAMINE/TYRAMINE RECEPTOR TO MULTIPLE SECOND MESSENGERS

Sandra Robb[1], Timothy R. Cheek[1], Frances L. Hannan[1], Vincenzina Reale[1], Linda Hall[2], John M. Midgley[3] and Peter D. Evans[1]

[1]AFRC Laboratory of Molecular Signalling, Department of Zoology, University of Cambridge, Downing Street, Cambridge CB2 3EJ, UK
[2]Dept. of Biochemistry and Pharmacology, School of Pharmacy, 317 Hochstetter, Buffalo, New York 14260, USA
[3]Dept. Pharmaceutical Sciences, University of Strathclyde, Royal College, 204 George Street, Glasgow G1 1XW, UK

The ability of single agonists to interact with more than one second messenger system is now well established. Initially, this was thought to be mediated via an agonist interacting with a range of different G-protein coupled receptor subtypes (Bylund, 1992; Hosey, 1992), but recently the use of cloned receptors has shown that single receptor subtypes can be directly linked to multiple second messenger systems (Thompson, 1992). Thus the cloned muscarinic-M2 (Lai et al. 1991) and α_2-adrenergic (Cotecchia et al. 1990) receptors both stimulate phosphoinositide hydrolysis and inhibit adenylate cyclase activity, whilst the cloned receptors for thyrotropin (Van Sande et al. 1990), calcitonin (Chabre et al. 1992), parathyroid hormone (Abou-Samra et al. 1992) and the three classes of tachykinin receptor (Nakajima et al. 1992; Mitsuhashi et al. 1992), all stimulate both pathways. The present paper reports that a cloned seven transmembrane spanning *Drosophila* octopamine/tyramine receptor (Arakawa et al. 1990; Sandou et al. 1990; Robb et al. 1991) permanently expressed in a Chinese Hamster Ovary K1 (CHO) cell line, both inhibits adenylate cyclase activity and leads to the elevation of intracellular Ca^{2+} levels by separate G-protein coupled pathways. However, tyramine is about two orders of magnitude more potent than octopamine, when assayed by direct binding or by depression of cyclic AMP levels, whereas octopamine is slightly more potent or faster than tyramine in causing a transient elevation of cytosolic Ca^{2+}. Thus agonists of this cloned receptor, differing by only a single hydroxyl group in the side chain, may be capable of differentially coupling it to different second messenger systems. The phenomenon of agonist specific coupling of cloned receptors to different second messenger systems could alter our current concepts of receptor pharmacology.

Radioligand binding studies using [³H]yohimbine, the α_2-adrenergic

0-8493-4591-X/94/$0.00 + $.50
© 1994 by CRC Press, Inc.

351

antagonist, were performed on membranes prepared from CHO-K1 cells transfected with a putative octopamine receptor cloned from *Drosophila* (Arakawa et al. 1990). We found that tyramine, the non-β-hydroxylated (i.e. side chain) precursor of octopamine, showed the highest agonist affinity for the receptor in these membranes of the range of related agonists that were tested, including β-phenylethylamine, β-phenylethanolamine, synephrine, octopamine and dopamine (Robb, Cheek, Hannan, Hall, Midgley and Evans, in preparation). This agrees with the findings obtained for an independent isolation of the same gene expressed in a different vertebrate cell line (Sandou et al. 1990). An important novel finding of our investigation was that the receptor showed little or no stereoselectivity between (+)- and (-)-enantiomers of the most potent structural isomers of octopamine or synephrine. This contrasts with previous studies on other insect octopamine receptors using physiological and biochemical studies where the (-)-isomers were much more potent than the (+)-isomers (Evans, 1980, 1981, 1993). These results suggest that despite sharing some pharmacological similarities with the OCTOPAMINE$_1$ subtype of receptor from locust skeletal muscle (Evans, 1981, 1993), this cloned *Drosophila* aminergic receptor represents a novel class of receptor.

A similar pharmacological profile has been obtained for [^{3}H]-yohimbine binding in membranes prepared from *Drosophila* heads, indicating that these novel receptor properties are not due to the expression of the *Drosophila* receptor clone in a vertebrate cell line. In addition, a similar pharmacological profile for the cloned *Drosophila* receptor was also obtained in studies on its ability to attenuate forskolin stimulated increases in cyclic AMP levels in transfected CHO cells (Robb, Cheek, Hannan, Hall, Midgley and Evans, in preparation).

Since OCTOPAMINE$_1$ receptors mediate their effects via an elevation of intracellular Ca^{2+} levels (Evans, 1984) we tested the ability of the cloned receptor to raise intracellular Ca^{2+} levels in single transfected CHO cells (Robb et al. 1991). Both the application of 100 μM DL-p-octopamine (Robb et al. 1991) and tyramine produce a transient ($\sim$30s) elevation in $[Ca^{2+}]_i$ despite the continued presence of the agonist. These effects were both blocked by yohimbine at concentrations below 5μM. Surprisingly, the dose response curves for the peak $[Ca^{2+}]_i$ responses achieved by tyramine and DL-octopamine were almost superimposable and in contrast to the binding studies and adenylate cyclase inhibition studies, tyramine was not two orders of magnitude more potent than octopamine, even allowing for the large variability and cell to cell heterogeneity typically encountered in single cell Ca^{2+} imaging studies (Robb, Cheek, Hannan, Hall, Midgley and Evans, in preparation). The kinetics of the $[Ca^{2+}]_i$ responses we observed suggests that DL-p-octopamine may be a more effective agonist than tyramine since it consistently produces effects with a shorter lag time and a faster time to peak than tyramine. Thus, the efficiency of coupling of this cloned *Drosophila* receptor to different second messengers varies with the agonist used.

The above results are unlikely to be due to genetic drift or to cell-cell heterogeneity in our cultures since the effects on cyclic AMP levels and calcium transients have now been observed in the same populations of cells at many different passage numbers. Equally they are unlikely to be due to the production of multiple receptor types due to alternative splicing since our binding studies indicate a single binding site for our ligand and since a PCR analysis of transfected CHO cell mRNA using closely spaced overlapping primer pairs gave no evidence for the production of multiple transcripts.

Our evidence suggests that this cloned *Drosophila* aminergic receptor is coupled to adenylate cyclase via a pertussis toxin sensitive G-protein coupled pathway, whilst it is coupled to the production of the Ca^{2+} signal via a pertussis toxin-coupled pathway (Robb, Cheek, Hannan, Hall, Midgley and Evans, in preparation). We have now successfully expressed this cloned *Drosophila* receptor in *Xenopus* oocytes (Reale, Hannan, Robb, Midgley and Evans, in preparation) where we are currently investigating its ability to couple to a range of specific G-proteins and the effects of agonist specific modulation on the coupling of this receptor to these G-proteins.

The results of this present study re-open the debate as to whether this cloned *Drosophila* aminergic receptor is really an octopamine (Arakawa et al. 1990) or a tyramine receptor (Sandou et al. 1990), but the resolution of this question must await the demonstration of the endogenous ligand at specific cellular locations. However, it could be a multifunctional receptor that is activated by octopamine at some locations and by tyramine at other locations. Equally, the receptor may be involved in a novel form of modulation, whereby agents that can alter the ratio of octopamine and tyramine released from octopaminergic neurones in insects could bias the post-synaptic responses of an effector cell to favour one second messenger over another. In general terms, agonists differing by only a single hydroxyl group in their side chain, may be able to bias the interactions of a single G-protein coupled receptor with multiple second messenger systems by inducing specific conformational changes which enable it to couple preferentially to separate G-proteins. Thus a single receptor may have a different pharmacological profile depending on which second messenger system is used to assay its efficacy.

REFERENCES

Abou-Samra, A.B., Juppner, H., Force, T., Freeman, M.W., Kong, X.F., Shipam, E., Urena, P., Richards, J., Bouventre, J.V., Potts, J.T., Kronenberg, H.M. and Segne, G.V., Expression cloning of a common receptor for parathyroid hormone and parathyroid hormone-related peptide from rat osteoblast-like cells - a single receptor stimulates intracellular accumulation of both cAMP and inositol trisphosphates and increases free calcium. *Proc. Natl. Acad. Sci. USA*, 89, 2732-2736, 1992.

Arakawa, S., Gocayne, J.D., McCombie, W.R., Urquhart, D.A., Hall, L.M., Fraser, C.M. and Venter, J.C., Cloning, localization and permanent expression of a *Drosophila* octopamine receptor. *Neuron*, 2, 343-354, 1990.

Bylund, D.B., Subtypes of α_1 and α_2 adrenergic receptors. *FASEB J.*, 6, 832-839 1992.

Chabre, O., Conklin, B.R., Lin, H.Y., Lodish, H.F., Wilson, E, Ives, H.E., Catanzariti, L., Hemmings, B.A. and Bourne, H.R., A recombinant calcitonin receptor independently stimulates cAMP and Ca^{2+}/inositol phosphate signalling pathways. *Mol. Endocrinol.*, 6, 551-556, 1992.

Cotecchia, S., Kobilka, B.K., Daniel, K.W., Nolan, R.D., Lapetina, E.Y., Caron, M.G., Lefkowitz, R.J and Regan, J.W., Multiple second messenger pathways of alpha-adrenergic receptor subtypes expressed in eukaryotic cells. *J. Biol. Chem.*, 265, 63-69, 1990.

Evans, P.D., Biogenic amines in the insect nervous system. **Adv. Insect Physiol.**, 15, 317-473, 1980.

Evans, P.D., Multiple receptor types for octopamine in the locust. *J. Physiol.*, 318, 99-122, 1981.

Evans, P.D., Studies on the mode of action of octopamine, 5-hydroxytryptamine and proctolin on a myogenic rhythm in the locust. *J. exp. Biol.*, 110, 231-251, 1984.

Evans, P.D., Molecular studies on insect octopamine receptors, in *Comparative Molecular Biology*, Pichon, Y., Ed., Birkauser Verlag AG, Basel, Switzerland, 1993, pp.286-296.

Hosey, M.M., Diversity of structure, signalling and regulation within the family of muscarinic cholinergic receptors. *FASEB J.*, 6, 845-852, 1992.

Lai, J., Waite, S.L., Bloom, J.W., Yamamura, H.I. and Roeske, W.R., The m2 muscarinic acetylcholine receptors are coupled to multiple signalling pathways via pertussis toxin-sensitive guanine nucleotide regulatory proteins. *J. Pharmacol. exp. Ther.*, 258, 938-944, 1991.

Mitsuhashi, M., Ohashi, Y., Shichyo, S., Christian, C., Sudduth-Klinger, J., Harrowe, G. and Payan, D.G., Multiple intracellular signalling pathways of the neuropeptide Substance P receptor. *J. Neurosci. Res.*, 32, 437-443, 1992.

Nakajima, Y., Tsuchida, K., Negishi, M., Ito, S. and Nakanishi, D., Direct linkage of three tachykinin receptors to stimulation of both phosphatidylinositol hydrolysis and cyclic AMP cascades in transfected Chinese Hamster Ovary cells. *J. Biol. Chem.*, 267, 2437-2442, 1992.

Robb, S., Cheek, T.R., Venter, J.C., Midgley, J.M. and Evans, P.D., The mode of action and pharmacology of a cloned *Drosophila* phenolamine receptor. *Pestic. Sci.*, 32, 369-371, 1991.

Sandou, F., Amlaiky, N., Plassat, J.-L., Borrelli, E. and Hen, R., Cloning and characterization of a *Drosophila* tyramine receptor. *EMBO J.*, 9, 3611-3617, 1990.

Thompson, E.B., Comment: Single receptors, dual second messengers. *Mol. Endocrinol.*, 6, 501, 1992.

Van Sande, J., Raspe, E., Perret, J., Lejeune, C., Maenhart, C., Vassart, G. and Dumont, J.E., Thyrotropin activates both the cyclic AMP and the PIP_2 cascades in CHO cells expressing the human cDNA of the TSH receptor. *Mol. Cell Endocrinol.*, 74, R1-R6, 1990.

LOCUST PROCTOLIN RECEPTOR :
BINDING STUDIES AND PARTIAL PURIFICATION

Jacques Puiroux, Anne Pédelaborde and Barry G. Loughton

Department of Biology, York University, 4700 Keele Street,
North York, Ontario, M3J 1P3   Canada

Proctolin (N-Arg-Tyr-Leu-Pro-Thr-OH) was the first insect neuropeptide to be purified (Brown and Starratt 1975). Since its isolation from Periplaneta americana, this molecule has been detected in the nervous system of several insect species and others arthropods (O'Shea and Adams 1986). The studies on distribution of proctolin revealed that nerve terminals containing proctolin were frequently associated with visceral and skeletal muscles. Despite the contractile activity of proctolin, this neuropeptide does not respond to the definition of a conventional neurotransmitter and is rather considered to function as a neuromodulator (Orchard et al. 1989).

The small size of proctolin, only 5 residues, renders this molecule a very attractive subject for structure/activity studies. Insect muscle preparations were used in bioassays and these investigations resulted in extensive information on the importance of specific moieties for proctolin activity. Unfortunately no antagonist molecule was identified and the lack of proctolin blocker restrains the results on proctolin receptor(s) obtained with bioassays. In an attempt to further elucidate the proctolin receptor, we developed an alternative method of investigation using tritiated proctolin. We began binding studies on locust hindgut and oviduct. The specificity of these binding sites was assessed by testing a variety of proctolin analogues in competition studies. Successful solubilization of receptors was achieved with digitonin and the partial purification was performed by isoelectric focussing (IEF) and by anion exchange HPLC.

METHODS

Locust hindgut and oviduct membranes were suspended in 10 mM HEPES buffer (pH = 7.3, 5mM MgCl2, 1mM PMSF, 1mM EGTA, 1mM 0-phenanthroline, 5% glycerol w/v). Membranes were incubated (100 µg protein/500 µl) with [3H]-proctolin (24.3 Ci/mmol) in Eppendorf. The reaction was stopped by centrifugation (20,000 g) and the radioactivity in the pellet was counted. A final concentration of 1 % (w/v) digitonin in

0-8493-4591-X/94/$0.00 + $.50
© 1994 by CRC Press, Inc.

membrane preparations stirred for 1 hour at 4°C was successful to solubilize proctolin receptors. After a 100,000 g centrifugation (45 minutes), supernatant (53.5 ml) was added with ampholytes (1.5 ml of a 40% w/v solution, pH range 3 to 10). The sample was loaded into a IEF preparative electrophoresis cell and the separation was carried out at 12 W (constant power) for 2 hours. The second step purification was performed by HPLC on a Waters protein pack anion exchange column (10x100 mm) with HEPES buffer used for binding (plus 0.1 % detergent) at a flow rate of 1.5 ml/min and monitored at 280 nm. Solubilized receptors were incubated in 500 μl HEPES buffer. The reaction was terminated by addition of 75 μl of a bovine gamma globuline solution (10 mg/ml) followed by 475 μl of a polyethylene glycol solution (PEG 8000, 40% w/v) and then centrifugation. Non specific binding was measured in presence of an excess (1000 fold that of labelled proctolin) of cold ligand.

RESULTS

A preliminary study was conducted to determine the degradation pathways of proctolin incubated with locust hindgut and oviduct membrane preparations. Several peptidase inhibitors were also tested for their capacity to prevent proctolin metabolization. O-phenanthroline revealed to be efficient at 1 mM (Puiroux and Lougthon 1992) and binding studies with tritiated proctolin could be performed in satisfactory conditions.

Proctolin binding to hindgut and oviduct membranes was specific, proportional to membrane protein concentration, time dependent, saturable and reversible. The apparent Kd of the hindgut binding sites was 96 ± 8 nM and the Bmax was 3.37 ± 0.44 picomols/mg protein (Puiroux et al. 1992a). The studies on oviduct membranes indicated the probable presence of two classes of binding sites (Puiroux et al. 1992b) with an apparent lower affinity for proctolin (Kd_1 = 400 nM; Kd_2 = 2.5 μM) although oviduct muscles are more sensitive to proctolin than those of the hindgut.

Peptides known to affect contractile activity of insect muscles as well as a variety of proctolin analogues were tested in competition studies for their ability to displace proctolin binding (Table 1). The analogues modified in position 3 by susbtitution of serine for leucine and in position 4 by substitution of alanine for proline revealed to possess high competitive activity. By contrast, the so called supra analogue (Arg-[O-Met]Phe-Leu-Pro-Thr) was the less potent competitor of proctolin binding. The molecules were also tested for their biological activity by using the locust oviduct bioassay and a partial antagonistic activity of the analogue with a serine in position 3 was deduced (Puiroux et al. in press).

Table 1
Amino acid sequences of proctolin analogues used for the competition studies of hindgut proctolin binding site and analysis of their displacement curve.

Peptide	Amino acid sequence	IC_{50}
Proctolin	Arg - Tyr - Leu - Pro - Thr	90 nM
	Orn - Tyr - Leu - Pro - Thr	760 nM
	Tyr - Leu - Pro - Thr	2.34 µM
	Arg - Phe - Leu - Pro - Thr	240 nM
	Arg - Phe - Leu - Pro - Thr $\vert$ O - Methyl	2.49 µM
	Arg - Tyr - Gly - Pro - Thr	1.28 µM
	Arg - Tyr - Ser - Pro - Thr	22.2 nM
	Arg - Tyr - Leu-dPro - Thr	1.57 mM
	Arg - Tyr - Leu - Ala - Thr	112.9 nM
	Arg - Tyr - Leu - Pro - Ala	1.65 µM
	Arg - Tyr - Leu - Pro	1.29 µM

The solubilization of proctolin receptors was achieved by addition of digitonin to membrane preparation as indicated by the measurement of high specific binding in supernatants from 100,000 g centrifugation. 75 hindgut or 40 oviduct membranes were solubilized and the supernatants were added with ampholytes, the separation was performed by IEF. Proctolin receptors were usually recovered in a single fraction (12 to 15 fold purified, Table 2) which pH was 4.8 corresponding to the isoelectric point (Pi) of proctolin receptor. Pi was identical for receptors of both types of membranes. The saturability of partially purified proctolin receptor revealed a single class of receptor in hindgut and oviduct samples with a similar Kd value of 15 ± 4 nM.

Six separate IEF runs were required to process 1000 hindgut or 500 oviduct samples. Fractions with high specific binding were pooled (15 ml) and dialyzed to remove the ampholytes. The pH of the sample was adjusted to 7.1 by addition of 15 ml HPLC buffer. The total volume (approx. 30 ml) was loaded onto the anion exchange column via one HPLC pump. When the baseline was stabilized, elution was performed by a gradient of 0 to 250 mM NaCl in 50 minutes. 1.5 ml fractions were collected. Proctolin binding activity was detected in 4/5 fractions eluted at about 120 mM NaCl. Hindgut and oviduct proctolin receptors were respectively 230 fold (Table 2) and 290 fold purified. The apparent molecular weight (MW) of partially purified proctolin receptor was determined by HPLC on size exclusion column (Beckman SEC 3000) previously calibrated with standard proteins. The MW is approximately 50,000. The mode of action of proctolin is mediated via phosphoinositol pathway in locust oviduct (Lange 1989) and implicates an interaction of proctolin with a membrane receptor which is regulated by G-proteins. The first estimation of proctolin receptor molecular weight ranging at 50 kDa corresponds to the size of G-protein associated receptors.

Table 2
Purification of locust hindgut proctolin receptor

	Protein	pmols/ mg protein	recovery	purification
1000 hindguts	90 mg	0.7	100 %	1
Solubilisation	53 mg	1.3	100 %	1.9
Rotofor (IEF)	3.6 mg	8.3	50 %	12.4
DEAE HPLC	80 µg	150	20 %	230

Partially purified hindgut and oviduct proctolin receptors exhibit great similarities (identical Kd, Pi and MW values) what is in contrast with their different binding characteristics on membranes. Thus enkephalins (Tyr-Gly-Gly-Phe-Leu and Tyr-Gly-Gly-Phe-Met) and the major proctolin metabolite Tyr-Leu-Pro-Thr, sharing the same N-terminal tyrosine residue, proved to induce a significant potentiation of proctolin specific binding exclusively on oviduct membranes. Investigations with several opioid molecules (μ and ∂ agonists and antagonists) lead us to consider a possible regulation of proctolin binding on the oviduct membranes by a ∂ type opioid receptor. Locust hindgut and oviduct proctolin receptors might reveal very similar molecules, however their activation would depend on distinct regulation systems.

Brown, B. E. and Starratt, A. N., Isolation of proctolin, a myotropic peptide, from *Periplaneta americana*, *J. insect Physiol.*, 21, 1879-1881, 1975.

Lange, A. B., Inositolphospholipid hydrolysis may mediate the action of proctolin on insect visceral muscle, *Arch. Insect Biochem. Physiol.*, 9, 201-209, 1989.

O'Shea, M. and Adams, M., Proctolin: from "gut factor" to model neuropeptide, *Adv. Insect Physiol.*, 19, 1-28, 1986.

Orchard, I., Belanger, J. H. and Lange, A. B., Proctolin: a review with emphasis on insects, *J. Neurobiol.*, 20, 470-496, 1989.

Puiroux, J. and Loughton, B. G., Degradation of the neuropeptide proctolin by membrane bound proteases of the hindgut and ovary of *Locusta migratoria* and the effects of different inhibitors, *Arch. Insect Biochem. Physiol.*, 19, 193-202, 1992.

Puiroux, J., Pédelaborde, A. and Loughton, B. G., Characterization of a proctolin binding site on locust hindgut membranes, *Insect Biochem. and Molecular Biol.*, 22 (6), 547-551, 1992a.

Puiroux, J., Pédelaborde, A. and Loughton, B. G., Characterization of proctolin binding sites on locust oviduct membranes, *Insect Biochem. and Molecular Biol.*, 22 (8), 859-865, 1992b.

Puiroux, J., Pédelaborde, A. and Loughton, B. G., The effect of proctolin analogues and other peptides on locust oviduct muscle contractions, *Peptides*, in Press.

CHARACTERIZATION OF ACHETAKININ-BINDING SITES ON CRICKET MALPIGHIAN TUBULE PLASMA MEMBRANES

Jum Sook Chung[1]*, Colin H. Wheeler[2], Graham J. Goldsworthy[1], and Geoffrey M. Coast[1]

[1] Department of Biology, Birkbeck College, London, WC1E 7HX;
[2] Sussex Centre for Neuroscience, University of Sussex, East Sussex, BN1 9QG

I. INTRODUCTION

The cockroach hindgut provides a versatile bioassay for the isolation of neuropeptides, which either increase or decrease the frequency and/or amplitude of spontaneous contractions (Holman *et al.*, 1990; Holman *et al.*, 1991). In at least two instances, peptides identified in this assay have been shown subsequently to have other sites of action. The C-terminus (Phe-Thr-Pro-Arg-Leu-NH$_2$) of leucopyrokinin (LPK) is identical to that of pheromone biosynthesis activating neuropeptide (PBAN) from *Pseudaletia separata* (Matsumoto *et al.*, 1992), and stimulates pheromone biosynthesis in *Bombyx mori* (Fonagy *et al.*, 1992). Likewise, myokinins from cockroaches (leucokinins), crickets (achetakinins), locusts (locustakinin) and mosquitoes (culekinins), stimulate ion transport and fluid secretion by isolated Malpighian tubules (Hayes *et al.*, 1989; Coast *et al.*, 1990). In both myokinins and LPK, biological activity resides in a C-terminal pentapeptide ("active core") sequence (Nachman and Holman, 1991). Myokinins have the core sequence Phe-X^1-X^2-Trp-Gly-NH$_2$, where X^1 is Asn, His, Tyr or Ser, and X^2 is Pro or Ser. The small size of this active core facilitates the design of analogues with which to investigate requirements for biological activity. To investigate further the structure-activity relationships of myokinins, we have developed a receptor-binding assay using an iodinated probe based on the core of achetakinin-II (AKII), and preparations of membranes from cricket Malpighian tubules.

II. RESULTS AND DISCUSSION

Preparation of a ^{125}I-labelled biologically active probe

AKI has a tyrosyl residue within the active core, but all biological activity is lost when this peptide is iodinated. AKII core (Phe-Ser-Pro-Trp-Gly-NH$_2$) conjugated to Bolton-Hunter (BH) reagent is a more potent diuretic than the core peptide, but iodination destroys biological activity. Reasoning this might be due to steric hindrance of receptor binding, the N-terminus

0-8493-4591-X/94/$0.00 + $.50
© 1994 by CRC Press, Inc.

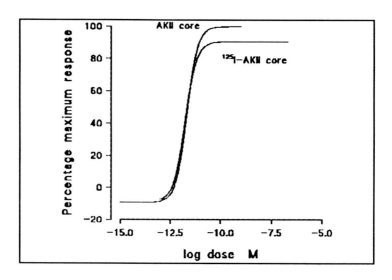

Figure 1. Dose-response curves for the stimulation of fluid secretion by labelled and unlabelled glycine-extended AKII core conjugated to BH reagent. Peptides were assayed on cricket Malpighian tubules isolated *in vitro*.

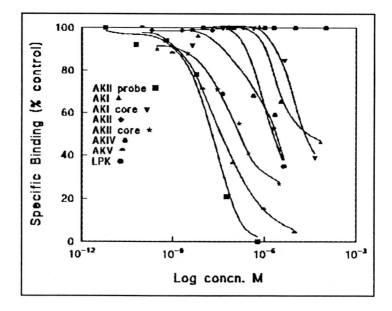

Figure 2. Displacement of ^{125}I-AKII specific binding from membrane preparations. Membranes were incubated with 4.5×10^{14} M ^{125}I-AKII in the presence of different concentrations of competing peptide. The results were analyzed by a computer program (EBDA, Biosoft), and are presented as a percentage of control binding.

was extended by 3 glycyl residues. The glycine-extended analogue, conjugated to BH reagent, was iodinated with IODO-GEN reagent (Pierce). Full activity and potency are retained (Fig. 1), and the previous loss of activity is therefore attributed to steric hinderance, and not to oxidation of the tryptophan residue. ^{125}I-AKII core was used as a probe for receptor-binding studies, and has a specific activity of *c*. 1000 Ci/mmol.

Membrane preparation
Malpighian tubules from adult male and female crickets were transferred to ice-cold 25 mM Tris-maleate buffer (pH 7.4) containing 0.6 mM EGTA and a cocktail of protease inhibitors (1 mM PMSF, 1 mM bacitracin, 1μM pepstatin and 10 μM leupeptin). They were disrupted with a Polytron (30 sec), and centrifuged at 1,000 rpm (10 min at 4°C). The pellet was discarded, and the supernatant centrifuged at 19,000 rpm (30 min at 4°C). The pellet was resuspended in Tris-maleate buffer and centrifuged again at 19,000 rpm. Membrane preparations resuspended in Tris-maleate buffer were stored -20°C in aliquots of 300 μl containing 3 mg protein.

Receptor-binding assay
The incubation medium for receptor-binding assays was 25 mM Tris-maleate buffer (pH 7.4), 5 mM $MgCl_2$, 0.6 mM EGTA, protease inhibitors and 1% BSA. Assays were started by the addition of membrane protein to tubes containing medium and radiolabelled probe. Free and bound ligand were separated by filtration on GF/C filters (Whatman) pre-soaked in buffer containing BSA. The filters were washed once with 5 ml ice-cold buffer. Bound ligand was counted in a gamma counter (LKB), and the data analyzed by a computer program (EBDA, Biosoft).

Characterization of ^{125}I-AKII receptor binding site(s)
Conditions were optimized to maximise specific binding. Optimum binding occurred between pH 6.6 and 7.6, and was decreased by the addition of NaCl. For the characterization of receptors, incubations were performed at room temperature in 100 μl buffer (pH 7.4) containing no added NaCl. Binding increased linearly with increasing concentration of membrane protein, and over the range 10-100 μg protein specific binding was *c*. 75% total binding. The binding of ^{125}I-AKII reached equilibrium after 45 min and was reversible with excess unlabelled probe. Specific binding was saturable and, with 100 μg membrane protein, reached a maximum at 100,000 cpm. Scatchard analysis indicates a single class of binding sites with a K_d of 0.51 nM and a B_{max} of 35.5 fmol/mg protein. In the fluid secretion assay, the EC_{50} of the probe is 0.002 nM, and there is a considerable receptor reserve.

Receptor specificity
Membranes (100 μg protein/assay tube) were incubated for 60 min with

^{125}I-AKII (100,000 cpm) and increasing concentrations of competing peptides. Results from competition studies are shown in Fig. 2. LPK has no diuretic activity, and at < 0.01 mM did not compete for ^{125}I-AKII binding sites. However, achetakinins and core peptides all displaced the iodinated probe (Fig 3). Their affinities (IC$_{50}$ values) ranked as follows: AKV > AKII core > AKII > AKIV ≥ AKI > AKI core. Each of these peptides doubles the rate of fluid secretion by isolated tubules, and in this bioassay their affinities are ranked AKV > AKII = AKII core = AKIV > AKI core > AKI (Coast et al., 1990). In the hindgut bioassay, affinities based on threshold concentrations for myotropic activity rank AKV = AKII > AKI > AKIV. Thus, in both assays, AKV and AKII are more potent than AKI and AKIV, which is in general agreement with their affinities for receptors on Malpighian tubule membranes. Intriguingly, AKII core conjugated to BH reagent has a higher affinity than AKV or AKII in the receptor-binding assay, and is an order of magnitude more potent in the fluid secretion assay. It appears that the hydroxyphenylpropionate group of BH reagent enhances receptor binding, possibly by interacting with some auxiliary site (Portoghese, 1989). This is currently being investigated.

This work was supported by the SERC and by a Collaborative Research Grant from NATO.

III. REFERENCES

Coast, G.M., Holman, G.M., and Nachman, R.J., The diuretic activity of a series of cephalomyotropic neuropeptides, the achetakinins, on isolated Malpighian tubules of the house cricket, *Acheta domesticus, J. Insect Physiol.* 36, 481-488,1990.

Fonagy, A., Schoofs, L., Matsumoto, S., De Loof. A., and Mitsue, T., Functional cross-reactivities of some locust myotropins and *Bombyx* pheromone biosynthesis activating neuropeptide. *J. Insect Physiol.* 38, 651-657, 1992.

Hayes, T.K., Pannabecker, T.L, Hinkley, D.J., Holman, G.M., Nachman, R.J., Petzel, D.H. and Beyenbach, K.W. Leucokinins, a new family of ion transport stimulators and inhibitors in insect Malpighian tubules. *Life Sci.* 44, 1259-1266, 1989.

Holman, G.M., Nachman, R.J., Schoofs, L., Hayes, T.K., Wright, M.S. and De Loof, A. The *Leucophaea maderae* hindgut preparation: a rapid and sensitive bioassay tool for the isolation of insect myotropins of other insect species. *Insect Biochem.* 21, 107-112, 1991.

Holman, G.M., Nachman, R.J., Wright, M.S. Comparative aspects of insect myotropic peptides. In *Progress in Comparative Endocrinology*. Epple, A., Scanes, C.C. and Stetson, M.H. Eds. Wiley-Liss, New York, pp. 35-39, 1990.

Matsumoto, S., Fonagy, A., Kurihara, M., Uchiumi, K., Nagamine, T., Chijimatsu, M., and Mitsui, T., Isolation and primary structure of a novel pheromonotropic neuropeptide structurally related to leucopyrokinin from the armyworm larvae, *Pseudaletia separata*. *Biochem. Biophys. Res. Commun.* 182, 534-539, 1992.

Nachman, R.J. and Holman, G.M. Myotropic insect neuropeptide families from the cockroach *Leucophaea maderae*, In *Insect Neuropeptides Chemistry, Biology and Action*, ACS Symposium Series 453, Menn, J.J., Kelly, T.J., and Masler, E.P. Eds., American Chemical Society, Washington, DC, 1991, pp. 194-214.

Portoghese, P.S., (1989) Bivalent ligands and the message-address concept in the design of selective opioid receptor antagonists. *TIPS* , 101, 230-235, 1989.

CHARACTERIZATION AND EXPRESSION OF A MEMBRANE GUANYLATE CYCLASE *DROSOPHILA* GENE

C. Malva, S. Gigliotti, V. Cavaliere, A. Manzi and F. Graziani

Istituto Internazionale di Genetica E Biofisica,
via Marconi 10, 80125, Napoli, Italy

During our study of maternal effect mutations contained within region 32 of the standard salivary gland chromosome map of the second chromosome of *D. melanogaster* we have isolated a gene coding for a putative membrane form of guanylate cyclase Guanylate cyclases (GCs) have been studied and described in detail in both the soluble and particulate cell compartments in sea urchins and mammals, including man (Garbers, 1992).

The *Drosophila* gene was first identified by isolation of cDNAs from embryo and head libraries. The gene structure was subsequently defined by sequence analysis of these cDNAs and a genomic region spanning about 7 kb. Due to the lack of any full length cDNA, we obtained the missing carboxyl-terminal amino acids of the protein from the genomic sequence. We deduced almost the entire protein up to the 50th NH2 terminal amino acid taking advantage of the strong conservation between the GC proteins.

The gene structure of the GC gene so far obtained is unusual for *Drosophila*: it is made of 21 small exons, each one coding for less than 50aa, separated by small introns. All the functional domains, including the small transmembrane and the cyclase catalytic regions, are separated by introns (Fig. 1). A canonical polyadenylation signal is present in the genomic sequence, 574 bp after the end of the coding region. We cannot perform any evolutionary analysis on GC gene structure because in sea urchin and mammalian species only cDNAs have been isolated and therefore no other genomic organisation is known.

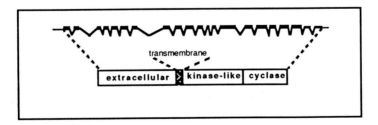

Figure 1. Schematic representation of the *Drosophila* Guanylate Cyclase gene and its putative product with the indication of the functional domains.

0-8493-4591-X/94/$0.00 + $.50
© 1994 by CRC Press, Inc.

The putative *Drosophila* GC protein has a protein kinase-like and a cyclase domain and is very similar to the other members of this receptor protein family. Hydropathic analysis demonstrates a major hydrophobic stretch of 22 amino acids in the middle of the protein. This suggests the presence of only one membrane-spanning segment dividing the protein into putative extra cellular and cytoplasmic catalytic regions, in agreement with the structure of the other membrane forms of GCs (Garbers, 1992).

From sequence comparison between the *Drosophila* GC amino acid sequence and that of other species present in the data base banks we found that the closest are the *Strongilocentrotus purpuratus* GC (38.1% identity in 1005 aa overlap) and the human atrial natriuretic peptide receptor A (33.4% identity in 1005 aa overlap). In the same relative position, also the *Drosophila* protein contains the protein kinase-like sequence previously reported in GCs. Of the 33 residues highly conserved in all reported protein kinases (Hanks et al., 1988), 25 are present also in the *Drosophila* protein.

The intracellular region, where the putative cyclase catalytic domain is located (Thorpe and Morkin, 1990) appears more highly conserved between the *Drosophila* and the other GCs (Fig. 2). We observed in this region about 60% identity with the membrane form of GCs. In the same part of the protein homology is also found with soluble GCs (Koesling et al., 1988) and with Adenylate Cyclases, including the *Drosophila* protein (Levin et al., 1992). The region that Beuve and Danchin (1992) have proposed as the one involved in recognition of the base by the Guanylate cyclase enzymes is the most conserved (99% identity).

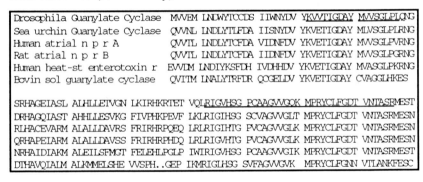

Figure 2. Sequence comparison of the *Drosophila* Guanylate Cyclase catalytic domain with the corresponding region of other members of the GC family. The proposed catalytic domain and the region proposed for the recognition of the base by GCs are underlined.

We analyzed the expression pattern of the GC gene by Northern blot and "in situ" hybridization with digoxigenin labelled probes on different wild type tissues. We found a very low level of expression and in Northern blots we obtained faint signals only in RNA prepared from larvae and adult heads (4.5 Kb transcript). "In situ" expression is detected in ovaries, first in germaria and later at stage 10 (Gigliotti et al., 1993). A signal is also present in very early embryos, immediately after fertilization and in larval imaginal discs (data not shown).

Despite the fundamental role played by the cyclic GMP in fertilization, cell growth, migration and differentiation and signal transduction in visual system, we are not yet able to correlate this receptor protein gene with any of the mutations mapped in region 32. We isolated the GC gene in an attempt to identify genes located near the insertion site of a blood transposon that we have found (Lavorgna et al., 1989) in correlation with the presence of the *abo* mutation. However, the GC transcript is detected "*in situ*" on *abo* mutant ovaries (data not shown). In the same 32 region, the tumor suppressor gene *1(2) giant disc (1(2)gd)* has been identified by Bryant and Schubiger (1971). Although the most conspicuous aspect of the *1(2)gd* phenotype is the imaginal disc overgrowth, it has been demonstrated that the gene plays an essential role for normal egg production in female germ line (szabad et al., 1991). We are using different approaches to identify the phenotype induced by disruption of the GC gene function and the potential ligands of this *Drosophila* receptor protein.

Acknowledgments

This work was supported by National Research Council of Italy Special Project on "Genetic Engineering".

REFERENCES

Beuve, A., and Danchin, A., From adenylate cyclase to guanylate cyclase: mutational analysis of a change in substrate specificity, *J. Mol. Biol.* 225, 933-938, 1992.

Bryant, P. J., and Schubiger, G., Giant and duplicated imaginal discs in a new lethanl mutant of *Drosophila melanogaster*, *Dev. Biol.* 24, 233-263, 1971.

Garbers, D. L., Guanylyl cyclase receptors and their endocrine, paracrine and autocrine ligands, *Cell* 71, 1-4, 1992.

Gigliotti, S., Cavaliere, V., Manzi, A., Tino, A., Graziani, F. and Malva, C., A membrane guanylate cyclase Drosophila homolog gene exhibits maternal and zygotic expression, *Devl. Biol.*, in press, 1993.

Hanks, K., Quinn, A. M. and Hunter, T., The protein kinase family: conserved features and deduced phylogeny of the catalytic domains, *Science* 241, 42-52, 1988.

Koesling, D., Herz, J., Gausepohl, H., Niroomand, F., Hinsch, K. D., Mulsch, A., Bohme, E., Schultz, G., and Frank, R., The primary structure of the 70 kDa subunit of bovine soluble guanylate cyclase, *FEBS Lett.* 239, 29-34, 1988.

Lavorgna, G., Malva, C., Manzi, A., Gigliotti, S., and Graziani, F., The *abnormal oocyte* phenotype is correlated with the presence of the *blood* transposon in *D. melanogaster*, *Genetics* 123, 485-494, 1989.

Levin, L. R., Hant, P. L., Hwang, P. M., Feinstein, P. G., Davis, R. L., and Reed, L. R., The *Drosophila* learning and memory gene *rutabaga* encodes a Ca2+/calmodulin-responsive adenylyl cyclase, *Cell* 68, 479-489, 1992.

Szabad, J., Jursnich, V. A., and Bryant, P., Requirement for cell-proliferation control genes in *Drosophila* oogenesis, *Genetics* 127, 525-533, 1991.

Thorpe, D. S., and Morkin, E., The carboxyl region contains the catalytic domain of the membrane form of guanylate cyclase, *J. Biol. Chem.* 265, 14717-14720, 1990.

CLONING, SEQUENCING AND TISSUE DISTRIBUTION OF A PUTATIVE LOCUST G-PROTEIN COUPLED RECEPTOR

Jozef Vanden Broeck, Peter Verhaert, and Arnold DeLoof

Zoological Institute, Catholic University of Leuven,
Naamsestraat 59, B-3000 Leuven (Belgium)

INTRODUCTION

G-protein coupled receptors form a large family of integral membrane proteins. They are involved in the specific perception of extracellular messengers (*e.g.* many neurotransmitters and hormones) or signals (*e.g.* light or olfactory). Via the action of G-proteins they also initiate the appropriate physiological responses. Several structural features are shared by all members of this important family of signal transducing proteins (Probst *et al.* 1992).

When comparing members of the family of G-protein coupled receptors, amino acid sequence preservation is clearly observed, especially within the seven transmembrane domains (TMDs). This observation was used to derive oligonucleotides corresponding to the consensus amino acid sequences of TMD3 and TMD6 for a number of known mammalian biogenic amine receptors (Libert *et al.* 1989). These oligonucleotides were then used as primers in PCR (Polymerase Chain Reaction) on locust (*Locusta migratoria*) brain and ventral nerve cord cDNA templates.

MATERIALS AND METHODS

A. Preparation of cDNA template

PCR-template DNA was synthesized from locust nervous system (brain-suboesophageal ganglion-ventral nerve cord) mRNA using a modified MMLV-reverse transcriptase (Superscript, BRL).

B. PCR primers

The oligonucleotide sequences were purchased from Eurogentec S.A. (Belgium): (TMD3)
5'GCAGATCTAGAGCCAT(CT)GC(CT)ITIGA(CT)(CA)G(CG)TAC3'
and (TMD6)
5'GGCTCGAGCTCG(TA)(GA)(GA)(TA)A(GC)GGCA(GA)CCA(GA)CAGA3'.

C. PCR reactions

The enzyme of choice was AmpliTaq (Perkin Elmer Cetus) which was used in combination with the recommended buffer (10 mM Tris-HCl, pH 8.3; 50 mM KCl; 1.5 mM $MgCl_2$; 0.01 %(w/v) gelatin). The final concentration of each nucleotide was 0.2 mM and both primers were used in a concentration of 0.5 μM each. Template contents varied from 1/100 to 1/10 of the cDNA derived from 2 μg of locust CNS mRNA.

0-8493-4591-X/94/$0.00 + $.50
© 1994 by CRC Press, Inc.

Thermal cycling consisted of about 38 cycles with a denaturation step of 1 min. at 94°C, an annealing step that varied from 53°C up to 65°C (for 70 seconds) depending on the conditions using and an extension time of 85 seconds at 72°C. Under optimal reaction conditions a specific PCR-product was obtained.

D. Sequence analysis

The PCR fragment was purified, cleaved with the appropriate combination of restriction enzymes and unidirectionally inserted into $M_{13}mp_{18}$ and $M_{13}mp_{19}$ for further sequencing (Sequenase, USB).

E. cDNA library preparation and screening

A locust CNS cDNA library was constructed in lambda gt22A (BRL). The library contained an estimated ratio of recombinant clones to non-recombinant ones of approximately 90% with a total titer of about 4.10^6 of primary plaque forming units.

About 6.10^5 pfu from the primary lambda phage mixture were screened by hybridization with the radio-labeled PCR fragment. Positive phage clones were further purified and the cDNA inserts were sequenced.

F. Northern blot analysis

RNAs derived from different tissues (adult locust CNS, skeletal muscle and fat body) and from the CNS of the first larval stages were separated on formaldehyde containing gels. Northern blot analysis was performed according to "Maniatis" (Sambrook *et al.* 1989).

RESULTS AND DISCUSSION

Our study demonstrates the presence of an mRNA encoding an evolutionary conserved G-protein coupled receptor in the locust central nervous system. The entire cDNA was cloned from a *Locusta* library and sequenced. The cDNA sequence was submitted to the "EMBL Data Library" and has been assigned the accession number X-69520. Clear sequence homology is found with known mammalian and fruitfly receptors for biogenic amines. From all amino acid sequences in the Swiss-Prot database the highest degree of similarity was observed with a fruitfly (*Drosophila melanogaster*) G-protein coupled receptor which was proposed to be a receptor for octopamine (OA) and/or tyramine (TA) (Arakawa *et al.* 1990, Saudou *et al.* 1990). TA is the natural precursor of OA, but some researchers suggest that in the insect CNS it might also play a role as a transmitter or modulator. Our locust (Orthoptera, Exopterygota) receptor is most probably an evolutionary homologue of this fruitfly (Diptera, Endopterygota) TA/OA receptor. This suggestion is mainly supported by the high degree of sequence conservation despite a long period of evolutionary divergence. In agreement with most other G-protein coupled receptors seven putative transmembrane regions are present (Fig. 1). Putative N-glycosylation sites are found in the extracellular parts of the molecule. Several possible phosphorylation sites are indicated. These are preferentially located in the intracellular parts of the protein, especially in the third loop and may be involved in the regulation of receptor functioning.

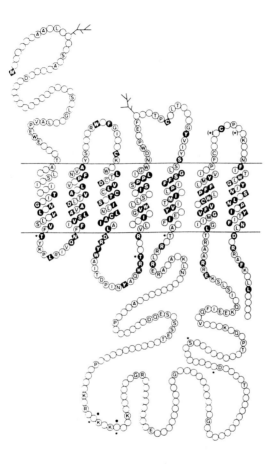

Figure 1. Representation of the predicted locust receptor protein showing transmembrane domains, conserved amino acid residues and putative glycosylation and phosphorylation sites. The amino acid positions that are shown as black squares were found to be conserved in more than 20 different biogenic amine receptors. The black circles represent conserved positions in a few related mammalian and insect receptors. The amino acids shown in one-letter code can be aligned with the fruitfly TA/OA receptor. cAMP/cGMP dependent protein kinase (PKA) phosphorylation sites are indicated as · and PKC sites as *.

Northern blot analysis (Fig. 2) revealed that the receptor encoding mRNA is mainly found in the central nervous system. The receptor-mRNA has an approximate full length of 4.5 kb. This 4.5 kb mRNA-band is already present in the central nervous system at the first larval stage of *Locusta migratoria*.

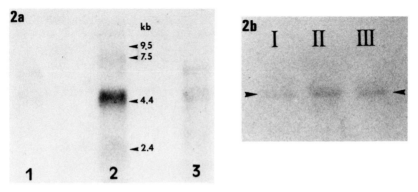

Figure 2. Northern blot analysis of the locust receptor mRNA in different adult tissues (2a) [1 = 10 μg of skeletal muscle poly-A+ RNA; 2 = 10 μg of CNS poly-A+ RNA; 3 = 20 μg of fat body poly-A+ RNA] and in the CNS of the first larval stages (2b) [approximately 1 μg of poly-A+ RNA of the first (I), second (II) and third (III) larval stage CNS].

Many signal substances as well as their receptors, in addition to the important key-proteins involved in the intracellular signal transduction, have been moderately to highly conserved during Evolution to preserve the conformational matching of important molecular interactions (Stefano 1988). The identification of receptor sequences in several insect species may become particularly useful to define insect-specific features of these molecular interactions. This will not only broaden the field and improve our understanding of the phylogenetic history of protein families, but it may also lead to additional insights in the molecular structure-function relationships. In combination with steadily progressing protein modelling abilities, this will probably lead to novel prospects for a more rationalized insecticide design.

REFERENCES

Arakawa, S., Gocayne, J.D., McCombie, W.R., Urquhart, D.A., Hall, L.M., Fraser, C.M. and J.C. Venter, Cloning, localization and permanent expression of a *Drosophila* octopamine receptor. *Neuron* 2, 343-354, 1990.

Libert, F., Parmentier, M., Lefort, A., Dinsart, C., Van Sande, J., Maenhaut, C., Simons, M.-J., Dumont, J.E. and Vassart, G., Selective amplification and cloning of four new members of the G protein-coupled receptor family. *Science* 244, 569-572, 1989.

Probst, W.C., Snyder, L.A., Schuster, D.I., Brosius, J. and Sealfon S.C., Sequence alignment of the G-protein coupled receptor superfamily. *DNA Cell Biol.* 11, 1-20, 1992.

Sambrook, J., Fritsch, E.F. and Maniatis, T., Molecular Cloning - A Laboratory Manual, Cold Spring Harbor laboratory Press (CSH), 1989.

Saudou, F., Amlaiky, N., Plassat, J.-L., Borrelli, E. and Hen, R., Cloning and characterization of a *Drosophila* tyramine receptor. *EMBO J.* 9, 3611-3617, 1990.

Stefano, G.B., The evolvement of signal systems: conformational matching a determining force stabilizing families of signal molecules. *Comp. Biochem. Physiol.* 90C, 287-294, 1988.

The authors wish to thank the N.F.W.O. of Belgium and the European Community (C.E.C., 'Science and Technology for Development' TS 2-0264-B and TS 2-0295-D) for financial support. Dr. F. Libert (Brussels) is gratefully acknowledged for his initial advices. J.VdB. and P.V. are respectively "senior research assistant" and "research associate" of the N.F.W.O. of Belgium.

Subject Index

SUBJECT INDEX

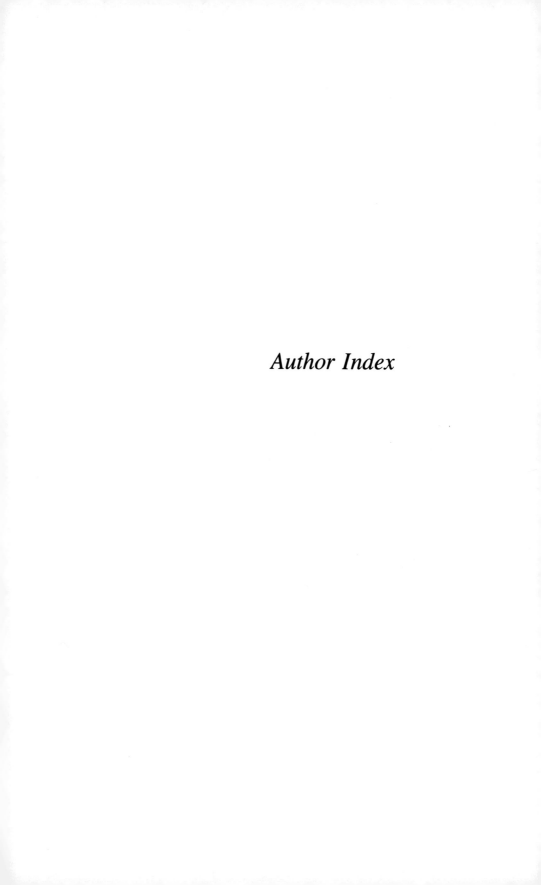

Author Index

AUTHOR INDEX

Conference Attendees

Attendees at ICINN '93

Hiromu AKAI, Tokyo University of Agriculture, Setagaya, Tokyo 156, JAPAN

Miriam ALTSTEIN, Institute of Plant Protection, Volcani Center, Bet Dagan 50250, ISRAEL

Richard A. BAINES, Sussex Centre for Neuroscience, University of Sussex, Falmer, Brighton BN1 9QG, UNITED KINGDOM

Nancy E. BECKAGE, Department of Entomology, University of California, Riverside, CA 92521, USA

Robert A. BELL, USDA, ARS, BA, PSI, Insect Neurobiology & Hormone Lab, 10300 Baltimore Ave. Bldg. 309, Rm. 214, BARC-East, Beltsville, MD 20705-2350, USA

Blanka BENNETTOVÁ, Institute of Entomology, Czech Academy of Sciences, Branišovská 31, 370 05 České Budějovice, CZECH REPUBLIC

Alexej A. BOŘKOVEC, USDA, ARS, BA, PSI, Insect Neurobiology & Hormone Laboratory, 10300 Baltimore Ave., Bldg. 306, Room 322, BARC-East Beltsville, MD 20705-2350, USA

Malcolm BURROWS, Department of Zoology, University of Cambridge, Downing Street, Cambridge CB2 3EJ, UNITED KINGDOM

Yang CHANSUNG, (CHINA) c/o Dr. F. Sehnal, Institute of Entomology, Czech Academy of Sciences, Branišovská 31, 370 05, České Budějovice, CZECH REPUBLIC

Jum S. CHUNG, Department of Biology, Birkbeck College, University of London, Malet Street, London WC1E 7HX, UNITED KINGDOM

Ornella CUSINATO, Department of Biology, Birkbeck College, University of London, Malet Street, London WC1E 7HX, UNITED KINGDOM

Carl H. DAHM, Department of Biology, Texas A&M University, College Station, TX 77843-3258, USA

Kenneth G. DAVEY, Department of Biology, York University, North York, Ontario M3J 1P3, CANADA

Roger G. H. DOWNER, Department of Biology, University of Waterloo, Waterloo, Ontario N2L 3G1, CANADA

Véronique DUBREIL, Laboratoire de Neurophysiologie, URA CNRS 611, Universite d'Angers, F-49045 Angers Cedex FRANCE

Maurice R. ELPHICK, Sussex Centre for Neuroscience, School of Biological Sciences, University of Sussex, Falmer, Brighton BN1 9QG, UNITED KINGDOM

Yasuhisa ENDO, Department of Applied Biology, Kyoto Institute of Technology, Matsugasaki, Sakyo, Kyoto 606, JAPAN

Peter D. Evans, Dept of Zoology, University of Cambridge, AFRC Laboratory of Molecular Signaling, Downing St., Cambridge CB2 3EJ, ENGLAND

Valerij FILIPPOV, (RUSSIA) c/o Dr. F. Sehnal, Institute of Entomology, Czech Academy of Sciences, Branišovská 31, 370 05, České Budějovice, CZECH REPUBLIC

Maria FILIPPOVA, (RUSSIA) c/o Dr. F. Sehnal, Institute of Entomology, Czech Academy of Sciences, Branišovská 31, 370 05, České Budějovice, CZECH REPUBLIC

Barry GANETZKY, Laboratory of Genetics, University of Wisconsin, 445 Henry Mall, Madison, WI 53706, USA

Ivan GELBIČ, Institute of Entomology, Czech Academy of Sciences, Branišovská 31, 370 05, České Budějovice, CZECH REPUBLIC

Dale B. GELMAN, USDA, ARS, BA, PSI, Insect Neurobiology & Hormone Lab, 10300 Baltimore Ave., Building 306, Room 317, BARC-East, Beltsville, MD 20705, USA

Lawrence I. GILBERT, Dept. of Biology, University of North Carolina, Wilson Hall 046A, Chapel Hill, NC 27599-3280, USA

Adrien GIRARDIE, Laboratoire de Neuroendocrinologie, Universite Bordeaux 1, 33405 Talence Cedex, FRANCE

Josiane GIRARDIE, Laboratoire de Neuroendocrinologie, Universite Bordeaux 1, 33405 Talence Cedex, FRANCE

Graham J. GOLDSWORTHY, Department of Biology, Birkbeck College, University of London, Malet Street, London WC1E 7HX, UNITED KINGDOM

Rosemary S. GRAY, Biology Department, University of North Carolina at Chapel Hill, Chapel Hill, NC 27599-3280, USA

Linda M. HALL, Dept. of Biochemical Pharmacology, SUNY at Buffalo, Buffalo, NY 14260, USA

Peter HARVIE (USA) c/o Dr. F. Sehnal, Institute of Entomology, Czech Academy of Sciences, Branišovská 31, 370 05, České Budějovice, CZECH REPUBLIC

Jon H. HAYASHI, Insecticide Discovery/Neurobiology, American Cyanamid, P.O. Box 400, Princeton, NJ 08543-0400, USA

John G. HILDEBRAND, ARL Division of Neurobiology, University of Arizona, 611 Gould-Simpson Science Bldg., Tucson, AZ 85721 USA

László HIRIPI, Balaton Limnological Research Institute of the Hungarian Academy of Sciences, H-8237 Tihany, HUNGARY

Magdalena HODKOVÁ, Institute of Entomology, Czech Academy of Sciences, Branišovská 31, 370 05, České Budějovice, CZECH REPUBLIC

Klaus H. HOFFMANN, Department of General Zoology, University of Ulm, D-7900 Ulm, GERMANY

Alica HŮCKOVÁ, Institute of Experimental, Phytopathology and Entomology, Slovak Academy of Sciences, 900 28 Ivanka pri Dunaji, SLOVAKIA

Blanka KALINOVÁ, Institute of Organic Chemistry and Biochemistry, Insect Chemical Ecology Unit, U Šalamounky 41, 158 00 Praha 5, CZECH REPUBLIC

Thomas J. KELLY, USDA, ARS, BA, PSI, Insect Neurobiology & Hormone Lab, 10300 Baltimore Blvd., Rm. 320, Bldg. 306, BARC-East, Beltsville, MD 20705-2350, USA

Dalibor KODRÍK, Institute of Entomology, Czech Academy of Sciences, Branišovská 31, 370 05, České Budějovice, CZECH REPUBLIC

Danuta KONOPIŃSKA, Institute of Chemistry, University of Wroclaw, ul. Joliet-Curie 14, 50383 Wroclaw, POLAND

Ladislav KOTĚRA, Institute of Organic Chemistry and Biochemistry, Insect Chemical Ecology Unit, U Šalamounky 41, 158 00 Praha 5, CZECH REPUBLIC

Vladimir LANDA, Institute of Entomology, Czech Academy of Sciences, Branišovská 31, 370 05, České Budějovice, CZECH REPUBLIC

Angela B. LANGE, Department of Zoology, University of Toronto, Erindale College, 3359 Mississauga Rd., Ontario L5L ICG, CANADA

Bruno LAPIED, Laboratoire de Neurophysiologie, Universite d' Angers, F-49045 Angers Cedex, FRANCE

Hans LAUFER, Dept. of Molecular and Cell Biology, University of Connecticut, U 125, 75 N. Eagleville Rd., Storrs, CT 06268 USA

Marcia J. LOEB, USDA, ARS, BA, PSI, Insect Neurobiology & Hormone Lab, 10300 Baltimore Ave., Bldg. 306, Room 319, BARC-East, Beltsville, MD 20705-2350, USA

Barry G. LOUGHTON, Department of Biology, York University, North York, Ontario M3J 1P3, CANADA

Carla MALVA, Instituto Internazionale, di Genetica e Biofisica, Via Marconi 10, 80125 Napoli, ITALY

Edward P. MASLER, USDA, ARS, BA, PSI, Insect Neurobiology & Hormone Lab, 10300 Baltimore Ave., Room. 309, Bldg. 306, BARC-East, Beltsville, MD 20705-2350, USA

Joanna MICHALIK, Institute of Biochemistry and Biophysics, Polish Academy of Sciences, Rakowiecka 36, 02-532 Warsaw, POLAND

David P. MUEHLEISEN, Biology Department, Wilson Hall, University of North Carolina, Chapel Hill, NC 27599-3280, USA

Malgorzta MUSZYŃSKA-PYTEL, Department of Invertebrate Physiology, Warsaw University, Zwirki i Wigury 93, PL 02-089 Warsaw, POLAND

Ronald J. NACHMAN, USDA, ARS, SPA, FAPRL, Veterinary Entomology Research Unit, Route 5, Box 810, College Station, TX 77845, USA

Voclav NĚMEC, Institute of Entomology, Czech Academy of Sciences, Branišovská 31, 370 05, České Budějovice, CZECH REPUBLIC

Vladimlr J. A. NOVÁK, Laboratory of Evolutionary Biology, Czech Academy of Sciences, Oldrichova 39, 128 00 Praha 2, CZECH REPUBLIC

Ian ORCHARD, Department of Zoology, University of Toronto, 25 Harbord Street, Toronto, Ontario, M5S 1A1, CANADA

M. Paul PENER, Department of Cell and Animal Biology, The Hebrew University of Jerusalem, Jerusalem 91904, ISRAEL

Maciej A. PSZCZÓLKOWSKI, Department of Invertebrate Physiology, Warsaw University, Zwirki i Wigury 93, 02-089 Warszawa, POLAND

Jacques PUIROUX, Department of Biology, York University, North York, Ontario, M3J 1P3, CANADA

Ada RAFAELI, Department of Stored Products, The Volcani Institute, P.O. Box 6, Bet Dagan 50250, ISRAEL

Richard C. RAYNE, Centre for Neuroscience, University of Sussex, Falmer, Brighton BN1 9QG, UNITED KINGDOM

Klaus RICHTER, Saxon Academy of Sciences of Leipzig, Erbertstrasse 1, Forschungsgruppe Jena, 07703 Jena, Postfach 100322, GERMANY

Sandra ROBB, AFRC Laboratory of Molecular Signalling, Department of Zoology, University of Cambridge, Downing Street, Cambridge CB2 3EJ, UNITED KINGDOM

František SEHNAL, Institute of Entomology, Czech Academy of Sciences, Branišovská 31, 370 05, České Budějovice, CZECH REPUBLIC

Veeresh L. SEVALA, Department of Biology, York University, North York, Ontario M3J 1P3, CANADA

P. SIVASUBRAMANIAN, Department of Biology, University of New Brunswick, Fredericton, NB E3B 6E1, CANADA

Karel SLÁMA, Institute of Entomology, Czech Academy of Sciences, Branišovska 31, 370 05 Ceske Budejovice, CZECH REPUBLIC

Mirko SLOVÁK, Institute of Experimental Phytopathology and Entomology, Slovak Academy of Sciences, 900 28 Ivanka pri Dunaji, SLOVAKIA

Hans M. SMID, Department of Entomology, Wageningen Agricultural University, P.O. 8031, 6700 EH Wageningen, THE NETHERLANDS

Wieslaw SOBOTKA, Institute of Industrial Organic Chemistry, Ul. Annopol 6, 03-236 Warszawa - Zeran, POLAND

Tomáš SOLDÁN, Institute of Entomology, Czech Academy of Sciences, Branišovska 31, 370 05 České Budějovice, CZECH REPUBLIC

John E. STEELE, Department of Zoology, The University of Western Ontario, London, Ontario, NGA 5B7, CANADA

J. STOUT, Biology Department, Andrews University, Berrien Springs, MI 49104, USA

Florence TRIBUT, Laboratoire de Neurophysiologie, URA CNRS 611, Universite d'Angers, F - 49045 Angers Cedex, FRANCE

James W. TRUMAN, Department of Zoology, University of Washington, Seattle, WA 98195, USA

Richard TYKVA, Czech Academy of Sciences, Institute of Organic Chemistry, 166 10 Praha 6, CZECH REPUBLIC

Jozef VANDEN BROECK, Zoological Institute, Catholic University of Leuven, Naamsestraat 59, B-3000 Leuven, BELGIUM

Peter VERHAERT, Zoological Institute, Catholic University of Leuven, Naamsestraat 59, B-3000 Leuven, BELGIUM

Henk G. B. VULLINGS, Department of Experimental Zoology, University of Utrecht, Padualaan 8, NL-3584 CH Utrecht, THE NETHERLANDS

Renee M. WAGNER, USDA, ARS, BA, LPSI, Livestock Insects Laboratory, 10300 Baltimore Ave., Rm. 120, Bldg. 307, BARC-East, Beltsville, MD 20705-2350, USA

Steven P. M. WARBURTON, Department of Life Sciences, University of Nottingham, University Park, Nottingham, UNITED KINGDOM

Li WEI WEI, (China) c/o Dr. F. Sehnal, Institute of Entomology, Czech. Academy of Sciences, 370 05 České Budějovice, CZECH REPUBLIC

Anne L. WESTBROOK, Department of Biology, University of North Carolina at Chapel Hill, CB# 3280, Coker Hall, Chapel Hill, NC 27599-3280, USA

Jan ŽĎAREK, Institute of Organic Chemistry and Biochemistry, Insect Chemical Ecology Unit, U Šalamounky 41, 158 00 Praha 5, CZECH REPUBLIC

Dušan ŽITŇAN, Dept. of Entomology, University of California, Riverside, CA 92521-0314, USA